PROPOSALS THAT WORK

5th EDITION

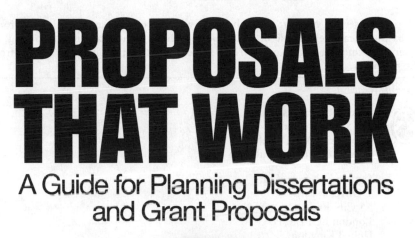

PROPOSALS THAT WORK

A Guide for Planning Dissertations and Grant Proposals

5th EDITION

Lawrence F. Locke
University of Massachusetts at Amherst
Waneen Wyrick Spirduso
The University of Texas at Austin
Stephen J. Silverman
Teachers College, Columbia University

For information:

Sage Publications, Inc.
2455 Teller Road
Thousand Oaks, California 91320
E-mail: order@sagepub.com

Sage Publications Ltd.
1 Oliver's Yard
55 City Road
London EC1Y 1SP
United Kingdom

Sage Publications India Pvt. Ltd.
B-42, Panchsheel Enclave
Post Box 4109
New Delhi 110 017 India

Printed in the United States of America

Library of Congress Cataloging-in-Publication Data

Locke, Lawrence F.
Proposals that work: A guide for planning dissertations and grant proposals / Lawrence F. Locke, Waneen Wyrick Spirduso, Stephen J. Silverman. — 5th ed.
p. cm.
Includes bibliographical references and index.
ISBN-13: 978-1-4129-2422-1 (cloth)
ISBN-13: 978-1-4129-2423-8 (pbk.)
1. Proposal writing in research—Handbooks, manuals, etc. 2. Dissertations, Academic—Handbooks, manuals, etc. 3. Research grants—Handbooks, manuals, etc. 4. Fund raising—Handbooks, manuals, etc. I. Spirduso, Waneen Wyrick. II. Silverman, Stephen J. III. Title.
Q180.55.P7L63 2007
001.4′4—dc22

2006029739

Printed on acid-free paper.

10 11 10 9 8 7 6 5 4 3

Acquiring Editor:	Lisa Cuevas Shaw
Assistant Editor:	Margo Beth Crouppen
Editorial Assistant:	Karen Margrethe Greene
Production Editor:	Sarah Quesenberry
Marketing Manager:	Stephanie Adams
Copy Editor:	Halim Dunsky
Proofreader:	Sue Irwin
Typesetter:	C&M Digitals (P) Ltd.
Cover Designer:	Michelle Kenny

Contents

Preface to the Fifth Edition

About the Nature of This Book

This book contains no direct instruction concerning how to do research. Rather, the material herein deals with the problem of how to write a research proposal. Although the two capacities—skill in conducting research and skill in writing about plans for that research—have a close relationship, they are far from coterminous. Individuals who have acquired considerable information about the mysteries of research methods and data analysis do not necessarily know how to undertake the task of planning and effectively proposing their own investigations.

The importance and, therefore, the perceived difficulty of the proposal task will seem greater under some conditions than under others. Even the novice in research is likely to appreciate fully the significance of the research proposal in preparing a grant application. Once the target for inquiry has been selected, it is obvious that the probability for obtaining funding will rest mostly on how clearly and persuasively the proposal displays the author's competence to design and execute the study.

In contrast, few graduate students initially recognize that preparation of the proposal will represent a major hurdle in gaining approval for their thesis or dissertation. In the context of graduate education, however, the research proposal plays a central role—and one that reaches beyond its simple significance as a plan of action. In most instances, the decision to permit the student to embark on a study is made solely on the basis of that first formal document. The quality of the proposal is likely to be used by advisors as a basis for judging the clarity of thought that has preceded the document, the degree of facility with which the study will be implemented if approved, and the adequacy of expository skills the student will bring to reporting the results. In sum, the proposal is the instrument through which the faculty must judge whether there is reasonable hope that the student can conduct research.

Given such an understanding of the importance invested in writing a proposal, our intention in writing this guide is clear and straightforward. Our purpose from start to finish is to assist in achieving one end—a fair and useful hearing for your proposed research. If the proposed study is rejected, it should be because the investigation lacks sufficient merit or feasibility, not because the presentation itself was flawed.

As you read, you will discover that in explanations and illustrations we treat a number of constructs (and their associated terminology) as though they were generic to all forms of research. We do that even though in the actual practice of particular research traditions different labels may be assigned or special meanings employed that are unique to a particular form of inquiry. Our preference for the use of general terminology is partly a matter of economy, in that referencing all of the nuanced variations in the language of research would be enormously cumbersome. It also is true, however, that despite the wide variety of available research formats there is a central set of recurring demands and concerns that cut across all types of inquiry. When proposals for research are being prepared, we think it is important to remember what is shared among, as well as what is particular to, the various strategies that might be employed.

As an illustration, proposals for historical or philosophic study must reflect canons unique to each of the two forms of inquiry. Likewise, plans for qualitative research will begin with several assumptions that are different from those that govern experimental investigations. Nevertheless, the planning stage within each of those divergent traditions requires that questions be formulated, that the prospectus be grounded in the current status of knowledge, that appropriate data sources be identified, that methodologies be rationalized in terms of the tradition used to frame the study, and that all of this be done in a manner that persuades the reader of the author's competence. In other words, different types of research may make distinctive requirements and yield particular virtues, but in this textbook we often will emphasize how much is shared in common among all forms of systematic inquiry.

About the Changes in This New Edition

A wide variety of considerations are involved in the decision to produce a new edition for an established textbook such as *Proposals That Work (PTW)*, and our fifth effort is no exception. We will not recount here all of the minutia that distinguish the latest iteration in the series. You can assume that we have accomplished all of the necessary but often invisible changes such as updating references, refreshing illustrative examples, and rewriting where greater clarity might be achieved. Instead, we want to draw your attention to the forces that shaped substantial parts of what you will find in this newest version of *PTW*.

Continuing feedback from readers has always encouraged revisions designed to improve utility, relevance, and ease of use. As an example, some of the users who were contemplating development of grant proposals reported that our previous treatment of the review procedures employed by the National Institutes of Health did little to serve the sorts of generally more modest applications they were likely to make. Accordingly, Chapter 8 has been reshaped so that *PTW* now provides examples and advice that better fits the needs of our readership.

In addition to that reorientation, Chapters 8 and 9 now include much closer attention to the preparation of grant proposals from students seeking financial support for theses and dissertations. In turn, that material is supported in Part III by the inclusion of a specimen proposal authored by a graduate student seeking internal funding for a dissertation. Finally, we have inserted a new section devoted to alternative dissertation formats that serve to more closely link the proposal with the subsequent dissertation and, in turn, with the production and publication of research reports.

Another group of considerations that worked with particular force to urge preparation of this new edition emerged from within the community of active researchers. Questions concerning ethics in the world of research have returned to our attention with renewed vigor. Both the public press and the pages of our own journals have made concern about the contaminations of dishonesty and carelessness unavoidable. Accordingly, in this new edition our discussion of "The Habit of Truth" in Chapter 2 has been updated and expanded to give this topic a new sense of urgency.

Although perhaps less weighty than matters of ethics, our own continuing experiences with the uses and misuses of PowerPoint in research presentations were the motivation for another addition. That will be found in Chapter 7 where we undertake to provide some guidelines to encourage uses that make Microsoft's powerful tool for creating graphics an adjunct to clarity rather than an enticing distraction.

Finally, as new tools for inquiry are refined through use, and as their utility is made more obvious through successful application, it is inevitable that they will have growing appeal to beginning researchers. Our readers, therefore, can reasonably expect that the problems to be encountered in proposing use of such "emerging" research instruments will be given proper consideration. As an example, the brief attention we gave to focus group studies in the fourth edition of *PTW* might have been reasonable in 1999, but it could not serve adequately in 2007. Research journals in all fields of social service reflect the extent to which the use of focus groups is an increasingly attractive option.

Likewise, although the obvious difficulties of writing proposals for mixed-method research may have previously made it reasonable to give limited attention to such designs, the realities of the world in which our readers will function make that benign neglect no longer prudent. Whatever our reservations, mixed method designs, like focus groups, are research strategies that are in general use—and a substantially revised Chapter 5 now reflects that fact.

Suggestions for Using the Book

This book has been divided into three major parts: Part I, Writing the Proposal; Part II, Money for Research; and Part III, Specimen Proposals. The chapters that constitute Part I represent the core of the book, serving both to present generic information that applies to all research proposals and to discuss some of the problems peculiar to the use of proposals in either graduate education or funding agencies. Chapter headings shown in the table of contents offer a self-explanatory outline of what is covered. We draw your attention to the fact that Chapter 5 deals with the particular demands of proposing studies framed by the assumptions established in the qualitative and critical research paradigms.

In Part II, we offer assistance to those who wish to seek funding for proposed studies. The general topic of locating sources and submitting applications is introduced in Chapter 8. Specific attention is given in that chapter to finding financial support for student research (projects, theses, and dissertations), and Chapter 9 is devoted to the specifics of preparing grant proposals. Collectively, the two chapters in Part II provide a sound starting place for anyone who wishes to locate and exploit the financial resources available from foundations and agencies (both governmental and private) that have an interest in research. Where required, more detailed information can be found in the books abstracted in the Appendix.

In Part III, four sample proposals are presented. The sections of the first document relate directly to each of the generic proposal tasks presented in Chapter 1. Our short commentaries inserted into the text are used to highlight those key elements. The next three proposals also are accompanied by critical evaluations intended to illuminate their individual strengths and limitations—both as plans for research and as written vehicles for presentation of those plans. Together, the four specimens include documents at both early and late stages of development, both quantitative and qualitative designs, and successful proposals for both graduate student dissertations and research funding.

Readers with different backgrounds will find it useful to employ different methods in using this book. Those who are completely inexperienced in writing proposals should begin with Chapter 1, which deals with the basic functions of the document. Some may find it helpful to couple each task introduced there with the parallel illustration found in the first specimen proposal (Part III). Skimming several of the other proposals to note how various functions are performed will allow the novice researcher to return to the following chapters with some concrete sense of the form actually assumed by proposals.

Readers with some previous experience in research may wish to turn directly to Chapter 3, in which we deal with specific problems in identifying research topics and initiating the proposal process. Those interested in the particular demands and problems presented by the grant proposal should begin with Chapters 8 and 9, subsequently turning to other sections of the guide as needed (including the grant application in Part III). The same pattern of use would be appropriate for those who are considering a qualitative study—in that case starting with Chapter 5 and the corresponding proposal in Part III, using other sections as required.

Acknowledgments

As in most complex writing endeavors, many persons whose names do not appear on the cover played important parts in its completion. First, we would like to acknowledge the contributions of students in our classes at the University of Massachusetts at Amherst, The University of Texas at Austin, and Teachers College, Columbia University who used earlier versions of this text and provided helpful advice for revision. Thanks also to Professor Lorraine Goyette at Elms College in Chicopee, Massachusetts, whose lively and articulate students provide a steady stream of commentary on both of our textbooks from Sage Publications, *Proposals That Work* and *Reading and Understanding Research*. Drs. Shannon Mihalko, Belinda Minor, Pamela Rothpletz-Puglia, and Jed Tucker wrote the excellent specimen proposals used in this edition of the guide. We particularly appreciate their generosity in allowing us to reprint and critique their work.

Our special appreciation goes also to Patricia Moran of Fordham University who again shared her insights into the inner workings of foundations and their complex roles in providing financial support. We also are indebted to our editors at Sage, Lisa Cuevas Shaw and Margo Crouppen, who got us started again and with gentle prodding kept us on track. Our deepest appreciation, however, must be reserved for Lori, Craig, and Pat, whose patience and understanding once more sustained our effort. As for so much else in our lives, we stand in your debt.

Acknowledgments

PART I

Writing the Proposal

1

The Function of the Proposal

The dissertation process begins with the development of a proposal that sets forth both the exact nature of the matter to be investigated and a detailed account of the methods to be employed. In addition, the proposal usually contains material supporting the importance of the topic selected and the appropriateness of the research methods to be employed.

Function

A proposal may function in at least three ways: as a means of communication, as a plan, and as a contract.

Communication

The proposal serves to communicate the investigator's research plans to those who provide consultation, give consent, or disburse funds. The document is the primary resource on which the graduate student's thesis or dissertation committee must base the functions of review, consultation, and, more important, approval for implementation of the research project. It also serves a similar function for persons holding the purse strings of foundations or governmental funding agencies. The quality of assistance, the economy of consultation, and the probability of financial support will all depend directly on the clarity and thoroughness of the proposal.

Plan

The proposal serves as a plan for action. All empirical research consists of careful, systematic, and preplanned observations of some restricted set of phenomena. The acceptability of results is judged exclusively in terms of the adequacy of the methods employed in making, recording, and interpreting the planned observations. Accordingly, the plan for observation, with its supporting arguments and explications, is the basis on which the thesis, dissertation, or research report will be judged.

The research report can be no better than the plan of investigation. Hence, an adequate proposal sets forth the plan in step-by-step detail. The existence of a detailed plan that incorporates the most careful anticipation of problems to be confronted and contingent courses of action is the most powerful insurance against oversight or ill-considered choices during the execution phase of the investigation. With the exception of plans for some qualitative research (see Chapter 5), the hallmark of a good proposal is a level of thoroughness and detail sufficient to permit another investigator to replicate the study, that is, to perform the same planned observations with results not substantially different from those the author might obtain.

Contract

A completed proposal, approved for execution and signed by all members of the sponsoring committee, constitutes a bond of agreement between the student and the advisors. An approved grant proposal results in a contract between the investigator (and often the university) and a funding source. The approved proposal describes a study that, if conducted competently and completely, should provide the basis for a report that would meet all standards for acceptability. Accordingly, once the contract has been made, all but minor changes should occur only when arguments can be made for absolute necessity or compelling desirability.

Proposals for theses and dissertations should be in final form prior to the collection of data. Under most circumstances, substantial revisions should be made only with the explicit consent of the full committee. Once the document is approved in final form, neither the student nor the sponsoring faculty members should be free to alter the fundamental terms of the contract by unilateral decision.

Regulations Governing Proposals

All funding agencies have their own guidelines for submissions, and these should be followed exactly. In the university, however, no set of universal rules or guidelines presently exists to govern the form or content of the research proposal. There may be, however, several sources of regulation governing the form and content of the final research report. The proposal sets forth a plan of action that must eventuate in a report conforming to these latter regulations; therefore, it is important to consider them in writing the proposal. As we discuss later in this chapter, understanding what the final report will look like may help you in completing the dissertation and submitting articles for publication.

Although it is evident that particular traditions have evolved within individual university departments, any formal limitation on the selection of either topic or method of investigation is rarely imposed. Normally, the planning and execution of student research are circumscribed by existing departmental policy on format for the final report, university regulations concerning theses and dissertation reports, and informal standards exercised by individual advisors or study committees.

Usually, departmental and university regulations regarding graduate student proposals are either so explicit as to be perfectly clear (e.g., "The proposal may not exceed 25 typewritten pages" or "The proposal will conform to the style established in the *Publication Manual of the American Psychological Association*") or so general as to impose no specific or useful standard (e.g., "The research topic must be of suitable proportions" or "The proposal must reflect a thorough knowledge of the problem area"). The student, therefore, should find no serious difficulty in developing a proposal that conforms to departmental and university regulations.

Some universities now allow students to elect alternative dissertation or thesis formats, such as a research paper (or series of papers) with an expanded literature review and supporting materials in the appendix. We discuss this in the last section of this chapter and urge you to consider such an option because the more compact research paper format can save considerable time in turning the completed dissertation into a publication. Alternative formats for the final report, however, do not alter the need for a complete proposal. A good study requires a sound plan, irrespective of the format used for reporting the results.

Another potential source of regulation, the individual thesis or dissertation committee, constitutes an important variable in the development of the thesis or dissertation proposal. Sponsoring committee members may have strong personal commitments concerning particular working procedures, writing styles, or proposal format. The student must confront these as a unique constellation of demands that will influence the form of the proposal. It always is wise to anticipate conflicting demands and to attempt their resolution before the collection of data and the preparation of a final report.

Committees are unlikely to make style and format demands that differ substantially from commonly accepted modes of research writing. As a general rule, most advisors subscribe to the broad guidelines outlined in this book. Where differences occur, they are likely to be matters of emphasis or largely mechanical items (e.g., inclusion of particular subheadings within the document).

General Considerations

Most problems in proposal preparation are straightforward and relatively obvious. The common difficulties do not involve the subtle and complex problems of design and data management. They arise instead from the most basic elements of the research process: What is the proper question to ask? Where is the best place to look for the answer? What is the best way to standardize, quantify, and record observations? Properly determining the answers to these questions remains the most common obstacle to the development of adequate proposals.

Simplicity, clarity, and parsimony are the standards of writing that reflect adequate thinking about the research problem. Complicated matters are best communicated when they are the objects of simple, well-edited prose. In the early stage of development, the only way to obtain prompt and helpful assistance is to provide advisors with a document that is easily and correctly understood. At the final stage, approval of the study will hinge not only on how carefully the plan has been designed but also on how well that design has been communicated. In the mass of detail that goes into the planning of a research study, the writer must not forget that the proposal's most immediate function is to inform readers quickly and accurately.

The problem in writing a proposal is essentially the same as in writing the final report. When the task of preparing a proposal is well executed, the task of preparing the final report is more than half done (an important consideration for the graduate student with an eye on university deadlines). Under ideal conditions, such minor changes as altering the tense of verbs will convert the proposal into the opening chapters of the thesis or dissertation, or into initial sections of a research report.

Many proposals evolve through a series of steps. Preliminary discussion with colleagues and faculty members may lead to a series of drafts that evolve toward a final document presented at a formal meeting of the full dissertation or thesis committee, or to a proposal submitted through the university hierarchy to a funding source. This process of progressive revision can be accelerated and made more productive by following these simple rules:

1. Prepare clean, updated copies of the evolving proposal and submit them to advisors or colleagues in advance of scheduled consultations.
2. Prepare an agenda of questions and problems to be discussed and submit them in advance of scheduled consultations.
3. Keep a carefully written and dated record of all discussions and decisions that occur with regard to each item on the consultation agenda.

General Format

Guidelines for the format of proposals, even when intended only as general suggestions, often have an unfortunate influence on the writing process. Once committed to paper, such guidelines quickly tend to acquire the status of mandatory prescription. In an attempt to conform to what they perceive as an invariant format, students produce proposal documents that are awkward and illogical as plans for action—as well as stilted and tasteless as prose.

Some universities and many funding agencies make very specific demands for the format of proposals. Others provide general guidelines for form and content. Whatever the particular situation confronting the writer, it is vital to remember that *no universally applicable and correct format exists for the research proposal.* Each research plan requires that certain communication tasks be accomplished, some that are common to all proposals and others that are unique to the specific form of inquiry. Taken together, however, the tasks encompassed by all proposals demand that what is written fit the real topic at hand, not some preconceived ideal. It is flexibility, not rigidity that makes strong proposal documents.

Specific Tasks

The following paragraphs specify communication tasks that are present in nearly all proposals for empirical research. Each proposal, however, will demand its own unique arrangement of these functions. Within a given proposal, the tasks may or may not be identified by such traditional section designations as "Background," "Importance of the Study," "Review of Literature," "Methodology," "Definitions," or "Limitations." Individual proposals are sure to demand changes in the order of presentation or attention to other tasks not specified below. This particularly will be the case with some of the tasks that are specific to grant proposals (see Part II). Finally, it is important to note that some of the adjacent tasks, shown as headings in the following paragraphs, often may be merged into single sections.

As you read each of the tasks below, an illustration can be found by turning to the first proposal in Part III of this guide. In that particular specimen, we have edited the proposal so the sections correspond to the discussion of each task. We have provided a critique preceding each section of the specimen proposal to summarize the suggestions presented in this chapter.

Introducing the Study 

Proposals, like other forms of written communication, are best introduced by a short, meticulously devised statement that establishes the overall area of concern, arouses interest, and communicates information essential to the reader's comprehension of what follows. The standard here is a "gentle introduction" that avoids both tedious length and the shock of technical detail or abstruse argument. A careful introduction is the precursor of three other tasks (purpose statement, rationale, and background). In many cases, it may be written simply as the first paragraph(s) of an opening proposal section that includes all three.

For most proposals, the easiest and most effective way to introduce the study is to identify and define the central construct(s) involved. In the sense that constructs are concepts that provide an abstract symbolization of some observable attribute or phenomenon, all studies employ constructs. Constructs such as "intelligence" or "teacher enthusiasm" are utilized in research by defining them in terms of some observable event, that is, "intelligence" as defined by a test score, or "teacher enthusiasm" as defined by a set of classroom behaviors. When the reader asks, "What is this study about?" the best answer is to present the key constructs and explain how they will be represented in the investigation. The trick in these opening paragraphs of introduction is to sketch the study in the bold strokes of major constructs without usurping the function of more detailed sections that will follow.

Relationships among constructs that will be of particular interest or about which explicit hypotheses will be developed should be briefly noted. Constructs with which the reader probably is familiar may be ignored in the introduction, for they are of less interest than the relationships proposed by the author.

The most common error in introducing research is failure to get to the point—usually a consequence of engaging in grand generalizations. For instance, in a proposed study of attributes contributing to balance ability, the opening paragraph might contain a sentence such as "The child's capacity to maintain balance is a factor of fundamental importance in the design of elementary school curriculum." The significance of the construct "balance" in accomplishing motor tasks may make it an attribute of some importance in elementary education, but that point may be far from the heart of a study involving balance. If, for example, the proposed study deals with the relationship of muscle strength to balance, observations about balance as a factor in the design of school curriculum belong, if anywhere, in a later discussion. What belongs up front is a statement that gets to the point: "The task of maintaining static balance requires muscular action to hold the pelvis in a horizontal position. When muscle strength is inadequate to accomplish this, performance is impaired."

Some indication of the importance of the study to theory or practice may be used to help capture the reader's interest, but in the introduction it is not necessary to explain completely all the study's significance. Present the basic facts first and leave the detail of thorough discussion until a more appropriate point. Use of unnecessary technical language is another impediment to the reader's ability to grasp the main idea. Similarly, the use of quotations and extensive references are intrusions into what should be a clear, simple preliminary statement. As a general rule, the first paragraph of the introduction should be free of citations. Documentation of important points can wait until a full discussion of the problem is launched.

Stating the Purpose 

Early in the proposal, often in the introductory paragraph(s), it is wise to set forth an explicit statement of your purpose in undertaking the study. We are using the word "purpose" in its general sense as a statement of why you want to do the study and what you intend to accomplish. Such statements can be divided broadly into those related to the desire to *improve* something and those reflecting a desire to *understand* something. In addition to such practical and theoretical purposes, Maxwell (2005) has pointed out that, in some instances, it may be wise to be explicit about more personal purposes as well, including interests related to simple curiosity, a sense of social responsibility, or career demands.

A statement of purpose need not be an exhaustive survey of your intentions, nor need it be written in the formal language of research questions (which are much more specific expressions of what you want to learn). An early and specific announcement of the primary target for the study, and your purpose in aiming at it, will satisfy the reader's most pressing questions—what is this all about, and why is this study being proposed? Succinct answers allow the reader to attend to your subsequent exposition without the nagging sense that he or she still is waiting to discover the main objective. Make your statement of purpose early, be forthright, keep it simple, and be brief.

Providing a Rationale

Once the reader understands the topic of the investigation and has at least a general sense of your purpose, the next task is to address the question "Why bother with that?" in terms that are more detailed and explicit. The development of a rationale that justifies the proposed study usually involves both logical argument and documentation with factual evidence. The intention is to persuade the reader not only that the investigation (with its component questions or hypotheses) is worthy of attention, but also that the problem has been correctly defined.

To that end, it often is helpful to diagram factors and relationships that support your formulation of the problem. Suppose that an assertion proposed for experimental testing is that older adults who had oxygen therapy for six months would show superior cognitive function when compared to subjects assigned to a control group. The implication of such an assertion is that there is a relationship between the level of oxygen provided to the brain and cognitive capacity in older adults. The reasons for such a complex supposition can be clarified by diagramming them in a simple form like the one shown in Figure 1.1. Assuming that the constructs have been defined, the rationale can be developed by documenting the information within each box, and then

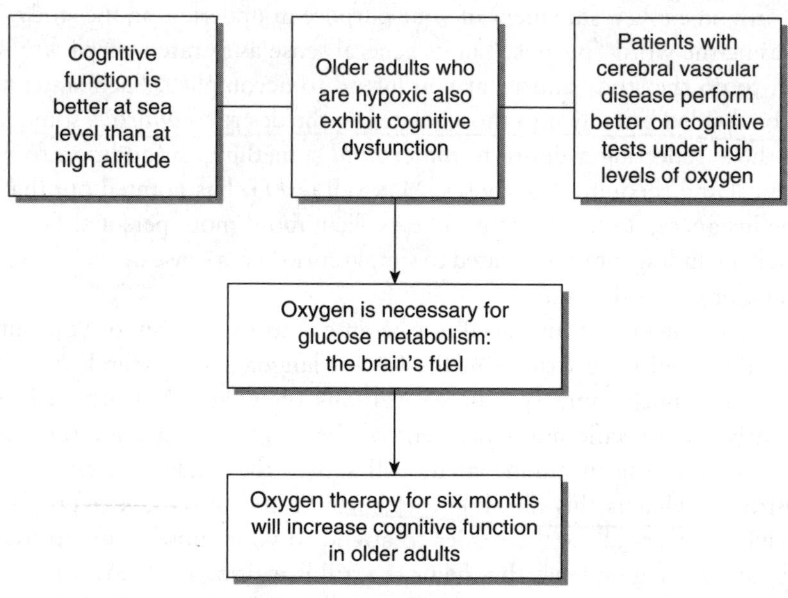

Figure 1.1 Example of Diagram of Logic for Rationale

explaining the enormous practical consequences that would attend a positive finding. A sound rationale is one that convinces the reader that you are raising the right question—and that the answer is worth finding.

In most cases, this early attention to justifying the proposed study should be limited to the basic matters of defining what is to be studied and why it is worth so doing. These reasons may be practical, theoretical, or both and should be presented economically. The detail of rationale for particular choices in methods of data collection and analysis can be deferred until such matters are discussed in subsequent parts of the proposal.

Formulating Questions or Hypotheses

All proposals must arrive at a formal statement of questions or hypotheses. These statements should be written in carefully constructed language that specifies each variable in explicit terms. A statement such as "Studying each day should result in improved learning" is better written as "Sixty minutes of studying each day will result in significantly increased scores on a standardized test of achievement." These statements of questions or hypotheses may be set aside as a separate section or simply included in the course of other discussion. Such statements differ from what was contained in the statement of the purpose in that (a) they are normally stated in formal terms appropriate to the design and analysis of data to be employed, and (b) they display, in logical order, all subsections of the research topic.

The question form is most appropriate when the research is exploratory, when it is impossible and inappropriate to state hypotheses, or for qualitative studies where the question format is much more appropriate. The researcher should indicate by the specificity of questions, however, that the problem has been subject to thorough analysis. By careful formulation of questions, the proposed study should be directed toward outcomes that are foreshadowed by the literature or pilot work, rather than toward a scanning of potentially interesting findings.

The hypothesis form is employed when the state of existing knowledge and theory permits formulation of reasonable predictions about the relationship of variables. Hypotheses ordinarily have their origin in theoretical propositions already established in the review of literature. Because the proposal must ensure that the reader grasps how the relationships expressed in theory have been translated into the form of testable hypotheses, it often is useful to provide a succinct restatement of the theoretical framework at a point contiguous to the presentation of formal research hypotheses.

The most common difficulty in formulating a research question is the problem of clarity. Students who have read and studied in the area of their topic for weeks or months often are distressed to discover how difficult it is to reduce all they want to discover to a single, unambiguous question.

The clarity of a research question hinges on adequate specificity and the correct degree of inclusiveness. The major elements of the investigation must be identified in a way that permits no confusion with other elements. At the same time, the statement must maintain simplicity by including nothing beyond what is essential to identify the main variables and any relationships that may be proposed among them. Questions for quantitative studies, for example, must meet three tests of clarity and inclusiveness:

1. Is the question free of ambiguity?
2. Is a relationship among variables expressed?
3. Does the question imply an empirical test?

Applying these standards to the question "Does a relationship exist between self-esteem and reading achievement in children?" might appear to identify the study's main elements in reasonably clear fashion. Self-esteem and reading achievement are variables, and children are the subject population. A relationship is suggested, and correlation of self-esteem and reading scores clearly is implied as an appropriate empirical test of the relationship. The constructs of self-esteem and reading achievement, however, are quite broad and might be taken by some readers to indicate variables different from those intended. These potential sources of ambiguity might be resolved without destroying the simplicity of the question by altering it to ask, "Does a relationship exist between scores on the Children's Test of Self-Esteem and scores on the reading portion of the Tri-State Achievement Test?" Whether it also might be important to provide more specificity for the generic word "children" would depend on whether the intent was to examine self-esteem and reading in a particular type of child. If not, the generic word would be adequate, but if so, the importance of that variable calls for more careful specification in the question.

In the case of qualitative research (discussed at length in Chapter 5), because pre-established hypotheses are seldom used, questions are the tool most commonly employed to provide focus for thesis and dissertation studies. Although there is disagreement among scholars about the use of formal questions in qualitative research, there is no escape from the need to have a question (whether explicit or implicit) that will serve to direct what is observed or who is interviewed—at least at the outset of the study.

The question(s) frequently are phrased in ways that make them appear very different from those used in the natural science model (and, thereby, discrepant with some aspects of the advice given in this chapter). Some, for example, will sound highly generalized, as in the following examples paraphrased from qualitative proposals.

1. What is going on in this urban school classroom?
2. How do professional wrestlers understand their work?
3. What does residence in a hospice mean to a patient?

Other question statements reflect the intention to use a particular theoretical framework in the study.

1. What perspective do medical students adopt to make sense of their experience in medical school?
2. How do gay and lesbian soldiers manage the presentation of their sexual preference within the social setting of their workplace?
3. How do social roles influence the interaction between teachers and students as they attempt to realize personal goals in the classroom?

In contrast with quantitative research, questions in a qualitative proposal often are treated as more tentative and contingent on the unfolding of the study. Nevertheless, their careful formulation is no less important. They must give initial direction to planning, bring the power of theoretical constructs to the process of analysis, and reflect the degree of sophisticated thought employed in determining the focus of inquiry.

Experienced qualitative researchers sometimes do, in fact, elect not to package their curiosity, interests, concerns, and foreshadowings into the form of explicit research questions. Graduate students, however, embarking on their first attempt within the qualitative paradigm, often find that their advisors are greatly reassured when the proposal contains a careful accounting of what the data are expected to reveal that is not already known. In other words, it is a good idea for the novice to explicate the questions that motivate their interest, thereby firmly grounding the study in the conventions of scholarly inquiry. How a qualitative investigator's assumptions about the world, and about research, serve to shape those questions will be addressed in Chapter 5.

Research *hypotheses* differ from research *questions* in that hypotheses both indicate the question in testable form and predict the nature of the answer. A clear question is readily transformed into a hypothesis by casting it in the form of a declarative statement that can be tested so as to show it to be either true or false. Getting precisely the hypothesis that is wanted, however, often is more exacting than it appears.

Unlike a question, the hypothesis exerts a direct influence on each subsequent step of the study, from design to preparation of the final report. By specifying a prediction about outcome, the hypothesis creates a bridge between the theoretical considerations that underlie the question and the ensuing research process designed to produce the answer. The investigator is limited to procedures that will test the truth of the proposed relationship, and any implications to be deduced from the results will rest entirely on the particular test selected. Because it exerts such powerful a priori influence, a hypothesis demands considerable attention at the start of a study but makes it easier to preserve objectivity in the later stages of design and execution.

Aside from specific impact on design of the study, the general advantage of the hypothesis over the question for quantitative studies is that it permits more powerful and persuasive conclusions. At the end of a study, a research question never permits the investigator to say more than "Here is how the world looked when I observed it." In contrast, hypotheses permit the investigator to say, "Based on my particular explanation of how the world works, this is what I expected to observe, and behold—that is exactly how it looked! For that reason my explanation of how the world works must be given credibility." When a hypothesis is confirmed, the investigator is empowered to make arguments about knowledge that go far beyond what is available when a question has been asked and answered.

We would be remiss here if we did not note the current debate among researchers about the value of hypotheses and statistical significance testing. It has been argued (Schmidt, 1996; Thompson, 1996, 1997; Thompson & Kieffer, 2000) that statistical significance testing (one step in the process of testing hypotheses) has certain technical limitations. For some studies, at least, other types of analyses, such as examining effect sizes, might provide greater benefits. That debate is beyond the scope of this text. What is certain, however, is that graduate students should discuss the matter with advisors and committee members until a consensus emerges that meets both their expectations and those (if any) of the graduate school. Whether hypotheses are tested or questions are used to guide the research, they should be written with the greatest care for precision and must be exactly appropriate to the purposes of the study.

A hypothesis can be written either as a null statement (conveniently called a null hypothesis), such as "There is no difference between . . . ," or as a directional statement indicating the kind of relationship anticipated (called a research or directional hypothesis), such as "When this, also that" (positive) or "When this, not that" (negative). Many arguments favor the use of directionality because it permits more persuasive logic and more statistical power. If a pilot study has been completed or the literature review provides strong reasoning for a directional result, then directional hypotheses are clearly appropriate. In some instances, particularly evaluation studies, practical matters may dictate use of a directional hypothesis. For instance, if a therapy program is being evaluated and the only practical consequence would be finding that therapy provides greater gains in stress reduction than the program in current use, a directional hypothesis would permit a direct test of this singular outcome.

Some of the technical debate about the form of hypotheses is beyond the scope of this guide, but a good rule of thumb for the novice is to employ directional hypotheses when pilot data are available that clearly indicate a direction, or when the theory from which the hypotheses were drawn is sufficiently robust to include some persuasive evidence for directionality. If the investigation is a preliminary exploration in an area for which there is no well established theory, and if it has been impossible to gather enough pilot data to provide modest confidence in a directional prediction, the format of the null hypothesis is the better choice. Ultimately, as a researcher pursues a line of questioning through several investigations, directional hypotheses become more obvious and the null format less attractive.

Hypotheses can be evaluated by the same criteria used to examine research questions (lack of ambiguity, expression of relationship, and implication of appropriate test). In addition, the statement must be formulated so that the entire prediction can be dealt with in a single test. If the hypothesis is so complex that one portion could be rejected without also rejecting the remainder, it requires rewriting.

Several small, perfectly testable hypotheses always are preferable to one that is larger and amorphous. For example, in the following hypothesis the word "but" signals trouble. "Males are significantly more anxious than females, but male nurses are not significantly more anxious than female teachers." The *F* test for the main effect of sex in the implied analysis of variance (ANOVA) will handily deal with males and females, but a separate test as a part of a factorial ANOVA would be required for professional status. Should the tests yield opposite results, the hypothesis would point in two directions at once.

Similarly, the presence of two discrete dependent variables foreshadows difficulty in the following example: "Blood pressures on each of five days will be significantly lower than the preceding day, whereas heart rate will not decrease significantly after Day 3." The implied multivariate analysis of variance (MANOVA) could not rescue the hypothesis by indicating whether we could accept or reject it. The required follow-up test might reject the blood pressure prediction while accepting it for heart rate. In all such cases, division into smaller, unitary hypotheses is the obvious cure.

When a number of hypotheses are necessary, as a result of interest in interaction effects or as a consequence of employing more than one dependent variable, the primary hypotheses should be stated first. These primary statements may even be separated from hypotheses that are secondary or confirmatory, as a means of giving prominence to the main intent of the study.

Finally, hypotheses should be formulated with an eye to the qualitative characteristics of available measurement tools. If, for example, the hypothesis specifies the magnitude of relationship between two variables, it is essential that this be supportable by the reliability of the scores for the proposed instrumentation. Returning to the earlier example of self-esteem and reading, the fact should be considered that the correlation between scores from two tests cannot exceed the square root of the product for reliability in each test. Accordingly, if reliability of the self-esteem test is .68 and that of the reading test is .76, then a hypothesis of a positive correlation greater than .80 is doomed to failure ($\sqrt{.68 \times .76} = .72$).

Delimitations and Limitations

In some cases, a listing of delimitations and limitations is required to clarify the proposed study. Delimitations describe the populations to which generalizations may be safely made. The generalizability of the study will be a function of the subject sample and the analysis employed. *Delimit* literally means to define the limits inherent in the use of a particular construct or population.

Limitations, as used in the context of a research proposal, refer to limiting conditions or restrictive weaknesses. They occur, for example, when all factors cannot be controlled as a part of study design, or when the optimal number of observations simply cannot be made because of problems involving ethics or feasibility. If the investigator has given careful thought to these problems and has determined that the information to be gained from the compromised aspect of the study is nevertheless valid and useful, then the investigator proceeds but duly notes the limitation.

All studies have inherent delimitations and limitations. Whether these are listed in a separate section or simply discussed as they arise is an individual decision. If they are few in number and perfectly obvious, the latter is desirable. Whatever format is used, however, it is the investigator's responsibility to understand these constraints and to assure the reader that they have been considered during the formulation of the study.

Providing Definitions

All proposals for research use systematic language that may be specific to that field of research or to that proposal. We discuss the use of definitions in greater detail in the section of Chapter 6 titled "Clarity and Precision: Speaking in System Language."

Discussing the Background of the Problem 

Any research problem must show its lineage from the background of existing knowledge or previous investigations, or, in the case of applied research, from contemporary practice. The author must answer three questions:

1. What do we already know or do? (The purpose here, in one or two sentences, is to support the legitimacy and importance of the question. Major discussions of the importance and significance of the study will come under the "rationale for the study" section.)
2. How does this particular question relate to what we already know or do? (The purpose here is to explain and support the exact form of questions or hypotheses that serve as the focus for the study.)
3. Why select this particular method of investigation? (The purpose here is to explain and support the selections made from among alternative methods of investigation.)

In reviewing the research literature that often forms the background for the study, the author's task is to indicate the main directions taken by workers in the area and the main issues of methodology and interpretation that have arisen. Particular attention must be given to a critical analysis of previous methodology and the exposition of the advantages and limitations inherent in various alternatives. Close attention must be given to conceptual and theoretical formulations that are explicit or implicit within the selected studies.

By devising, when appropriate, a theoretical basis for the study that emerges from the structure of existing knowledge, by making the questions or hypotheses emerge from the total matrix of answered and unanswered questions, and by making the selection of method contingent upon previous results, the author inserts the proposed study into a line of inquiry and a developing body of knowledge. Such careful attention to background is the first step in entering the continuing conversation that is science.

The author should select only those studies that provide a foundation for the proposed investigation, discuss these studies in sufficient detail to make their relevance entirely clear, note explicitly the ways in which they contribute to the proposed research, and give some indication of how the proposal is designed to move beyond earlier work. The second section of Chapter 4 provides guidelines for preparing the literature review.

It is important for students and novice proposal writers to resist the impulse to display both the extent of their personal labors in achieving what they know and the volume of interesting, but presently irrelevant, information accumulated in the process. The rule in selecting studies for review is exactly the same as that used throughout the proposal—limit discussion to what is essential to the main topic. A complete list of all references used in developing the proposal (properly called a bibliography as distinct from the list of references) may be placed in an appendix, thereby providing both a service to the interested reader and some psychological relief to the writer. We should note, however, that many dissertation committees will think the references are all that is needed and including both a reference section and a bibliography would be overkill.

Whenever possible, the author should be conceptually or theoretically clear by creating organizing frameworks that encompass both the reviewed studies and the proposed research. This may take the form of something as obvious and practical as grouping studies according to certain methodological features (often for the purpose of examining divergent results), or something as esoteric as identifying and grouping the implicit assumptions made by various researchers in formulating their statement of the problem (often for the purpose of clarifying the problem selected in the present proposal).

In many proposals, creating an organized conceptual framework represents the most important single opportunity for the application of original thought. In one sense, the organizing task is an extension of the need to achieve clarity in communication. A category system that allows division of diverse ideas or recondite events into easily perceived and remembered subsets is an organizational convenience for the author, as well as for the reader. Beyond convenience, however, organizing frameworks identify distinctive threads of thought. The task is to isolate the parallel ways by which researchers, working at different times and in varying degrees of intellectual isolation, have conceived of reality. In creating a schema that deals meaningfully with similarities and dissimilarities in the work of others, the author not only contributes to the body of knowledge but also deals with the immediate needs of communicating this research to others.

Even relatively simple organizing or integrating systems demand the development of underlying conceptual plans and, often, new ways of interpreting old results and presumed relationships. The sequence of variables in the study may provide a simple and generally adequate place to begin arranging the review. Such questions as "What is the relationship between social class and school achievement when ability is held constant?" consist of concepts placed within a convenient sequential diagram. In turn, such conceptual schemata often contain useful assumptions about causal relationships and thus can serve as effective precursors to explanatory theory. The most elegant kind of research proposals achieve exactly that sort of linkage, using the framework for organizing the review of literature as a bridge connecting existing knowledge, a proposed theory, and the specific, theory-based hypotheses to be empirically tested.

Explaining Procedures

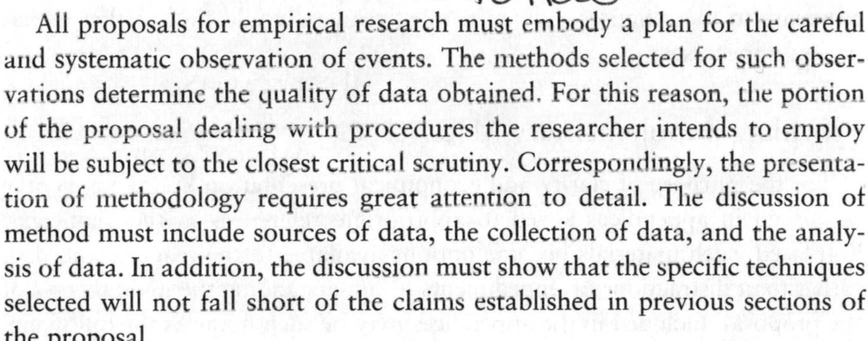

All proposals for empirical research must embody a plan for the careful and systematic observation of events. The methods selected for such observations determine the quality of data obtained. For this reason, the portion of the proposal dealing with procedures the researcher intends to employ will be subject to the closest critical scrutiny. Correspondingly, the presentation of methodology requires great attention to detail. The discussion of method must include sources of data, the collection of data, and the analysis of data. In addition, the discussion must show that the specific techniques selected will not fall short of the claims established in previous sections of the proposal.

The section(s) dealing with methodology must be freely adapted to the purpose of the study. Whatever the format, however, the proposal must provide a step-by-step set of instructions for conducting the investigation. For example, most studies demand explication of the following items:

1. Identification and description of the target population and sampling methods to be used
2. Presentation of instruments and techniques for measurement
3. Presentation of a design for the collection of data
4. Presentation of procedures for collecting and recording data
5. Explanation of data analysis procedures to be used
6. Development of plans for contingencies such as subject mortality

Many justifications for particular method selections will emerge in the development of background for the problem. The rationale for some choices, however, will most conveniently be presented when the method is introduced as part of the investigation plan.

In describing such elements, proposals can include pages of description that fatigue and frustrate the reader without yielding a clear picture of the overall pattern. In many cases, this problem can be avoided by the use of diagrams. Although Figure 1.2 displays a counterbalanced treatment design of moderate complexity, it would require no more than a brief paragraph of accompanying text to provide a clear account of the procedure.

Diagrams are helpful when presenting statistical models that will be tested later, once the data are collected. Note how clearly the interrelations of a hypothetical statistical model appear in Figure 1.3. In the figure, the ovals represent clusters of variables, the boxes show the variables in each cluster, and the various arrows represent interrelationships. Imagine how many words it would take to describe all of those relationships! Given a brief exposure to these figures, however, most readers would find further explanation unnecessary.

Providing Supplementary Material

For the purpose of clarity and economical presentation, many items may be placed in appendices keyed to appropriate references in the main text. So placed, such materials become options available to the reader as needed, rather than distractions or impediments to understanding the main themes of the proposal. Included in the appendices may be such items as the following:

1. Specifications for equipment
2. Instructions to subjects
3. Letters and other relevant documents
4. Subject consent forms
5. Raw data or tabular material from pilot studies
6. Tabular materials from related research
7. Copies of paper and pencil instruments
8. Questions for structured interviews
9. Credentials of experts, judges, or other special personnel to be employed in the study
10. Diagrammatic models of the research design
11. Diagrammatic models of the statistical analysis
12. Schematics for constructed equipment
13. Chapter outline for the final report
14. Proposed time schedule for executing the study
15. Supplementary bibliographies

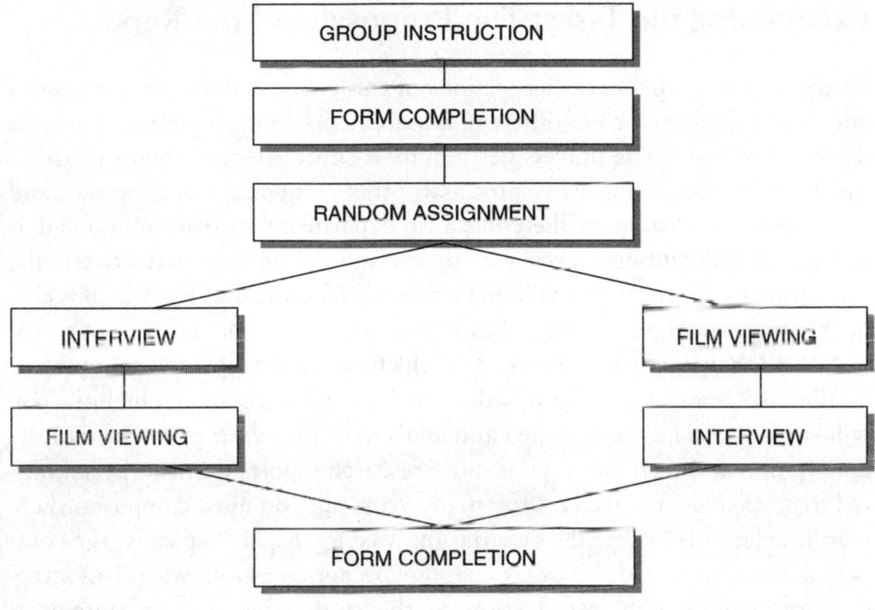

Figure 1.2 Example of Method Flow Chart

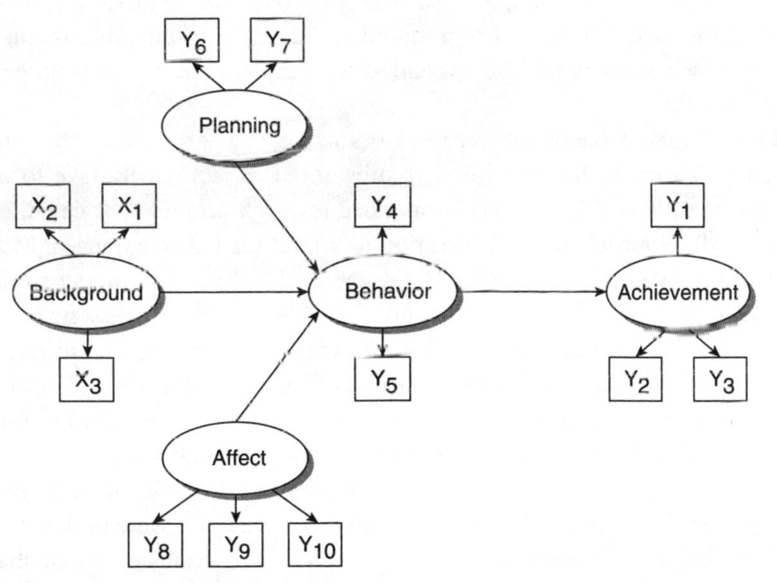

Figure 1.3 Example of Statistical Model

Completing the Tasks: The Proposal and the Report

As we have indicated in earlier sections of this chapter, different universities and funding sources have different requirements for completing the tasks discussed above. Some universities require a short pre-proposal, a prospectus, prior to completing a full proposal; others require a full proposal; and still others leave it up to the college or department to determine what is acceptable. We strongly urge you to get the documents that govern the requirements to which you will be held—and read them as early as possible in this process. There is no need to create an initial structure for your proposal and then convert to the one for which you will be held accountable.

We also think it is to your advantage to think ahead about how you will—once the data are collected and analyzed—turn your proposal into the dissertation and then into one or more research reports that can be submitted to academic journals. Not too many years ago, on most campuses, there was little latitude in how the dissertation was organized. Typically, the completed dissertation had five or six chapters organized as follows: (a) an introductory chapter with introduction to the study, the purpose statement, the rationale, the questions or hypotheses, and, in some cases, limitations, delimitations, and definitions; (b) the review of the literature (i.e., the background); (c) method; (d) results; and (e) discussion and conclusions (the latter may be divided into separate chapters). Ideally, when this format is used, the first three chapters of the proposal—introduction, literature review, and method—would be updated and then the results, discussion, and conclusions simply would be appended to complete the dissertation or thesis document.

One of the problems inherent in organizing a dissertation in the five- or six-chapter format, however, is that substantial revisions will have to occur to turn the dissertation into one or more research articles that can then be submitted to journals for consideration as a published article (Jensen, Martin, & Mann, 2003; O'Brien, 1995; West, 1992). We have seen many students, our own included, who have taken months or years to turn a dissertation into an article. Others, faced by the substantial task of preparing an entirely new document for publication, have not been willing to revisit their dissertation to take this next step. In some cases the result of that reticence must be counted as a genuine loss to both the author and the body of knowledge.

The purpose of this book is to help you navigate the tasks of planning and executing a dissertation. Beyond that, however, we have come to believe that early planning can both expedite and encourage the vital process of sharing what is learned. Getting the task done and graduating with a degree is the first priority, but to stop there is to leave yourself unfinished.

Many universities now permit—and some even encourage—dissertations that deviate from the traditional format. As with proposal regulations, we believe you should understand what options are available to you and discuss them with you advisor early in the process. If there is a format available that will expedite turning your dissertation into research articles, we urge you to give it close consideration.

Table 1.1 shows how the proposal can be converted to a traditional dissertation and then to a format that makes it easier to revise the dissertation into research articles. (The particular format used here is only one among a number now in growing use in higher education.) In this alternative configuration the first chapter or section is a general introduction that sets up the complete study. The next chapter or chapters are potential articles representing individual parts of the study, in this case reports containing the review of literature, the design and methodology, and the findings. The decision as to the actual number of publishable articles would be a function of the particular dissertation or thesis, as well as the availability of appropriate venues for dissemination.

Following the chapters that appear in the form of individual articles, the dissertation would present a general discussion and conclusions, followed by references and appendices. Where the content of articles does not include all of the material that would be essential to the proposal, as, for example, might be the case for a review of literature, a complete version simply can be placed in an appendix.

The article chapters, with one last edit and the addition of references, can be quickly converted to article format and submitted for publication.

Table 1.1 Moving From the Proposal to the Dissertation

Chapter or Section	*Dissertation Proposal*	*Traditional Dissertation*	*Dissertation Prepared to Facilitate Publishing*
1	Introduction Introduction Purpose Rationale Questions or hypotheses Limitations/delimitations/ definitions (if included as separate sections)	Introduction Introduction Purpose Rationale Questions or hypotheses Limitations/delimitations/ definitions (if included as separate sections)	Introduction Introduction Purpose Rationale Questions or hypotheses Limitations/delimitations/ definitions (if included as separate sections)
2	Review of Literature (complete)	Review of Literature (complete)	Article 1[1]
3	Method	Method	Article 2
4		Results	Article 3
5		Discussion and Conclusions	
6	References	References	References
7	Appendices—including some or all of the information found in the list on page 20.	Appendices—including some or all of the information found in the list on page 20.	Appendices—including a *complete review of literature* and some or all of the information found in the list on page 20.

[1] Complex designs may involve a set of related inquires, each of which represents a legitimate and reportable research investigation. Thus, there may be one or several articles, the number depending on the nature of the study. Each article consists of material drawn from the seven chapters or sections and each constitutes a chapter in the dissertation document. In some cases, the review of literature also may be included as one of the publishable reports.

2

Doing the Right Thing

"The Habit of Truth"

The foundation of scholarship as a collective human enterprise is neither intellect nor technical skill. It is simple honesty. If scholars did not have what Jacob Bronowski (1965) called "the habit of truth," there could be no accumulation of reliable knowledge, and thus no science. The rules for this habit of conscience are absolute: no compromises, no evasions, no shortcuts, no excuses, and no saving face. Planning, conducting, and reporting research make sense only so long as the social contract among scholars is honored—*everyone tells the truth as well as he or she can know it.*

Certainly, there is no reason to believe that researchers are paragons of virtue. Nor should we expect that by some magic of nature or nurture they have been exempted from the human frailties of temptation. Pride, selfishness, greed, sloth, and vindictiveness are possibilities for them as much as for anyone else, and the opportunities to succumb are fully as plentiful in the course of inquiry as they are in any other human undertaking.

Why the habit of truth is sustained in any individual scholar is a complex, personal matter. Intellectual integrity, like love, may defy complete analysis. Of this much we can be certain. Researchers understand that there is no way to cheat "just a little." Any dishonesty, large or small, corrodes and contaminates both the process and the products of inquiry—doing harm to guilty and innocent alike by denying truth to all. To this we add another fact that should be noted by anyone proposing to do research: To be discovered, to be caught at cheating in scholarship, brings swift and awful consequences. In this game, the spoilsport is never forgiven.

There may once have been a time when science, as the avocation of people with the leisure to pursue their curiosity, did not invite unethical conduct, but our world has changed, and so has research as a social process. In business, the defense industry, medicine, and the university, research often is a deadly serious contest, played for big stakes and fraught with the tension of competition.

In commercial laboratories with astonishing budgets, large staffs of investigators race around the clock to beat other research teams to lucrative patents and new contracts. In university departments, professors struggle to attract external funding, keep the favor of the institution's managers, and court the possibility of academic honors. Meanwhile, young assistant professors grind out publications in a sometimes desperate pursuit of tenure and promotion. Everywhere, graduate students hurry to complete dissertations so they can win a faculty appointment and rise at last above the enforced penury of teaching assistantships. Anyone who does not see fertile ground for unethical conduct in all of that simply does not understand human nature.

As with any social statistic, it is difficult to know whether the incidence of scientific cheating has undergone an increase that is truly out of proportion to the growth of the research industry itself. What is certain, however, is that we now hear and read regularly about senior investigators, assistants, and graduate students who have been caught doctoring the data, plagiarizing information, or otherwise violating the canons of good scholarship by failing to tell the truth about their work (Altman & Broad, 2005; Bell, 1992; Payne, 2005; Safrit, 1993; Swazey, Anderson, & Lewis, 1993; Taubes, 1995). In their haste to complete the project and reap the rewards, they have broken the fragile habit of truth.

In some cases, of course, that habit never was adequately established. In the preparation of researchers, many universities fail to socialize students into the ethical norms of honest scholarship with the same thoroughness and rigor employed to induct them into the mysteries of sound data analysis. Courses in the philosophy of science and seminars on ethical issues in scholarship have been crowded out of the curriculum. Professors, busy in the frantic pace of their own studies, fail to give time to mentoring students in the complexities of what is acceptable and unacceptable behavior among scholars. We believe that the academic community will be living with the consequences of that neglect for many years to come.

This chapter is no substitute for thorough, on-the-job indoctrination into the ethical standards of inquiry in your field. What we can accomplish here is only to flag with warnings those areas in which the proposal writer is likely to encounter problems related to the ethics of scholarship. One set of those issues attends the process itself (acquiring the cooperation of subjects, gaining access to research sites, gathering data, writing reports, and publishing results). A second category contains ethical problems encountered in relationships, personal and professional, with mentors (notably in the case of professors and graduate students) and colleagues.

Throughout the chapter, we will note sources for learning more about the topic of ethics in research. We urge, however, that you use every opportunity to initiate informal discussions about ethics with active researchers, that you seek out courses and seminars dealing with the moral dimensions of scholarship, and that you attend symposia and workshops on ethical issues whenever they are offered at research conferences in your own field. You also may wish to do some self-study; the cases presented by Penslar (1995) and Macrina (2005) are good places to start. In scholarship, as elsewhere, proper behavior is shaped less by good intentions and far more by healthy doses of forewarning and forearming.

Ethics and the Research Process

What constitutes ethical behavior in research, as elsewhere in life, is the product of social definition. As a social construct, it is relative to a particular place, time, and set of people. In the instance of research-based scholarship (what is commonly called "science"), however, there is an important difference. Because science must be cumulative and self-correcting over time, and uniform across cultures (and national boundaries), standards for ethical behavior have evolved that constitute a widely agreed-to social contract among the participants.

In North America, for example, members of the National Academy of Sciences (NAS) have defined what constitutes scientific misconduct (breaches of ethical behavior) in explicit detail (Panel on Scientific Responsibility and the Conduct of Research, 1992). These have been divided into three broad categories (not all the individual items deal directly with research activity): (a) scientific misconduct (e.g., falsifying data, plagiarism), (b) questionable research procedures (e.g., keeping inadequate records, careless data collection), and (c) other misconduct (e.g., violation of government regulations, sexual harassment in the workplace).

Our own experience as well as the results from studies and reports (Altman & Broad, 2005; Payne, 2005; Swazey et al., 1993) confirm that all the NAS categories of ethical concern deal with problems that have a very real presence in the places where research is done. In the particular context of planning and proposing a study, however, we have found that inadequate provision for the protection of human subjects is among the most common defects in research procedures that fail the test of ethical behavior. In part, this arises from the fact that the risks to participants are not always obvious. It also is true that devising safe, equitable, and respectful treatment for the people who make themselves available for study can sometimes be difficult. Nevertheless, observance of the social contract for ethical research begins with the proposal. That ethical failures as often involve acts of omission as acts of commission within that document makes them no less dangerous—to both people and the enterprise of scholarship.

The Protection of Human Subjects

All universities that receive funding from a federal agency must have in place a system of mandatory review of all research proposals that involve the use of human subjects. The purpose of the review is to ensure protection of the participants' health and welfare. The process may operate at one or several levels, in some cases functioning through a department or college committee and in others by means of a university board (in some institutions, both are employed). Whatever the structure of review procedures, most institutions have published guidelines that lay out the process for obtaining approval and specify the ethical issues to be addressed in the proposal. We strongly recommend that you obtain that document and digest its contents *before* beginning to prepare your proposal. The humane treatment of human beings is far easier to build in than it is to add on!

In addition to the institutional review of protection given human participants, some research journals now have ethical guidelines that must be met if a manuscript is to be considered for review. Professional and scholarly organizations have published standards that should govern research conducted by members. In all of these, attention is given to how the researcher must think and act concerning the people who make themselves available for scrutiny. We believe this body of writing reflects a vital dialogue to which every novice researcher should be a witness—and in which, over the course of their careers, they may become participants.

The proposal for a thesis or a dissertation is not intended just to prepare graduate students to undertake and finish a study. For all, it is basic instruction about what science must be if it is to serve us well and wisely, and for some it is the first rite of initiation into a career of scholarship. Accordingly, all students should invest time in learning and thinking about human rights and the ways in which they become entangled with research.

If you have no convenient vehicle for that exercise, a valuable first step can be achieved by reading *Ethical Standards of the American Educational Research Association* (American Educational Research Association, 1992), *Ethical Principles of Psychologists and Code of Conduct* (American Psychological Association, 2002), *Ethics in Research With Human Participants,* published by the American Psychological Association (Sales & Folkman, 2000), and *On Being A Scientist* (1995), prepared by the Committee on Science, Engineering, and Public Policy, National Academy of Science, National Academy of Engineering, and Institute of Medicine for students about to begin their first research study. Also available are a number of books on the topic of ethics in research (for example, Farrell, 2005; Kimmel, 1988; Oliver, 2003; Reece & Siegel, 1986, Sieber, 1992; Simons & Usher, 2000; Stanley, Sieber, & Melton, 1996; and Zeni, 2001). Finally, for vivid examples of the complex tensions between our desire to protect participants against unethical use in research investigations and our need to preserve the integrity of research designs, we suggest a review of the problem by Kroll (1993).

It is easy for the novice to feel overwhelmed by all the responsibilities that seem to come with the direct study of other human beings. It is true that in recent years the domain of concerns about the health and welfare of participants has expanded. In large part, this reflects the fact that a greater volume and diversity of research activity has necessitated a wider definition of what constitutes ethical behavior for the investigator and reasonable protection for the subjects of investigation. Indeed, even at a great distance from the physical intrusions required to study human physiology, there is much to consider in writing that required proposal section on "Protection of Human Subjects."

An Ethical Benchmark

In the thickets of ethical complexity faced by the novice researcher, it sometimes is difficult to keep one's moral bearings. Our suggestion is to keep sight of something basic in your value system and use it as a benchmark to test each decision. For us, the starting place is simpler and far more encompassing than any of the now traditional concerns about the physical and psychological safety of participants. We believe that the right to protection begins with the right of free and informed choice.

Every human has the right not to be used by other people. That means the investigator's need for a human source of data is always subordinate to the other person's right to decide whether to provide it or not. The right not to be used applies with equal force to fifth graders, college sophomores, trash collectors, professional athletes, and residents in retirement communities. People who are asked to participate in a study have a right to know what they are getting into and the right to give or withhold their cooperation on the basis of that information.

The Self-Interest of Respect

The fact that an investigator uses volunteers responding to a newspaper advertisement, or paid subjects from a prison population, does not alter the participant's right not to be used as chattel. Research workers have both a responsibility and a special interest in protecting that right. That the rule of informed consent has so often been ignored or compromised in the past accounts in large measure for the difficulties, both obvious and subtle, in obtaining cooperation from prospective participants. Subjects do not always miss (or choose to ignore) the fact that they have suffered the disrespect of having been used by a researcher. They may remember such violation of their person for a long time and on the next occasion may protect themselves by not cooperating. Worse still, they may take a measure of revenge by finding a way to make mischief in the study.

The Danger of Seeming Innocent

What constitutes an unethical use of people is not always clearly established. Social scientists, for example, disagree on the degree to which the use of psychometric instruments, questionnaires, and survey interviews should be circumscribed by procedures to protect respondents' right to informed consent. At the least, this is an issue to be discussed with advisors during development of a proposal that involves such methods.

Our own position on such questions is unequivocal. Concern for the rights of participants should accompany the use of research tools such as questionnaires and interviews just as it does any other form of data collection. The procedures used to protect the participant's rights may be much less elaborate than those used in an experiment involving physical discomfort or some degree of risk, but they should nonetheless be designed with care and applied with scrupulous uniformity.

It is the ubiquity of questionnaires, the conversational friendliness of interviews, the seeming innocuousness of a Likert-type scale, and the innocence of jotting field notes in social settings that present the greatest danger. Where hazards may be obvious, as for example in medical studies, careful attention to subject protection is the norm. Where risks are less obvious, and where they attach to intangibles such as "the right to full disclosure," the need for protection may be brushed aside as a mere formality, and abuse itself then becomes the insidious norm.

Each time people are treated as though they have no right to the privacy of their bodies, their thoughts, or their actions, the implication is that such treatment is right and acceptable. It is not. In contrast, when an investigator treats entry into the private world of a participant as a special privilege, granted by a fellow being as an act of informed cooperation, the opposite instruction is given—and it will be learned. From that lesson, we all profit.

A Protocol for Informed Consent

At minimum, you should consider the following standards for any proposed study that involves the use of human subjects.

1. Participants should be informed of the general nature of the investigation and, within reasonable limits, of their role in terms of time and effort. A script may be used when such information is transmitted orally. (However this is accomplished, a complete, word-for-word account of the information provided should be placed in the appendix of the proposal.)

2. Participants should be informed of procedures used to protect their anonymity. It should be made clear that anonymity cannot be guaranteed. All that should be promised is the use of rigorous procedures for protection.

3. Participants should, after reasonable consideration (including the right to ask questions), sign a document affirming that they have been informed of the nature of the investigation and have consented to give their cooperation. (A copy of this should be included in the proposal.) Many universities now have their own guidelines for informed consent letters. We encourage you to understand the local requirements and look at the examples provided on your campus before designing the informed consent letters for your research.

4. In experimental studies, participants should receive an explanation of all treatment procedures to be used and any discomforts or risks involved. If risks, physical or psychological, are more than minimal, procedures that will be taken to protect the well-being of the subject should be fully explained.

5. Participants should be told what benefits they will receive by participating in the study and what alternative benefits are being made available to other subjects in the study.

6. Participants should be explicitly instructed that they are free to withdraw their consent and to discontinue participation in the study at any time. In some circumstances it is essential that they also be assured that no reprisal will attend such a decision. (Complicated issues, such as setting a time limit

on this right or establishing the right to withdraw data already provided if cooperation is withdrawn, require careful attention in the proposal.)

7. Participants should be provided with the name of the person responsible for the study, to whom they can direct questions related to their role or any consequence of their participation. In the case of research conducted by graduate students, this should include the name, address, and telephone number of both the student and the supervising professor.

8. Participants should be offered the opportunity to receive feedback about the results of the study. This should be in a form that is appropriate to the needs and interests of the participant.

The Reactive Effects of Honesty

Arguments to the effect that procedures for informed consent introduce unknown bias effects are, for the most part, spurious. All contacts between investigators and participants hold the potential for generating unknown effects. The procedures for ensuring the rights of subjects do no more to produce response biases than any other interaction. On the other hand, respectful observance of these protections, quite aside from their ethical import, can exercise a measure of steadying influence over the most fundamental (and sometimes the most capricious) of all subject-related variables—the participant's cooperation.

Omission of Information

If it seems essential to withhold some specific item of information from participants, then the omission may be considered. This can be done, however, only after a careful search for any harm or substantial disadvantage that might occur. The use of placebo treatments and control groups sometimes involves exactly this form of limitation on full disclosure. The fact that such procedures are commonplace, however, does not make them less ethically problematic.

The seriousness of any decision to limit what the participant knows is underscored by the fact that deception in the form of giving false information about the nature of the study or the risks of participation (actual deception) is absolutely prohibited in every field of inquiry. As everyone knows, the boundary between sins of omission and commission is dangerously thin!

When participants are not given complete disclosure, they should know, in virtually every instance, that this is the case before making a decision about joining the study. When any information is withheld from the participants, the proposal must present explicit detail about how and when they will be debriefed and how any disadvantage created by the omission will be rectified.

Cooperation by Coercion

There is a spirit as well as a letter to be observed in rules about informed consent. When individuals are under the supervision or control of another individual, it may be difficult for them to refuse an invitation to participate in a study. In a sense, it may truly be impossible for them to freely give their cooperation. Thus, a graduate student may be loath to refuse an advisor's request to serve as a subject, or an undergraduate may feel reluctant to decline a teaching assistant's suggestion that he or she become a participant in a dissertation study. Secretaries and classified staff may be too intimidated to refuse a chairperson's invitation. Any of these individuals may sign the consent form and participate but inwardly feel resentful. People's rights are violated just as much by coercion that is unintended as by that which is so designed.

Many researchers feel that it is good ethical practice not to use students who are in their classes or under their supervision. Some institutions now have regulations that forbid such practice. It may be wise to use the media to solicit volunteers or, in the case of students, to invite only those from other departments or from places outside the purview of your authority.

Reciprocity

If participants give cooperation, time, effort, and access to what is by right theirs to control (not least of all their privacy), what does the researcher give in return? Once that question is allowed legitimacy, satisfying answers may prove more scarce than the novice might expect. It is difficult to establish what is proper compensation for serving as a study participant. In any case, few researchers could afford the purchase price for what their subjects contribute. In some measure, serving as a participant in a research study always has to take the form of a gift, which makes the demand that the gift be freely given and fully acknowledged all the more important.

Among the means of reciprocity are small symbols that show appreciation: a handwritten note, a free lecture or question-and-answer session on a topic of interest to the participants, or even just a sincere "thank you!" Pride also is its own reward when participants feel that they have contributed to the search for knowledge (or to finding solutions for practical problems), though that wears thin when the required investment of time or effort is large. Qualitative researchers often find that giving a respectful, nonjudgmental ear to participants' accounts of their experiences can have important value as a return for cooperation. An opportunity to satisfy curiosity, or for participants to learn something about themselves (or even about research), also may have some place in an honest exchange. In the end, however, the researcher always is left to answer two hard questions: For whose benefit is this work being done, and where are the interests of my subject in the calculus by which I find the answer?

The Ethics of Writing

Plagiarism

Ideas, and the words that express those ideas, are the valuables of scholarship. The people who create them have a right to receive credit for them. When one person appropriates the ideas or text of another person and presents them as his or her own, it is theft. The technical term is plagiarism, but it is theft, and the ethics of scholarship make no provision for petty crime—it is all grand theft.

Yes, gray areas exist where ideas, phrases, and even entire ways of thinking about things pass into the public domain where acknowledging their origin no longer is obligatory. And yes, the distinction between a paraphrase in your words and one that derives too much from the original author is a judgment call. For problems of that kind, we leave you to other resources—and your conscience.

What we are talking about here is thievery, and that includes failure to use quotation marks where they belong, omitting citations that credit material found in someone else's work, carelessness in preparing the list of references, and failure to obtain permission for the use of figures, tables, or even illustrations from another document—whether published or not. In each of those, doing the right thing does not involve knowing the niceties of custom or the precise reading of an obscure rule. You know exactly what is ethical without any coaching. All you need to remember is what your mother said: "*Don't cheat!*" If you need more motivation than that, you might consider the proposition that it is prudent to treat the property of others as you hope they will treat yours.

Providing All the Facts

The purpose of a proposal, you will remember, is to help other people understand what you plan to do. To accomplish that, they need all the relevant facts, not just those favorable to your inclinations. Deliberate omission of information hostile to some part of your proposal is dishonest. Accidental omission of important information (that includes simply not having learned it) is incompetence. Either way, an incomplete account defeats one of the main purposes in writing the proposal—to provide the basis for obtaining good advice.

Manufacturing the Facts

Once the study is under way, frequent opportunities will occur to fabricate data (creating it rather than collecting it) and falsify data (tampering with it to make it appear to be what it is not). Although obviously a violation of ethical practice, it is surprising how reasonable such fraud can seem. If you are collecting data on a subject over five days, and she happens to miss the third day, and your final record sheet shows scores of 10, 20, (?), 40, and 50, what is so unreasonable about inserting the 30 *except that it is a lie?*

Temptations like that are not rare in research. In fact, such situations are more the rule than the exception. In the rush to finish before a deadline, how reasonable will a small lie seem to you? In your response will reside the answer to another question: What kind of scholar do you want to become? (In case you wondered, in many forms of data analysis there are appropriate ways to deal with missing data. All you need to do is use the correct procedure and then be honest—describing in your report exactly how data were handled.)

Inappropriate use of data is no different from making it up. Using raw scores when you know they should be transformed before analysis is just another kind of falsification. Data dredging—repeatedly changing your analysis when results do not come out as anticipated—is another way to fudge. Use the data you have, use it correctly, and tell the reader exactly what you did. With those simple rules, it is hard to go wrong.

Other falsehoods do not directly involve the data. For example, it may be tempting to describe testing procedures as having more rigor than you actually were able to maintain or to report taking the precautions against contamination that you had proposed, when, in fact, it proved impossible to do so. It may be easy to allow the reader to infer that a standard protocol was used unchanged, when, in fact, it had to be modified, or to give the impression that your selection of observation sites was random, when, in fact, you had to pick the ones closest to your office. All are breaches of ethics, all undercut the value of your research, and all serve to create an insidious pollution in the atmosphere of science.

Discussion of your findings provides one last temptation to tilt the facts in a way that favors your personal interests but falls short of an honest accounting. The researcher has a perfect right to emphasize the results that seem most important. That power is abused, however, when disconfirming data are left out of the discussion, or are included without a clear notation of their implications.

Graduate students can be particularly vulnerable to the natural attraction of selective reporting. There often is subtle pressure to present the results that advisors expect. If having to find a way to deal with equivocal or counterintuitive findings is likely to delay graduation, the temptation to simply "not have them" grows ever more attractive. The choice to do the right thing can be inconvenient, expensive, and even painful. No experienced investigator would claim that research is a rose garden.

Ethics in Publication

Sending off the manuscript to a journal, mailing your abstract to the review committee for a national conference, finding your name in the table of contents—those are golden moments, and any researcher who no longer finds them so should consider a long vacation. The route to those wonderful moments, however, is strewn with ethical pitfalls for the unwary novice. There is a lot to know about the rules that govern ethics in publication, and much of that is beyond the purview of this handbook.

As an example, in the case of research studies to which a number of people contribute, there are rules (both written and unwritten) about whose names appear on the report, in what order they appear, and who should be acknowledged but not given status as an author. Unfortunately, these rules come in versions established by universities, journals, scholarly organizations, government agencies, laboratories, and even individual professors. The best we can do is to advise that you make it a point to find out the regulations (or local conventions) that apply in your circumstance. In such matters as whose name will appear on publications and who (or what entity) will have ownership of research products, it is prudent that you locate this information *early* in the proposal development process—and for this purpose, formal documents are always more appropriate than informal conversations.

We will note here only three of the many ethical concerns related to publication, one of them because it is among the first and most obvious, and two because they are well-disguised pitfalls that may not be recognized for the dangers they present. First, can you submit your manuscript to more than one journal at the same time? The short answer is: No, don't ever do it. If you have a fertile mind, you will recognize at once that there are dozens of questions that surround that simple rule. For example, can you submit a manuscript to a journal after the abstract for the same study was published in a conference proceedings? (The answer is yes.) For each of these subsidiary problems, you will have to find what is considered ethical in the particular situation. Just do not forget the basic rule—duplicate submissions are absolutely prohibited, and the event of duplicate publication brings swift and draconian consequences.

A second temptation will come early in your study, and it is that timing that gives it both great attraction and serious danger. Almost every researcher we know has been tempted to submit an abstract to a prestigious conference (to be held well after the proposed conclusion of the study), when the analysis has not been finished, or even when all the data are not yet in hand. The motive is understandable. Researchers are excited about their work and always eager to join the great conversation that represents scholarship in their field. Writing up an abstract that is incomplete or that projects anticipated results seems a minor transgression, but one thing can lead to another. What if the final analysis contradicts your confident projection? What if the work slows down and still is incomplete when the conference date arrives? Among the embarrassing alternatives are withdrawing the paper or making a public confession. Worse, however, there will be temptation into serious violation of ethics, such as covering your tracks by faking results or downplaying findings that do not harmonize with the earlier abstract. Cooking up an abstract so you can reserve a place on the program is risky business. Our advice is to do what you know is the right thing—just wait until next year.

A third pitfall lies waiting beyond presentations and journal articles, in a seemingly risk-free document—your résumé. Most beginners in academe do a bit of padding by including fairly minor accomplishments that later will be omitted as publication lists begin to lengthen, but this practice has limitations and points at which it becomes dishonesty rather than innocent enthusiasm. Employers and promotion committees have become increasingly restive about the kind of unethical padding that has proliferated in academic résumés. Among the truly dishonest inclusions are abstracts from presentations that are given a title different from an ensuing published report—so that the two appear to be different studies when they are not. Another is listing research articles or grant proposals in a way that allows the reader to assume you were a coinvestigator, when in fact you contributed only a minor part to the submission. Contributions to the "Brief Research Notes" sections of journals are found displayed as though they were major achievements.

All such forms of padding are not innocent enthusiasm; they are dishonesty. Happily, a simple procedure can be adopted that will keep you safe from any complaint about your résumé—and still allow you to take credit for everything. Put your accomplishments in categories headed by labels that clearly describe what each contains, and cross-reference where citations represent the same piece of work.

Personal and Professional Relationships

A very fine distinction separates violations of research ethics (which include most of what has been discussed to this point) and ethical misconduct. For a helpful discussion that defines categories in the domain of ethics, see volume 1 of *Responsible Science,* a publication prepared by the Panel on Scientific Responsibility and the Conduct of Research (1992). It is the latter category of misconduct, which includes the ethics of how scholars treat one another in their personal and professional relationships, to which we will now attend.

Relationships With Faculty Mentors

As has been described in a number of publications (Guston, 1993; National Academy of Science, National Academy of Engineering, and Institute of Medicine, 1997; Roberts, 1993), in the ideal situation, the relationship between graduate student and advisor gradually matures until at the completion of the degree the student and professor have become colleagues. Unfortunately, what is ideal is not always real. Some dysfunctions in professor/student relationships are no more than the collision of personality styles, the abrasion of conflicting intellectual interests, or the tensions of differing personal politics. Those are painful, and even harmful if not addressed, but they do not necessarily involve unethical behavior by anyone. Certain types of interpersonal behavior, however, are not acceptable because they are dangerous, or violations of what is equitable, or both.

First among these are instances in which pressure is put on a student to go beyond the expected academic or professional relationship. In our judgment, any attempt by a faculty member to coerce a student into an expanded relationship is inappropriate, and most universities now have policies that specifically address relationships between students and faculty members.

Perhaps not obvious as a form of ethical misconduct is pressure from faculty that urges students to perform tasks that are not part of the academic program—and that they do not wish to do. These subtle abuses include requesting errand service, assigning attendance at social functions, and even the expectation of jogging together. In most collegial relationships, such activities are normal. People who work closely together often exchange favors and socialize. The problem is that the relationship between professors and students is not truly collegial. There is an enormous imbalance of power. In the most innocent and well-intended invitations, students often find the hidden force of coercion.

The point at which inappropriate pressures shade into reasonable requirements is a gray area. An important part of any graduate program is gaining research experience by hands-on engagement. Working with faculty members on their projects can be a valuable source of such training. There are limits to be observed, however, and care needs to be taken so that apprenticeship does not become servitude in disguise.

When the task performed is so narrow as to yield no new skill or knowledge, when the student's own work has to be unreasonably delayed, and when the topic of inquiry is not remotely related to the student's specialization, the line between education and abuse has been crossed. By then, it may be too late to extricate yourself comfortably. For that reason, we suggest prior discussion of all such research opportunities with other faculty advisors, the prospective research mentor, and other graduate students.

Much more serious examples of abusing the faculty/student imbalance of power are all the forms of sexual harassment. The form taken may range from subtle comments to direct demands for sex. Although it is obvious that what one person considers offensive another person may regard as repartee or humor, that understanding does not make remarks with sexual connotations any more acceptable or any less serious as breaches of ethical conduct. Students have an absolute right to be free of such discomforts. Subtle or glaring, intended or inadvertent, sexual harassment is wrong and not to be tolerated.

If you encounter such problems, we suggest confronting them immediately, rather than allowing them to fester. Some faculty members may not understand that you find something troubling, and a private discussion may set everything straight. Repeated incidents or any invitation or pressure for a sexual relationship should be brought to the attention of someone you can trust—another faculty member, the department chair, the dean, or perhaps the best choice of all, the university ombudsperson.

It is to everyone's advantage that student/professor relationships stay at the formal level of teacher and pupil. If you want a mentor/mentee relationship to turn into something more, we urge that you acquire a new mentor before allowing that to happen. Human relationships are complex, but those that attempt to bridge stations of unequal power are doubly so.

Where goodwill reigns and a modicum of sensitivity is the norm, the relationship between professor and student is not a mine field of hidden risks; it is the place for a rare and invaluable kind of human exchange. Given even reasonable nurture and protection, it will flourish and grow—into collegiality.

Discovery and Obligation

What if it really happens? What if you discover or have strong reason to suspect unethical behavior on the part of a colleague or faculty member? Our first advice is to think carefully, very carefully, about how to protect yourself from involvement or reprisal. Also consider the terrible consequences of an accusation that proves false. Whether or not you take action will be a matter of conscience, and your conscience should tell you that cheating in research can never be condoned or tolerated. What course of action you choose, however, will be a function of circumstance, prudence, and courage.

Talking with the person(s) involved is one course of action. If you judge that to be unreasonable, then you must find other options. In most universities, appropriate authorities are available, and again the ombudsperson will know who they are. Any institution that receives federal research funds will have procedures for investigating violations of research ethics. In some instances, there are provisions to protect the "whistle-blower" (Miceli & Near, 1992), though that is not always possible.

In the end, anyone accused of a violation has a proper claim to due process, and even well-intentioned reports of ethical violations that cannot be sustained when investigated will rebound to the discomfort (and often disadvantage) of the accuser (for an interesting account see Sprague, 1998). Again, take the time to review the pages of *Responsible Science* (Panel on Scientific Responsibility and the Conduct of Research, 1992) before you act. Doing the right thing may not be easy, but at least it can be done the right way.

None of us can decline to take responsibility. Every study is part of a shared enterprise—the work of scholarship. When anyone does not do what is right, everyone is harmed.

3

Developing the Thesis or Dissertation Proposal

Some Common Problems

The general purposes and broad format of the proposal document have now been presented. There remain, however, a number of particular points that cause a disproportionate amount of difficulty in preparing proposals for student-conducted research. In some cases, the problems arise because of real difficulty in the subtle and complex nature of the writing task. In other cases, however, the problems are a consequence of confusion, conflicting opinions, and ambiguous standards among research workers themselves and, more particularly, among university research advisors.

As with many tasks involving an element of art, it is possible to establish a few general rules to which most practitioners subscribe. Success in terms of real mastery, however, lies not in knowing, or even following, the rules but in what the student learns to do within the rules.

Each student will discover his or her own set of special problems. Some will be solved only through practice and the accumulation of experience. While wrestling with the frustrations of preparing a proposal, you should try to remember that the real fascination of research lies in its problematic nature, in the search for serviceable hypotheses, in selecting sensitive means of analyzing data, and in the creative tasks of study design.

Some of the problems graduate students face cannot be solved simply by reading about them. What follows, however, is an effort to alert you to the most common pitfalls, to provide some general suggestions for resolution of the problems, and to sound one encouraging note: consultation with colleagues and advisors, patience with the often slow process of "figuring out," and scrupulous care in writing will overcome or circumvent most of the problems encountered in preparing a research proposal. In the midst of difficulty, it is useful to remember that problems are better encountered when developing the proposal than when facing a deadline for a final copy of the report.

The problems have been grouped into two broad sections: "Before the Proposal: First Things First" and "The Sequence of Proposing: From Selecting a Topic to Forming a Committee." Each section contains a number of specific issues that may confront the student researcher and provides some rules of thumb for use in avoiding or resolving the attendant difficulties. You should skim through the two sections selectively, because not all the discussions will be relevant to your needs. Chapter 4 ("Content of the Proposal: Important Considerations"), Chapter 6 ("Style and Form in Writing the Proposal"), and Chapter 7 ("The Oral Presentation") deal with specific technical problems and should be consulted after completing a review of what follows here.

Before the Proposal: First Things First

Making Your Decision: Do You Really Want to Do It?

The following idealized sequence of events leads to a thesis or dissertation proposal.

1. In the process of completing undergraduate or master's level preparation, the student identifies an area of particular interest in which he or she proposes to concentrate advanced study.
2. The student selects a graduate institution that has a strong reputation for research and teaching in the area of interest.
3. The student identifies an advisor who has published extensively and regularly chairs graduate student research in the area of interest.
4. Based on further study and interaction with the advisor, the student selects and formulates a question or hypothesis as the basis for a thesis or dissertation.

Because we do not live in the best of all possible worlds, few students are able to pursue the steps of this happy and logical sequence. For a variety of reasons, most students have to take at least one of the steps in reverse. Some even find themselves at the end of several semesters of study just beginning to identify a primary area of interest, in an institution that may be less than perfectly appropriate to their needs, and assigned to an advisor who has little or no experience in that particular domain. For this unfortunate state of affairs, we offer no easy solution. We do believe that one significant decision is, or should be, available to the student—the decision to do, or not to do, a research study. Faced with conditions such as those described above, if the option is available, the more rational and educationally profitable course may be to elect not to undertake a research study. You can determine whether this option is available before the school is selected, or at least before the program of study is selected.

There are sound reasons to believe that experience in the conduct of research contributes to graduate education. There also are good and substantial reasons to believe that other kinds of experiences are immeasurably more appropriate and profitable for some students. The question is, "Which experience is right for you?"

If you are, or think you might be, headed for a career in scholarship and higher education, then the decision is clear. The sooner you begin accumulating experience in research activities, the better. If you are genuinely curious about the workings of the research process, interested in combining inquiry with a career of professional service, or fascinated by the problems associated with a particular application of knowledge to practice, again the decision is clear. An experience in research presents at least a viable alternative in your educational plans.

Lacking one of these motives, the decision should swing the other way, toward an option more suited to your needs. Inadequately motivated research tends not to be completed or, worse, is finished in a pedestrian fashion far below the student's real capacity. Even a well-executed thesis or dissertation may exert a powerful negative influence on the graduate experience when it has not been accepted by the student as a reasonable and desirable task.

One problem touches everyone in graduate education, faculty and students alike—the hard constraints of time. Students want to finish their degree programs in a reasonable period of time. The disposition or circumstances of some, however, may define reasonable time as "the shortest possible time." Others find the thought of any extension beyond the standard number of semesters a serious threat to their sense of adequacy. For students such as these, a thesis or dissertation is a risky venture.

Relatively few research studies finish on schedule, and time requirements invariably are underestimated. Frequent setbacks are almost inevitable. This is one aspect of the research process that is learned during the research experience: Haste in research is lethal to both quality of the product and worth of the experience. If you cannot spend the time, deciding to initiate a research project endangers the area of inquiry, your advisor, your institution, your education, your reputation, and any satisfaction you might take in completing the task. In short, if you can't afford the time, then don't do it at all.

Choosing Your Turf: Advisors and Areas

Once a firm decision has been made to write a thesis or dissertation, the choice of an advisor presents a less difficult problem. Here, area of interest dictates selection because it is essential to have an advisor who is knowledgeable. Further, it always is preferable to have one who is actively publishing in the domain of interest.

Competent advisement is so important that a degree of student flexibility may be required. It is far better for students to adjust their long-range goals than to attempt research on a topic with which their advisor is completely unfamiliar. It may be necessary for the thesis or dissertation to be part of the advisor's own research program. As long as the topic remains within the broad areas of student interest, however, it is possible to gain vital experience in formulating questions, designing studies, and applying the technology and methods of inquiry that are generic to the domain.

It is desirable for student and advisor to interact throughout the development of the proposal, beginning with the initial selection and formulation of the question. On occasion, however, the student may bring an early stage proposal to a prospective advisor as a test of his or her interest or to encourage acceptance of formal appointment as advisor. Experience suggests that this strategy is most likely to produce immediate results if the proposal is in the primary interest area of the advisor. If the proposal involves replication of some aspect of the advisor's previous research, the student may be amazed at the intensity of attention this attracts.

Finding Your Question: What Don't We Know That Matters?

Before launching into the process of identifying a suitable topic for inquiry, we suggest a short course of semantic and conceptual hygiene. The purpose of this small therapy is to establish a simple and reliable set of terms for thinking through what can sometimes be a difficult and lengthy problem—what do I study?

All research emerges from a perceived problem, some unsatisfactory situation in the world that we want to confront. Sometimes the difficulty rests simply in the fact that we don't understand how things work and have the human itch to know. At other times, we are confronted by decisions or the need for action when the alternatives or consequences are unclear. Such perceived problems are experienced as a disequilibrium, a dissonance in our cognition. Notice, however, they do not exist out in the world, but in our minds.

That may sound at first like one of those "nice points" of which academics are sometimes fond, but for the purposes of a novice researcher, locating the problem in the right place and setting up your understanding of exactly what is unsatisfactory may represent much more than an arbitrary exercise. Thinking clearly about problems, questions, hypotheses, and research purposes can prevent mental logjams that sometimes block or delay clear identification of what is to be investigated.

The novice will encounter research reports, proposals, and even some well-regarded textbooks that freely interchange the words "problem" and "question" in ways that create all sorts of logical confusion (as in "The question in this study is to investigate the problem of . . ." or "The problem in this study is to investigate the question of . . ."). The problem is located alternately in the world or in the study, the distinction between problems and questions is unclear, and what is unsatisfactory in the situation is not set up as a clear target for inquiry.

We suggest that you be more careful as you think through the question of what to study. Define your terms from the start and stick with them, at least until they prove not to be helpful. The definitions we prefer are arbitrary, but it has been our experience that making such distinctions is a useful habit of mind. Accordingly, we suggest that you use the following lexicon as you think and begin to write about your problem.

> *Problem*—the experience we have when an unsatisfactory situation is encountered. Once carefully defined, it is that situation, with all the attendant questions it may raise, that can become the target for a proposed study. Your proposal, then, will not lay out a plan to study the problem but will address one or several of the questions that explicate what you have found "problematic" about the situation. Note that in this context neither situation nor problem is limited to a pragmatic definition. The observation that two theories contradict each other can be experienced as a problem, and a research question may be posed to address the conflict.

> *Question*—a statement of what you wish to know about some unsatisfactory situation, as in the following: What is the relation between . . . ? Which is the quickest way to . . . ? What would happen if . . . ? What is the location of . . . ? "What is the perspective of . . . ? As explained below, when cast in a precise, answerable form, one or several of these questions will become the mainspring for your study—the formal research question.

Purpose—the explicit intention of the investigator to accumulate data in such a way as to answer the research question posed as the focus for the study. The word "objective" is a reasonable synonym here. Although only people can have intentions, it is common to invest our research design with purpose (as in "The purpose of this study is to determine the mechanism through which . . . ").

Hypothesis—an affirmation about the nature of some situation in the world. A tentative proposition set up as a convenient target for an investigation, a statement to be confirmed or denied in terms of the evidence.

Given this lexicon, the search for a topic becomes the quest for a situation that is sufficiently unsatisfactory to be experienced as a *problem.* The proposal has as its *purpose* the setting up of a research *question* and the establishment of exactly how (and why) the investigator intends to find the answer, thereby eliminating or reducing the experience of finding something problematic about the world. Problems lead to questions, which in turn lead to the purpose of the study and, in some instances, to hypotheses. Table 3.1 shows the question, purpose, and hypotheses for a study. Note that the hypotheses meet the criteria established in Chapter 1 and are the most specific.

The research process, and thus the proposal, begins with a question. Committed to performing a study within a given area of inquiry and allied with an appropriate advisor, students must identify a question that matches their interests as well as the resources and constraints of their situation. Given a theoretically infinite set of possible problems that might be researched, it is no small wonder that many students at first are overwhelmed and frozen into indecision. The "I can't find a problem" syndrome is a common malady among graduate students, but fortunately one that can be cured by time and knowledge.

Research questions emerge from three broad sources: logic, practicality, and accident. In some cases, the investigator's curiosity is directed to a gap in the logical structure of what already is known in the area. In other cases, the investigator responds to the demand for information about the application of knowledge to some practical service. In yet other cases, serendipity operates and the investigator is stimulated by an unexpected observation, often in the context of another study. It is common for several of these factors to operate simultaneously to direct attention to a particular question. Personal circumstance and individual style also tend to dictate the most common source of questions for each researcher. Finally, all the sources depend on a more fundamental and prior factor—thorough knowledge of the area.

Table 3.1 Problem, Question, Purpose, and Hypotheses

Problem—Extensive teacher planning of lessons requires large investments of time and energy, and often must compete with other important responsibilities—both professional and personal.

Question—Is the amount or kind of lesson planning done by teachers positively related to student in-class learning behaviors such as time-on-task?

Purpose—The purpose of this study is to examine the relationships between several categories (types) of teacher lesson planning and student time on-task in a high school automobile mechanics class.

Hypotheses (Note that directional hypotheses are used for Hypotheses 1-3 and that even Hypothesis 4, stated in the null form, could be based on data from a pilot study.)

1. The number of teacher lesson planning decisions that relate to design and use of active learning strategies will be positively related to student time-on-task when those lessons are implemented.
2. The number of class management planning decisions related to particular lesson components will be positively related to student time-on-task when those components are implemented.
3. Teacher lesson planning decisions that require students to wait for the availability of tools or work sites will be negatively related to student time-on-task when those lessons are implemented.
4. The total number of teacher planning decisions (irrespective of category) will not be related to student time-on-task when those lessons are implemented.

It is this latter factor that accounts for the "graduate student syndrome." Only as one grasps the general framework and the specific details of a particular area can unknowns be revealed, fortuitous observations raise questions, and possible applications of knowledge become apparent. Traditional library study is the first step toward the maturity that permits confident selection of a research question. Such study, however, is necessary but not sufficient. In any active area of inquiry, the current knowledge base is not in the library—it is in the invisible college of informal associations among research workers.

The working knowledge base of an area takes the form of unpublished papers, conference speeches, seminar transcripts, memoranda, dissertations in progress, grant applications, personal correspondence, telephone calls, and electronic mail communications, as well as conversations in the corridors of conference centers, restaurants, hotel rooms, and bars. To obtain access to this ephemeral resource, the student must be where the action is.

The best introduction to the current status of a research area is close association with advisors who know the territory and are busy formulating and pursuing their own questions. Conversing with peers, listening to professorial discussions, assisting in research projects, attending lectures and conferences, exchanging papers, and corresponding with faculty or students at other institutions are all ways of capturing the elusive state of the art. In all of these, however, the benefits derived often depend on knowing enough about the area to join the dialogue by asking questions, offering a tangible point for discussion, or raising a point of criticism. In research, as elsewhere, the more you know, the more you can learn.

Although establishing a network of exchange may seem impossible to young students who view themselves as novices and outsiders, it is a happy fact that new recruits generally find a warm welcome within any well-defined area of intensive study. Everyone depends on informal relationships among research colleagues, and this rapport is one source of sustaining excitement and pleasure in the research enterprise. As soon as you can articulate well-formulated ideas about possible problems, your colleagues will be eager to provide comment, critical questions, suggestions, and encouragement.

In the final process of selecting the thesis or dissertation problem, there is one exercise that can serve to clarify the relative significance of competing questions. Most questions can be placed within a general model that displays a sequence of related questions—often in an order determined by logic or practical considerations. Smaller questions are seen to lead to larger and more general questions, methodological questions are seen necessarily to precede substantive questions, and theoretical questions may be found interspersed among purely empirical questions. The following is a much simplified but entirely realistic example of such a sequential model. It begins with an everyday observation and leads through a series of specific and interrelated problems to a high-order question of great significance.

OBSERVATION: Older adults generally take longer than young adults to complete cognitive tasks, but those who are physically active seem to be quicker mentally, especially in tasks that demand behavioral speed.

1. What types of cognitive function might be related to exercise?
2. How can these cognitive functions be measured?
3. What are the effects of habitual exercise on one of these types of cognitive function—reaction time?

4. Are active older adults faster on a simple reaction time task than sedentary older adults?
5. Are active older adults faster on a more complex reaction time task, such as choice reaction time, than older sedentary adults?

QUESTION: What effect does habitual exercise have on choice reaction time in older adults?

By making the twists and turns of speculation visible in the concrete process of sequential listing, previously unnoticed possibilities may be revealed or tentative impressions confirmed. In the simple example given above, the reader may immediately see other questions that could have been inserted or alternative chains of inquiry that branch off from the main track of logic. Other diagrammatic lists of questions about exercise and cognitive function might be constructed from different but related starting points. One might begin, for example, with the well-established observation that circulation is superior in older individuals who exercise regularly. This might lead through a series of proximal experiments toward the ultimate question, "What is the *mechanism* by which exercise maintains cognitive function?"

Building such diagrams will be useful for the student in several other ways. It is a way of controlling the instinct to grab the first researchable question that becomes apparent in an area. Often such questions are inferior to what might be selected after more careful contemplation of the alternatives. A logical sequence can be followed for most questions, beginning with "What has to be asked first?" Once these serial relationships become clear, it is easier to assign priorities.

In addition to identifying the correct ordering and relative importance of questions, such conceptual models also encourage students to think in terms of a series of studies that build cumulatively toward more significant conclusions than can be achieved in a one-shot thesis or dissertation. The faculty member who has clear dedication to a personal research program can be a key factor in attracting students into the long-term commitments that give life to an area of inquiry.

Researchable questions occur daily to the active researcher. The problem is not finding them but maintaining some sense of whether, and where, they might fit into an overall plan. Although this condition may seem remote to the novice struggling to define a first research topic, formulating even a modest research agenda can be a helpful process. The guidance of a sequential display of questions can allow the student to settle confidently on the target for a proposal.

The Sequence of Proposing: From Selecting a Topic to Forming a Committee

A Plan of Action: What Follows What?

Figure 3.1 presents a plan of action for developing a proposal. It can be useful for the novice if one central point is understood. A tidy, linear sequence of steps is not an accurate picture of what happens in the development of most research proposals. The peculiar qualities of human thought processes and the serendipity of retrieving knowledge serve to guarantee that development of a proposal will be anything but tidy. Dizzying leaps, periods of no progress, and agonizing backtracking are more typical than is a continuous, unidirectional flow of events. The diagram may be used to obtain an overview of the task, to establish a rough time schedule, or to check retrospectively for possible omissions, but it is not to be taken as a literal representation of what should or will happen.

To say that development of a proposal is not a perfectly predictable sequence is not to say, however, that it is entirely devoid of order. Starting at the beginning and following a logical sequence of thought and work has some clear advantages. When the proposal has been completed, a backward glance often indicates that a more orderly progression through the development steps would have saved time and effort.

For instance, although the mind may skip ahead and visualize a specific type of measure to be used, Step 11 ("Consider alternative methods of data collection") should not be undertaken until Step 6 ("Survey relevant literature") is completed. Many methods of measurement may be revealed and noted while perusing the literature. Sometimes suggestions for instrumentation materialize in unlikely places or in studies that have been initially categorized as unlikely to yield information concerning measurement. Additionally, reported evidence of the reliability and validity of the scores from alternative procedures will be needed before any final selection can be made. Thus, a large commitment of effort to consideration of alternative methods can be a waste of time if it precedes a careful survey of the literature.

For simplicity, many important elements have been omitted from Figure 3.1. No reference is made to such pivotal processes as developing a theoretical framework, categorizing literature, or stating hypotheses. Further, the detailed demands that are intrinsic to the writing process itself, such as establishing a systematic language, receive no mention. What are presented are the obvious steps of logic and procedure—the operations and questions that mark development toward a plan for action. Finally, the reader who begins to make actual use of the diagram will find that the sequence of steps at several junctures leads into what appear to be circular paths. For example, if at Question F a single form of inquiry does not present itself as most appropriate, the exit line designated "NO" leads back to the previous procedural step of considering alternative forms of inquiry. The intention in this arrangement is not to indicate a trap in which beginning researchers are doomed forever to chase their tails. In each case, the closed loop suggests only that when questions cannot be answered, additional input is required (more study, thought, or advice), or that the question itself is inappropriate to the case and must be altered.

For the most part, Figure 3.1 is self-explanatory. We have assumed that students will be working with, and obtaining advice from, their advisor as they navigate the various steps. In the pages that follow, however, we have selected a few of the steps and questions for comment, either because they represent critical junctures in the proposal process or because they have proven particularly troublesome for our own advisees. It will be helpful to locate in the diagram sequence each of the items selected for discussion so that the previous and succeeding steps and questions provide a frame for our comments.

Step 3: Narrow down. "What do I want to know?" Moving from general to specific is always more difficult for the beginner than is anticipated. It is here that the student first encounters two of the hard facts of scientific life: logistic practicality and the perverse inscrutability of seemingly simple events. Inevitably, the novice must learn to take one small step, one manageable question, at a time. In other words, the proposal must conform in scope to the realistic limitations of the research process itself. At their best, research tools can encompass only limited bits of reality; stretched too far, they produce illusion rather than understanding.

It may be important to think big at first, to puzzle without considering practicality, and to allow speculation to soar beyond the confines of the sure knowledge base. From such creative conceptual exercises, however, the researcher must return to the question, "Where, given my resources and the nature of the problem, can I begin?" Delimiting questions such as "In which people?" "Under what conditions?" "At what time?" "In what location?" "By observing which events?" and "By manipulating which variables?" serve the necessary pruning function.

Step 5: Identify reasons answer is important. This step places the proposed research in scientific-societal perspective. The study should contribute to the generation or validation of a theoretical structure or subcomponent or relate to one of the several processes by which knowledge is used to enhance

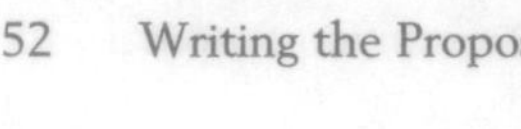

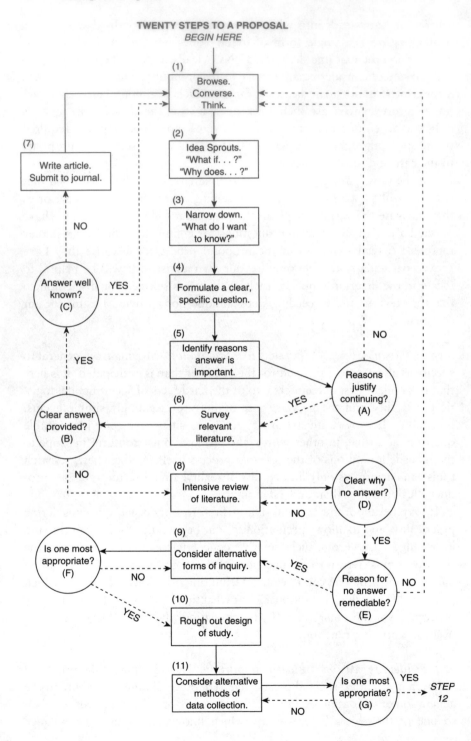

TWENTY STEPS TO A PROPOSAL

BEGIN HERE

(1) Browse. Converse. Think.

(2) Idea Sprouts. "What if. . . ?" "Why does. . . ?"

(3) Narrow down. "What do I want to know?"

(4) Formulate a clear, specific question.

(5) Identify reasons answer is important.

Reasons justify continuing? (A)

NO

YES

(6) Survey relevant literature.

Clear answer provided? (B)

YES

NO

Answer well known? (C)

YES

NO

(7) Write article. Submit to journal.

(8) Intensive review of literature.

Clear why no answer? (D)

NO

YES

Reason for no answer remediable? (E)

NO

YES

(9) Consider alternative forms of inquiry.

Is one most appropriate? (F)

NO

YES

(10) Rough out design of study.

(11) Consider alternative methods of data collection.

Is one most appropriate? (G)

NO

YES

STEP 12

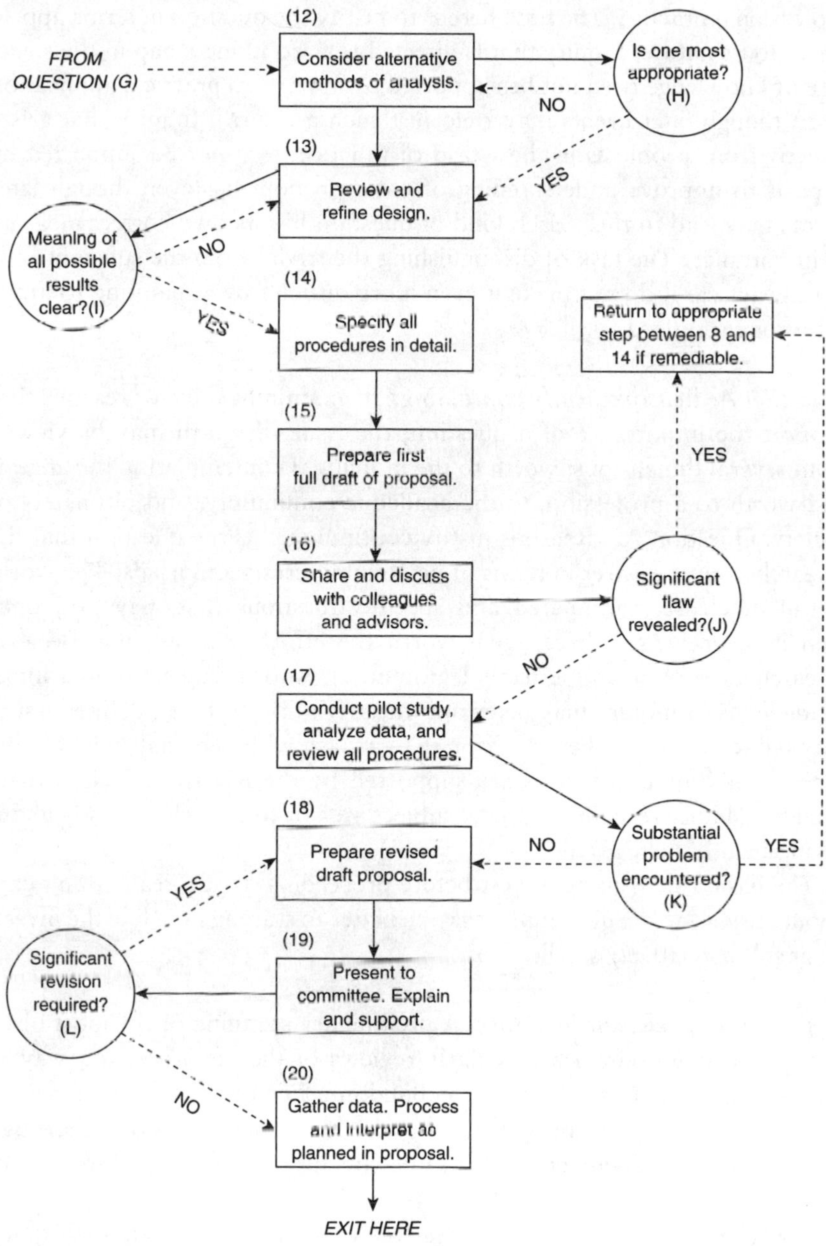

Figure 3.1 Twenty Steps to a Proposal

NOTE: Boxes represent major procedural steps, and unbroken lines trace the main sequence of those steps. Circles represent the major questions to be confronted, and broken lines lead to the procedural consequences of the alternative YES or NO answers.

professional practice. The trick here is to justify the question in terms appropriate to its nature. Inquiry that is directed toward filling a gap in the structure of knowledge need not be supported by appeals to practical application (even though later events may yield just such a return). Inquiry that arises directly from problems in the world of practice need not be supported by appeals to improve understanding of basic phenomena (even though later events may lead to this). Each kind of question has its own correct measure of importance. The task of distinguishing the trivial from the substantive is not always easy; do not make it even more difficult by attempting to apply the wrong standard.

Question A: Reasons justify continuing? In examining a list of reasons that support the importance of a question, the issue of worth may be viewed from several dimensions: worth to the individual contemplating the answer and worth to a profession, to the academic community, and ultimately to society. Question A, "Reasons justify continuing?" is the question that the researcher must answer in terms of personal interests and needs. The world is full of clearly formulated and specific questions that may not, once seen in their formal dress, seem worth the effort of answering. Because researchers are human, perfectly legitimate questions may seem dull, interesting veins of inquiry may peter out into triviality, and well-defined issues may fail to suit for no better reason than a clash with personal style. On the other hand, some questions are supported by the researcher's immediate need to enhance teaching in a vital subject area or to quench curiosity about a long-held hunch.

The basic rule is to be honest before proceeding. If you really don't care about answering the question, it may be better to start again while the investment still is relatively small.

Step 6: Survey relevant literature. A preliminary scanning of the most obvious, pertinent resources, particularly reviews of the literature, is a way of husbanding time. It is far better to abandon a line of thought after several weeks of selective skimming than to work one's way via slow, thorough digestion of each document to the same conclusion after several months of effort.

Conscientious students sometimes feel vaguely guilty about such quick surveys. Keeping in mind the real purpose, which is to identify questions that already have satisfactory answers, is one way of easing such discomfort.

Question E: Reason for no answer remediable? In some cases, the literature contains an empty area because the state of technology, the available knowledge framework, ethical considerations in completing the study, or the logistic demands peculiar to the question have made it impossible or unreasonable to conduct appropriate forms of inquiry. So long as the gap in knowledge seems to exist because no one has yet defined the question or become interested in pursuing the answer, it is reasonable to proceed. There are other reasons for empty or ambiguous areas in the literature, however, and they signal caution before proceeding.

Question I: Meaning of all possible results clear? The tighter the logic, the more elegant the theoretical framework, the more closely the design is tailored to produce clarity along one dimension—in short, the better the quality of the proposal—the greater the risk that the proposer will be lured into an unfortunate presumption: that the result of the study is known before the data are in hand. That student researchers sometimes are confronted by the stunning news that their treatment produced a reverse effect is in itself neither surprising nor harmful. Being unable to make an intelligent interpretation of such a situation, however, is unfortunate and in most cases avoidable.

Unanticipated results raise a fundamental question that the investigator must confront. Does the finding truly reflect what is resident in the data, or is it only an artifact of the analysis? If there is any doubt about the appropriateness of the analysis, particularly if the procedures were not perfectly aligned with the research question, the latter possibility must be considered. If reexamination of the analysis provides no accounting for findings that are sharply incongruent with expectations, another explanation must be sought. All of this is made more difficult if the possibility of discrepant findings has never been contemplated. A strong proposal, constructed in an orderly, step-by-step sequence, will enhance the likelihood that you can manage the unexpected with at least a degree of dignity.

Through serious consideration of alternative outcomes at the time of constructing the proposal, it may be possible to include elements in the study that will eliminate ambiguity in some of the most likely results. One method of anticipating the unexpected is to follow through the consequences of rejecting or failing to reject each hypothesis of the study. If the hypothesis was rejected, what is the explanation? How is the explanation justified by the rationale for the study? What findings would support the explanation? Conversely, if the findings of the study fail to provide a basis for rejection, what explanations are to be proposed? At the least, some careful preliminary thought about alternative explanations for each possible result will serve as a shield against the panic that produces such awkward post hoc interpretations as "no significant differences were observed because the instruments employed were inadequate."

Step 16: Share and discuss with colleagues and advisors. There is a well-known syndrome displayed by some who attempt research, characterized by the inclination to prolong the period of writing the final report—indefinitely. Some people simply cannot face what they perceive to be the personal threat implied in opening their work to challenge in the public arena. These individuals are terribly handicapped and only rarely can become mature, productive scholars. An early sign of this is seen in students who cannot bring themselves to solicit advice and criticism for their proposals.

Sometimes students experience severe criticism because they present their ideas before they have been sufficiently developed into a conceptual framework that represents careful preparation. Many professors avoid speculative conversations about "half-baked" ideas that have just arrived in a blinding flash of revelation to the student. Few professors, however, refuse a request for advice concerning a proposal that has been drafted as the culmination of several weeks of hard thought, research, and development. Even at that, having one's best effort devastated by pointed criticism can be an agonizing experience. Nevertheless, the only alternative is to persist in error or ignorance, and that is untenable in research.

If you are fortunate enough to be in a department that contains a vigorous community of inquiring minds, with the constant give and take of intellectual disputation, the rough and tumble soon will be regarded as a functional part of producing good research. The novice will solicit, if not always enjoy, the best criticism that can be found.

The notion that it is vaguely immoral to seek assistance in preparing a proposal is at best a parody of real science and at worst, as in the form of an institutional rule, it is a serious perversion arising from ignorance. Research may have some game-like qualities, but a system of handicaps is not one of them. The object of every inquiry is to get the best possible answer under the circumstances, and that presumes obtaining the best advice available. It is hoped that the student will not be held to any lesser standard.

It should be obvious that students, after digesting and weighing all the criticism received, must still make their own choices. Not all advice is good, and not all criticism is valid. There is only one way to find out, however, and that is to share the proposal with colleagues whose judgments one can respect, if not always accept.

The process of proposal development is enhanced if you obtain advice at various steps and do not wait until the end to solicit feedback. We strongly recommend working with your advisor and committee in ways that help you move steadily forward. For example, at Step 4 consulting your advisor about possible research questions may help you refine them and may assist you in finding relevant literature. At Steps 9 through 12, short, focused meetings where you are prepared to discuss specifics may be particularly beneficial. Advice and constructive criticism are best when received in small doses and integrated throughout the proposal process.

Step 19: Present to committee. Explain and support. Presentation of your proposal may take place before a thesis or dissertation committee on an occasion formally sanctioned by the graduate school, or at an informal gathering in the advisor's office. In either instance, the purpose served and the importance assumed will depend on both local traditions and the relationships that have evolved to that point among the chairperson, committee members, and student.

If, for example, the chairperson has closely monitored the developing proposal and is satisfied that it is ready for final review and approval, the nature of the meeting is shaped accordingly. In addition, if other committee members have consulted on the proposal at various stages of writing, the meeting may serve primarily as an occasion for final review and a demonstration of presentation skills, rather than evaluation, extensive feedback, and judgment. When these conditions do not apply, the meeting assumes far greater significance, in itself, and the length and nature of the presentation will be affected.

Whatever the circumstances, both a prudent respect for the important function the committee members must perform and a proper desire to demonstrate the extent to which the efforts of your advisors have been effective make careful preparation and a good presentation absolutely necessary. Much of our advice about that is contained in Chapter 7. For the present purpose, we want to underscore the following points.

1. The more you can work with committee members before an official meeting, the more that meeting can focus on improving (and appreciating) your proposal—rather than just on understanding it.

2. As committee members talk with you and with each other at the meeting, it is natural that new insights and concerns will surface. So long as those are accurately recorded, and so long as there is clear provision for how the committee will manage subsequent revisions in the proposal, that process is all to your advantage. The object is not simply to get the proposal (as it stands) accepted; it is to create the best possible plan for your dissertation or thesis.

3. Where you have had to make difficult choices, accept compromises in method for pragmatic reasons, or leave some final decision(s) for a later point in time, it is best to bring such matters directly to the attention of your committee. Don't wait to be questioned. Take the initiative and lay out the problematic aspects for your advisors as you go through the presentation. You need not

make the proposed study appear to be mired in difficulty. Propose solutions and give your rationale, but never ignore or gloss over what you know requires more attention—and the help of your committee.

4. If the proposal is approved with the understanding that certain revisions or additions will be made, the best procedure is to obtain signatures on documents while at the meeting. The signed forms can then be held by your chairperson until he or she has approved the final draft.

Step 20: Gather data. Process and interpret as planned in proposal. This is the payoff. A good proposal is more than a guide to action, it is a framework for intelligent interpretation of results and the heart of a sound final report. The proposal cannot guarantee meaningful results, but it will provide some assurance that, whatever the result, the student can wind up the project with reasonable dispatch and at least a minimum of intellectual grace. If that sounds too small a recompense for all the effort, consider the alternative of having to write a report about an inconsequential question, pursued through inadequate methods of inquiry, and resulting in a heap of unanalyzable data.

Originality and Replication: What Is a Contribution to Knowledge?

Some attention already has been given to considerations that precede the proposal, the critical and difficult steps of identifying and delimiting a research topic. One other preliminary problem, the question of originality, has important ramifications for the proposal.

Some advisors regard student-conducted research primarily as an arena for training, like woodchopping that is expected to produce muscles in the person who holds the axe, but not much real fuel for the fire. Whatever may be the logic of such an assumption, students generally do not take the same attitude. Their expectations are more likely to resemble the classic dictum for scholarly research, to make an original contribution to the body of knowledge.

An all-too-common problem in selecting topics for research proposals occurs when either the student or an advisor gives literal interpretation to the word "original," defining it as "initial, first, never having existed or occurred before." This is a serious misinterpretation of the word as it is used in science. In research, the word "original" clearly includes all studies deliberately employed to test the accuracy of results or the applicability of conclusions developed in previous studies. What is not included under that rubric are studies that proceed mindlessly to repeat an existing work either in ignorance of its existence or without appropriate attention to its defects or limitations.

One consequence of the confusion surrounding the phrase "original contribution" is that misguided students and advisors are led to ignore one of the most important areas of research activity and one of the most useful forms of training for the novice researcher—replication. That replication sometimes is regarded simply as rote imitation, lacking sufficient opportunity for students to apply and develop their own skills, is an indication of how badly some students misunderstand both the operation of a research enterprise and the concept of a body of knowledge.

The essential role of replication in research has been cogently argued (Gall, Gall, & Borg, 2007). What has not been made sufficiently clear, however, is that replication can involve challenging problems that demand creative resolution. Further, some advisors do not appreciate the degree to which writing proposals for replicative studies can constitute an ideal learning opportunity for research trainees.

In direct replication, students must not only correctly identify all the critical variables in the original study but also create equivalent conditions for the conduct of their own study. Anyone who thinks that the critical variables will immediately be apparent from a reading of the original report has not read very widely in the research literature. Similarly, an individual who thinks that truly equivalent conditions can be created simply by "doing it the same way" just has not tried to perform a replicative study. Thorough understanding of the problem and, frequently, a great deal of technical ingenuity are demanded in developing an adequate proposal for direct replication.

As an alternative to direct replication, the student may repeat an interesting study considered to have been defective in sample, method, analysis, or interpretation. Here the student introduces deliberate changes to improve the power of a previous investigation. It would be difficult to imagine a more challenging or useful activity for anyone interested in both learning about research and contributing to the accumulation of reliable knowledge.

In writing a proposal for either kind of replicative study, direct or revised, the student should introduce the original with appropriate citation, make the comments that are needed, and proceed without equivocation or apology to the proposed study. Replicative research is not, as unfortunate tradition has it in some departments, slightly improper or something less than genuine research.

Given the limitations of research reports, it often is useful to discuss the source study for the replication with the original author. Most research workers are happy to provide greater detail and in some instances raw data for inspection or reanalysis. In a healthy science, replication is the most sincere form of flattery. A proposal appendix containing correspondence with the author of the original report, or data not provided in that report, often can serve to interest and reassure a hesitant advisor.

Getting Started: Producing the First Draft

The student who has never written a research proposal commonly sits in front of a desk and stares at a blank piece of paper or an empty video monitor for hours. The mind is brimming with knowledge gleaned from the literature, but how does one actually get started? The concept of "a research proposal" conjures up ideas of accuracy, precision, meticulous form, and use of a language system that is new and unpracticed by the neophyte researcher. The demands can suddenly seem overwhelming. The student should realize that this feeling of panic is experienced by nearly everyone, not only those who are new to the writing endeavor but those who are skilled as well. Fanger (1985) expressed it beautifully: "I have come to regard panic as the inevitable concomitant of any kind of serious academic writing" (p. 28). For anyone temporarily incapacitated by a blank page or empty monitor, the following suggestions may be helpful.

Make an outline that is compatible with the format selected to present the communication tasks listed in Chapter 1. An initial approval of the outline by the advisor may save revision time later. Gather the resource materials, notes, and references, and organize them into groups that correspond to the outline topics. For instance, notes supporting the rationale for the study would be in one group, and notes supporting the reliability of an instrument to be used would be in another group.

Once the outline is made and the materials gathered, tackle one of the topics in the outline (not necessarily the first) and start writing. If the section to be written is labeled "The Purpose," try imagining that someone has asked, "What is the purpose of this study?" Your task is to answer that question. Start writing. Do not worry about grammar, syntax, or writing within the language system. Just write. In this way you can avoid one of the greatest inhibitions to creativity—self-criticism so severe that each idea is rejected before it becomes reality. Remember, it is easier to correct than to create. If all the essential parts of a topic are displayed in some fashion, they can later be rearranged, edited, and couched within the language system. With experience, the novice will begin thinking in the language system and forms of the proposal. Until that time, the essential problem is to begin. Awkward or elegant, laborious or swift, there is no substitute for writing the first draft.

One way to approach writing is to use the outline feature on your word processor, which allows you to develop your outline and then go back and progressively fill in the detail under each heading. Learn to use this feature. The effort needed to learn its use will be repaid many times over. Word processing programs provide the opportunity for writers to edit, rearrange text easily, and store manuscript copy for future revisions. There is a significant psychological advantage in the ease with which revised drafts can be produced. This encourages the author to make revisions that might otherwise be set aside under the press of limited time, and has greatly enhanced the ability of proposal writers to revise and polish their work.

Selecting Your Thesis or Dissertation Committee

Master's thesis committees vary in number from one professor to a committee of five or six faculty members. A doctoral dissertation committee typically consists of four to six members. In some instances, all committee members are from within the department of the student's major. In other instances, the committee is multidisciplinary, with faculty representing other departments on campus.

At most universities, students have some opportunity to request specific faculty members for their committee. If the student does have some freedom to exercise choice, committee membership should be designed to maximize the support and assistance available. A student interested in the study of behavioral treatment of drug abuse in young upwardly mobile women could tap the value of different faculty perspectives and skills by blending members from several departments. For this purpose, individuals with multiple interests are particularly useful. For example, a faculty member in the psychology department might be selected for both statistical competence and interest in behavior modification, someone in the school of social work might bring epidemiological expertise regarding drug usage, and a faculty member in the school of public health might be a part of the committee because of expertise in both experimental design and therapeutic compliance techniques.

Because students know from the beginning of the graduate program that faculty eventually will have to be selected for such a committee, it behooves them to be thinking about these matters during the selection of elective courses throughout the program. If a choice has to be made between two professors for an elective course, and one of them is more interested in the student's probable area of research, that may carry the day in determining which course to take. Although it is not essential that students have taken their committee members' classes, it is easier to ask a known faculty member to serve on your committee. That person is likely to take a greater interest in your work, and you have a good idea of his or her standards and methods of scholarship.

4

Content of the Proposal

Important Considerations

The topics covered in this chapter are designed to assist the true beginner. Most experienced proposal writers may want to go directly to Chapter 7 or simply skim this chapter for review.

Reviewing the Literature: Finding it First

Areas of inquiry within the disciplines exist as ongoing conversations among those who do the work of scholarship. The published literature of an area constitutes the archival record of those conversations: research reports, research reviews, theoretical speculation, and scholarly discourse of all kinds. You join the long conversation of science as you join any other, by first listening to what is being said, and only then formulating a comment designed to advance the dialogue.

The metaphor of scholarship as an extended conversation works well at a variety of levels—because at heart it is an accurate representation. The process of locating the voices of individual conversants, for example, is called *retrieval.* That involves searching through the accumulated archive of literature to find out what has been said (when, by whom, and on the basis of what evidence). The process of listening carefully to the ongoing discourse about a topic of inquiry is called *review.* That involves studying items previously retrieved until both the history and the current state of the conversation are understood. It does not stretch the metaphor too far to observe that writing the proposal is a step in preparing yourself to have your own voice be heard—to do research and enter what you learn into the long conversation.

Retrieval, review, proposing and conducting research, and even report writing are tasks that have their own sets of requisite technical skills. Each also has a place for art and instinct as well as intelligence and accumulated knowledge. This chapter deals with what goes into the proposal (content) once the topic and terms of discourse have been defined. It follows, then, that it must begin with what you have learned by listening in on the conversation—a review of the literature. In your review, you will establish what has been said to this point as the basis for proposing your own contribution. Retrieval, however, comes first in the order of things. You can't review what you haven't found. That brings us to the technical skill and fine art of searching the literature.

We will not burden you with the specifics of a particular search procedure. The demands made by proposals differ widely, as do the background and skills of each proposal writer. Further, the facilities for retrieval vary enormously at different institutions and, of course, each discipline and subspecialty has its own peculiar mechanisms for searching the literature. What we can do here is set forth the small number of general rules that, if observed from the outset, have the power to make any retrieval effort more efficient.

Knowing what you need to know is the obvious first step in formulating a retrieval strategy. Knowing how much you really need to know, however, is a vital second step—and one not always properly appreciated by the novice in research. Discussion with your advisor, consultation with colleagues who have written proposals, inspection of proposals previously accepted by the graduate school, and the preliminary reading already done during the process of identifying the topic of your proposed study will all serve to identify what you need to know—and thereby the literature you will seek. Normally that includes research reports and reviews related to your questions or hypotheses. This literature provides information about research methodology in the area and items dealing with both theory and application as they are related to your study.

Deciding how much you need to know is a more complex decision—in part because you often cannot answer that question until after some retrieval and review already have been accomplished. This is a matter of defining the purpose to be served by what you retrieve. Acquiring a broad overview of previous work in an area leads to one kind of retrieval strategy. If your purpose is to know what a small set of senior scholars have reported in the last two years, the strategy will be different, as will be the case if your purpose is to do an exhaustive search in which every scrap of fugitive literature is doggedly pursued until acquired.

What we suggest is that you talk with your advisor about the question of "how much?" and that you stay in touch with him or her on that topic as you proceed. The notion that every search of the literature needs to be exhaustive is one of the more destructive mythologies that persists in graduate student culture. When you have enough sense of the conversation to argue persuasively that the target for your proposed study is sound, and that the methods of inquiry are correct, you know enough for the purpose of the proposal.

Over a longer period of time, you may have plenty of motivation to read more widely and deeply than anything demanded by your proposal, but that is a different matter. (For a recent and provocative discussion of the place of literature reviews in preparation for dissertations, see Boote & Beile, 2005.) For the present, it is wise to acquire some sense of how much you need to know so that you can shape the size of your review task accordingly—and thereby know what and how much to retrieve. Until you have done some retrieval and review, it is likely that neither you nor your advisor will be able to set that target with precision, but some careful preliminary thought will serve you well.

Retrieval Rule 1. Do not begin by going to the library or your computer and starting to search for literature. Talk first to your advisor (or entire thesis or dissertation committee) and research colleagues who have some familiarity with the area of your proposal. Make a list of what they think you should read. Locate the items, skim them, and record full citations for all that appear to be appropriate. These form the core of your retrieved literature base. Go over the reference lists in the items that appear to be most directly relevant to your needs and make those citations the priority for retrieval when you return to the library.

Retrieval Rule 2. When you go to the library, do not begin by starting to search for literature. Talk first to the reference specialists who can identify the retrieval systems that are most likely to be productive for your topic. Then, we urge you to take advantage of seminars and classes on using specific retrieval systems that are offered at many academic libraries. Retrieval is one part sweat and two parts knowing where to look.

For every scholar, the marvels of computerized retrieval are available at your university library, through the university computer network, or on the Internet. The databases that any given system can access, however, differ in important and subtle ways. Attending introductory seminars and asking for expert help can save hours. Swift, accurate, flexible, and powerful beyond anything we could have dreamed when the first edition of this book appeared, computer retrieval systems are, nevertheless, only as good as their search structures, and those are just as full of limitations and idiosyncratic quirks as any of the printed indexes that scholars used in the past.

That observation leads to two items of advice. First, plan to devote a considerable amount of retrieval time to learning how each system works. Second, by all means use computerized systems, but do not automatically assume that manual search is without value. In other words, all but a very small number of highly technical topics can profit by a visit to whatever is the equivalent to the *Reader's Guide to Periodical Literature*—on CD-ROM or in the printed version—in your particular subject area.

Retrieval Rule 3. From the outset, think of your retrieval effort as consisting of a series of stages. It is unlikely that you will (or should) march through them in perfect sequence; it is more a matter of moving back and forth among the stages in ways that will make best use of your time.

Stage 1: Identification—Find and record citations that seem potentially relevant. This is work done with indexes, bibliographies, reference lists, and, most often, the computer.

Stage 2: Confirmation—Determine that the items identified can be obtained for use. This is work done with the library holdings of serials (electronic and hardcopy) and books, reprint services, interlibrary loan personnel, microfiche files, and the telephone.

Stage 3: Skimming and Screening—Assess each item to confirm that it actually contains content to be reviewed (to be read and studied with care). This is work that demands enough mastery of the system language and the constructs related to your topic to recognize what is and is not of potential use. Much of this (though not all) can be accomplished without taking the resource item into your physical possession or downloading it to your hard drive. This means time at the computer or in the stacks and time at the microfiche reader. The most important retrieval skill here is the ability to resist the temptation to stop the work of skimming and screening and immerse yourself in the conversation.

Stage 4: Retrieval—Acquire the literature. This is work done by checking out books, downloading or copying articles from journals, ordering microfiche and reprints, and initiating requests for interlibrary loans. Not everything must be (or should be) retrieved. There is a strong argument for not having every article immediately at hand when you are drafting the review of literature, and one way to ensure that is to take notes from references that then stay in the stacks or in a folder on your computer.

Stage 5: Review—Read and study the literature that records the conversation about your topic. Subsequent sections of this chapter will deal with how to use what you learn to build your proposal.

Retrieval Rule 4. From the first moment of your search, keep a log of all the words used to name what you have to learn about. These will become the keywords used by indexing systems (often they are specifically designated as keywords by the authors) for accessing their holdings. Building a keyword list is like acquiring a set of master keys to a large building. They can open doors in a variety of locations, and without them you can wander for hours without gaining entry to anywhere you want to be. Although most databases now provide the option of searching titles, authors, abstracts, and keywords—or all of them—knowing what phrases produce the greatest yield will help now and in the future. In almost every case, the novice will be astonished at the variety of words and phrases employed to categorize items that appear to be identical.

Retrieval Rule 5. Always take maximum advantage of other people's work. For that reason, research reviews in your area always should have the highest priority in your search plan, as should annotated bibliographies and the reference list at the back of every article and book you retrieve. For the same reason, your first stop in the library should be ProQuest Dissertations and Theses. What could be a better search strategy than reading the reviews of literature crafted by students who have worked on similar problems? Dissertations are the *Yellow Pages* of research retrieval. From the start, let your fingers do a lot of the walking.

Retrieval Rule 6. Record a complete citation for every item you identify. Whether with index cards or a computer program that alphabetizes and sorts by keywords, keep a complete running record of what you find—whether immediately reviewed or not. No frustration can match that of having to backtrack to the library for a missing volume or page number. We strongly recommend you use a computerized bibliographic note taking and retrieval program (e.g., EndNote, ProCite, or Reference Manager, among others). Employing such a program will aid in retrieving citations and developing your own searchable database. That, in turn, will greatly reduce the time later required to prepare the proposal's reference list. Many universities now have site licenses that provide for downloading such bibliographic programs without charge or at a very small cost.

Retrieval Rule 7. Whatever notes you may take during the stages of Skimming and Screening or Review, never write anything down in which there could be the slightest confusion at a later date as to whether the words are your own—or those of another author. If you take down a quotation, take it verbatim and attach the proper page citation. If you write anything other than a direct quotation, make absolutely sure it is a paraphrase in your own words. There is no place for anything in between those two species of notes.

Retrieval Rule 8. Be cautious about computerized systems that seem too good to be true. There are reference retrieval systems, for example, with which you can simply highlight a phrase and the computer then automatically imports the phrase to a designated point in your own document, complete with citations in the text and a reference at the back. Writers using such a system don't have to read the full article or do the intellectual work required to truly understand how it fits into the wider conversation among investigators in the area. What seems quick and efficient may serve to undercut the ability to make informed decisions about source material and, ultimately, to write a sound proposal.

If you follow these eight rules, if you build a reasonable sense of how much you really need to know, and if you persist, you may have one of the most wonderful epiphanies a scholar can experience. As you look down a page of references, you will recognize all the names, and the voices of their conversation will fill your ears. At that moment, you are current and you are ready to take part. Given the pace of work in many areas of science, that moment will be brief, but savor it! That is the sweet fruit of retrieval.

Reviewing the Literature: Writing the Right Stuff

By much deserved reputation, the reviews of literature in student research proposals are generally regarded as consisting of clumsy and turgid prose, written as pro forma responses to a purely ceremonial obligation in the planning format. Even when carefully crafted with regard to basic mechanics, they make dull reading, and when not so prepared they are excruciating torture for most readers. Much of this problem arises from a misunderstanding of the task served by reviewing the literature, and none of it need be true.

To begin, the common designation used in proposals, "review of the literature," is a misleading if not completely inappropriate title. A research proposal is not the place to review the body of literature that bears on a problematic area, or even the place to examine all the research that relates to the specific question raised in the proposal. A variety of methods for "reviewing the literature" do exist, such as best evidence synthesis, critical reviews, and even meta-analysis, but they are rarely appropriate for proposals. Analyses of that kind may be useful documents publishable in their own right. Indeed, some journals such as the *Review of Educational Research* are exclusively devoted to such critical retrospectives on scholarship. The task to be performed in the proposal, however, is different. It is not inferior to the true review, it simply is different.

In writing a research proposal, the author is obligated to place the question or hypothesis in the context of previous work in such a way as to explain and justify the decisions made. That alone is required. Nothing more is appropriate, and nothing more should be attempted.

Although the author may wish to persuade the reader on many different kinds of points, ranging from the significance of the question to the appropriateness of a particular form of data analysis, sound proposals devote most of the literature review to explaining (a) exactly how and why the research question or hypothesis was formulated in the proposed form and (b) exactly why the proposed research strategy was selected. What is required to accomplish these tasks is a step-by-step explanation of decisions, punctuated by reference to studies that support the ongoing argument. In this, the writer uses previous work, often some critique of previous work, and sometimes some exposition of the broad pattern of knowledge as it exists in the area to appeal for the reader's acceptance of the logic represented in the proposed study.

Whatever particular arguments must be sustained in the review of the literature, there is no place for the "Smith says this . . ." and "Jones says that . . ." paragraph-by-paragraph recital that makes novice proposals instruments for dulling the senses. This is the place to answer the reader's most immediate questions: What is it the author wants to know, and why has this plan been devised to find the answer? In a good review, the literature is made to serve the reader's query by supporting, explicating, and illuminating the logic now implicit in the proposed investigation.

It follows, then, that where there is little relevant literature, or where decisions are clear-cut and without substantial issues, the review should be brief. In some cases, the examination of supporting literature may best be appended or woven into another section of the proposal. To write a review of literature for the sake of having a review in the document is to make it a parody and not a proposal.

Remember, the writer's task is to employ the research literature artfully to support and explain the choices made *for this study,* not to educate the reader concerning the state of science in the problem area. Neither is the purpose of the section to display the energy and thoroughness with which the author has pursued a comprehensive understanding of the literature. If the author can explain and support the question, design, and procedures with a minimum demand on the reader's time and intellect, then that reader will be more than sufficiently impressed with the applicant's capabilities and serious purpose.

None of this is intended to undervalue the task that every researcher must face, that of locating and thoroughly assimilating what is already known. To do this, the student must experience what Fanger (1985) described as "immersion in the subject" by reading extensively in the areas that are either directly or indirectly related to the topic of study. This may lead at first to a sense of frustration and confusion, but perseverance usually leads out of the wilderness to the point at which what is known about the topic can be seen in the light of what is not known. The goals of the proposed study can be projected against that backdrop.

The proposal is the place to display the refined end products of that long and difficult process. It is not uncommon, for example, for the study's best support to emerge from a sophisticated understanding of gaps in the body of knowledge, limitations in previous formulations of the question, inadequate methods of data collection, or inappropriate interpretation of results. The review of the literature section then becomes a vehicle for illustrating why and how it all can be done better. What readers need, however, is not a full tour retracing each step the author took in arriving at the better mousetrap, but a concise summary of the main arguments properly juxtaposed to the new and better plan for action.

Most students will agonize over the many studies discovered that, although fascinating and perhaps even inspiring during the immersion process, in the final stages of writing turn out to fail the test of critical relevance and therefore merit exclusion from the proposal. It is tempting to see discarded studies, unused note cards, and bibliographic entries on your computer as wasted time, but that misses the long view of learning. The knowledge gained through synthesis and evaluation of research results builds a knowledge base for the future. The process of immersion in the literature provides not only the information that will support the proposal but also the intellectual framework for future expertise. What may appear in the crush of deadlines and overload stress to have been pursuit down blind alleys, ultimately may provide insights that will support new lines of thought and future proposals.

Writing the section on related literature often is no more complex than first describing the major concepts that led you to your research question or hypothesis and then describing the supporting research findings already in the literature. It may be as simple as hypothesizing that A is greater than C. Why do you hypothesize that A is greater than C? Because evidence suggests that A is greater than B, and B is greater than C; therefore, it is reasonable to hypothesize that A must be greater than C.

In the review of the related literature, you would present those conceptual relationships in an organized fashion and then document each with previously reported studies. For example, the first section would include the most important studies indicating that A is greater than B, and the second section would present similar evidence supporting the proposition that B is greater than C. The literature section would then conclude with the argument that given such information, it is reasonable to hypothesize that A is greater than C. In addition, either interwoven throughout or in separate sections, material from the literature would be presented in support of decisions about design and measurement in the proposed study.

Look at the example in Table 4.1, which also is represented diagrammatically in Figure 4.1. In this table, the general research question is posed, followed by the specific hypothesis through which the question will be answered. They are shown here merely to establish the frame of reference for the outline. In this example of the development of the related literature, three major concepts are necessary to support the legitimacy of this hypothesis.

Table 4.1 Preparing the Related Literature Section

QUESTION: Is physical fitness related to cognition in older adults? More specifically, can an aerobic exercise program increase cognitive processing speed in older adults?

HYPOTHESIS: Maintenance of physical fitness through a physical training program will significantly decrease (make faster) reaction time in older individuals.

First Stage Outline: Develop the Concepts That Provide the Rationale for the Study

I. Reaction time is related to physical fitness level.
II. Maintenance of cognitive function is dependent on maintenance of aerobic capacity of the brain.
III. The aerobic capacity of brain tissue is affected by physical activity-related regional cerebrovascular changes.

Second Stage Outline: Development of Subtopics for Each Major Concept

I. Reaction time is related to physical fitness level.
   A. Comparisons of the reaction time of physically active and inactive subjects.
   B. Training effects on reaction time.
   C. Reaction time of those in poor physical condition (cardiovascular disease, hypertension).

II. Maintenance of cognitive function is dependent on maintenance of aerobic capacity of the brain.
   A. Relationship of cognitive function and brain aerobic capacity in aging individuals.
   B. Relationship of the neurological measure of brain function, electroencephalography (EEG), to cerebral blood flow and cerebral oxygen uptake in older subjects.

III. The aerobic capacity of brain tissue is affected by physical activity-related regional cerebrovascular changes.
   A. Increased metabolism in specific regions leads to cerebral blood flow shifts to those regions.
   B. Regional blood flow shifts in motor areas of the brain are related to physical movement.
   C. Exercise is related to changes in brain capillarization.

(Continued)

Table 4.1 (Continued)

Third Stage Outline: Add the Most Important References That Support Each Subtopic

I. Reaction time is related to physical fitness level.
   A. Physically active individuals have faster reaction times than do sedentary individuals (Clark & Addison, 2003; Cohen, 1993, 1995; Jones, 1998, 1999; Jones & Johnson, 1991; Lloyd, 1994).
   B. Reaction time is faster after a physical training program (Black, 1992, 1997; Dougherty, 1987; Morgan & Ramirez, 2006; Ramirez, 2003; Richards, 1995, 1997; Richards & Cohen, 1989; Roe & Williams, 1995; Walters, 1991).
   C. Cardiovascular-diseased patients have slower reaction time than normal individuals (Brown, 1991; Brown, Mathews, & Smith, 1998; Miller, 1991, 1992; Miller & Roe, 2005; Smith, Brown, & Rodgers, 1999; Smith & Rodgers, 1998).

II. Maintenance of cognitive function is dependent on maintenance of aerobic capacity of the brain.
   A. Both cognitive function and aerobic capacity decrease with age (Gray, 1988; Petty, 2006).
   B. EEG, cerebral blood flow, and cerebral oxidative capacity decrease with age and are related (Doe & Smith, 1999; Doe, Smith, & Snyder, 1997; Goldberg, 1998; Smith & Doe, 1991; Waters, 1989, 1993; Waters & Crosby, 1992).

III. The aerobic capacity of brain tissue is affected by physical activity-related regional cerebrovascular changes.
   A. Increased regional metabolism leads to blood flow shifts (Green & Neil, 1966; Lewis, 1979; Thomas, 2001).
   B. Regional blood flow shifts to motor areas of the brain are related to physical movements being controlled (Caplan, Myerson, & Morris, 1991; Goldsmith, 1993, 1994; Johnson, Goldsmith, & Rodriguez, 1990).
   C. Exercise is related to changes in brain capillarization (Meyers & Templeton, 1991; Patrick, 1993; Patrick & Stone, 1995; Robinson & Spencer, 1997).

In Table 4.1, the question suggests that the way physical fitness and cognitive function are related is through a change in brain aerobic capacity as a result of training. If this is a reasonable question to ask, one would have to show that there have been some prior studies in which physical fitness level has been related to some measure of cognitive function (Concept I). Second, some evidence that cognitive function might be altered by aerobic functional capacity of the brain should be shown (Concept II). Finally, some evidence should exist that physical movement can alter blood flow shifts in the brain, and that blood flow shifts are related to aerobic capacity (Concept III).

Generally, the major concepts are supported by two or three subtopics, all of which lead to the formalization of the main concept. For example, the concept that reaction time and physical fitness level are related (I) can be supported in three different ways, by showing (a) that reaction time is faster in physically fit persons than in sedentary persons, (b) that physical training enhances reaction time, and (c) that reaction time in those on the lowest end of the physical fitness continuum is the slowest of all. Each of these subtopics is supported by the findings from several studies, as shown in the third stage outline section of the table.

Writing the related literature section is much easier if an outline is developed in stages of increasing detail, as shown in Table 4.1, prior to the actual writing. Once the outline is developed, this section of the proposal can be written in a straightforward manner, with little backtracking necessary. If meticulous care is taken in selecting each reference, an enormous amount of time will be saved in the long run.

Another easy way to conceptualize the organization of the related literature is to diagram it, as shown in Figure 4.1. In this figure, the question (Q) is shown in the first box, and then the major components of the rationale are shown as they relate to one another. Component I refers to the behavioral observations that have been reported in the research literature. Component II refers to the literature in which the effects of aging on cerebral aerobic capacity and cognitive function are described. Because these relationships have negative outcomes, they are shown with a negative sign. Component III refers to all literature that provides support for a relationship between physical activity and cerebral aerobic activity. In this case, these are all positive relationships, which are shown with plus signs. Finally, the last box depicts how all these relationships lead to the hypothesis (H) of the study.

Both the outline and diagram format can be very helpful in conceptualizing the related literature. The entire process can be summarized in the 15 steps in Table 4.2. Table 4.3 contains guidelines for evaluating the related literature section.

Spadework: The Proper Use of Pilot Studies

The pilot study is an especially useful form of anticipation, and one too often neglected in student proposals. When it comes to convincing the scholarly skeptic (sometimes your own advisor), no argument can be so effective as to write: "I tried it and here is how it worked."

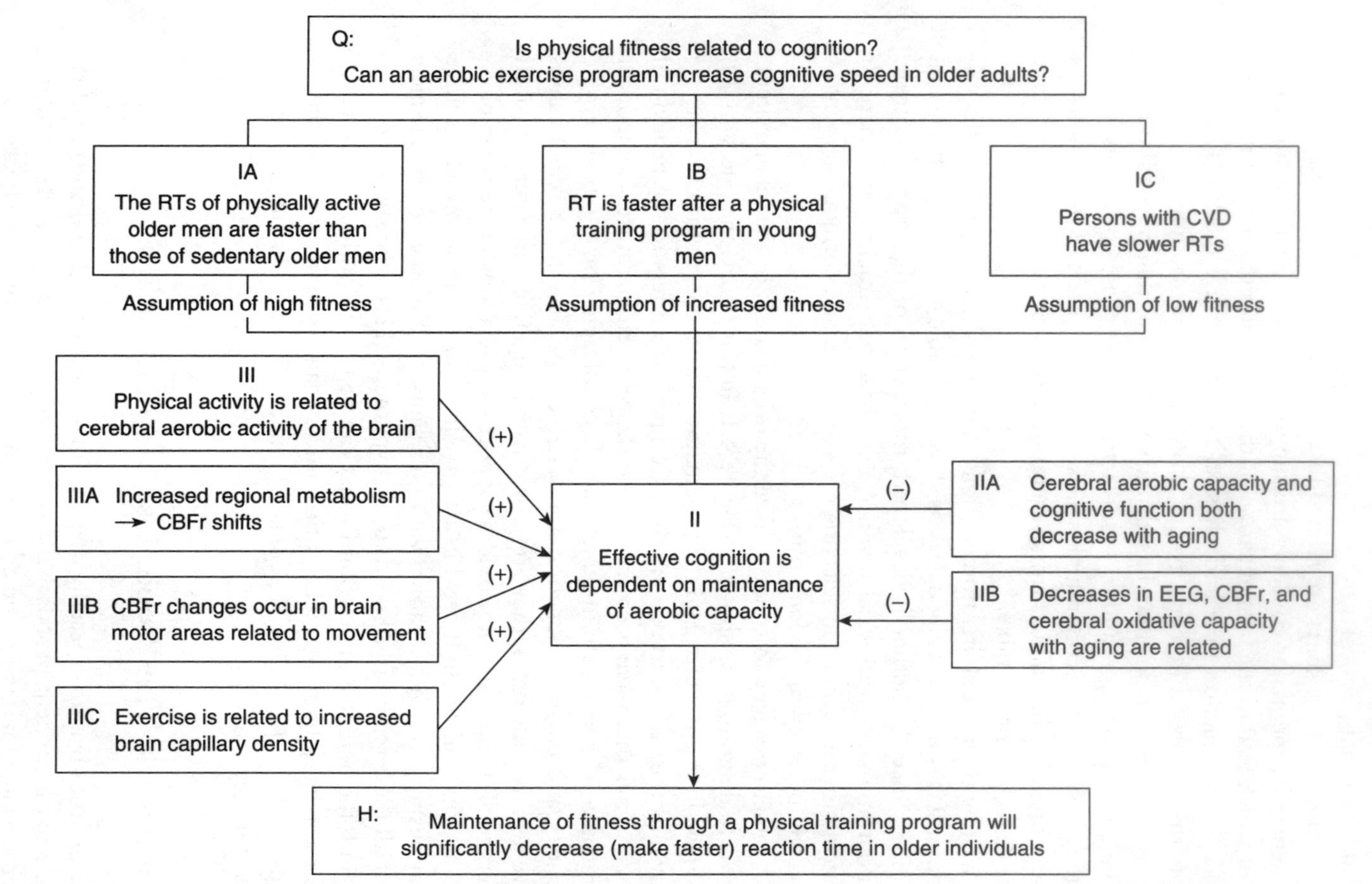

Figure 4.1 Example of a Diagrammatic Overview of the Related Literature

Table 4.2 Steps in Writing the Related Literature Section

1. Determine the major concepts (generally no more than two or three) that are pertinent to the proposed research question. That is, what are the concepts that must be true for your question to be appropriate or hypotheses tenable?
2. List concepts either in descending order of importance or in terms of logical presentation. That is, does one concept have to be understood before another can be introduced?
3. Prepare an outline with these major concepts as the major headings (such as the one in Table 4.1, Concepts I, II, and III).
4. Under each major heading, list the articles that are most directly related (authors and dates only).
5. If the articles under a major heading cluster themselves and suggest a subheading, then arrange the clusters under the major topics in logical order. For example, you might note that of the nine studies pertaining to the notion of a relationship between reaction time and physical fitness, in five of these reaction times of animals were reported, whereas in the other four studies, the reaction times were from humans. The interpretation of these studies, when clustered in terms of type of subject, might be different and have substantial bearing on the potential outcome of the proposed research.
6. Without referring to the details in the articles, summarize in one paragraph the combined findings of each cluster of studies. For example, in Concept I.A of Table 4.1, the summary might be that reaction time of physically active men and women is faster than that of sedentary individuals, as long as the subjects are over 60 years old. The summary of Concept I.B might be that aerobic training improves reaction time, but strength training in older individuals does not improve reaction time. At some point, you will have to discuss the interpretation to be made from different results of physical fitness on reaction time, depending on the way physical fitness as an independent variable is measured.
7. Write an introductory paragraph explaining what the two or three major areas are and in what order they will be discussed. Explain why the order used was selected, if that is important. Explain why some literature may be omitted if it might seem logical to the reader that it would be included.
8. Write a statement at the end of each section summarizing the findings within each cluster of studies. Show how this summary of findings relates to those in the cluster of studies described in the following paragraphs.
9. Write a paragraph at the end of each major topic (I, II, and III in Table 4.1), with a subheading if appropriate, that summarizes the major points, supports the cohesiveness of the subtopics, and establishes the relevance of these concepts to the proposed research question.

(Continued)

Table 4.2 (Continued)

10. Write a paragraph or short section (with the appropriate heading) at the conclusion that draws together all the major summarizing paragraphs.
11. Read the paragraphs and subject them to Steps 1-7 in the "Guidelines for Evaluating the Related Literature Section" (Table 4.3).
12. After all these concepts and subtopics have been carefully introduced, described, and summarized, return to the beginning and insert the documentation for each of the concepts in the proper location. That is, document the statements made in each of the paragraphs by describing the studies leading to them or verifying them.
13. Each time a reference is inserted, place the complete citation in a special file for eventual compilation of a reference list.
14. After a week has passed, reread the related literature section and use the complete "Guidelines for Evaluating the Related Literature Section" that are provided in Table 4.3. Make whatever revisions seem necessary and wait one more week.
15. Read the entire related literature section for coherence, continuity, and smoothness of transition from one concept to another. Check carefully for accuracy of all citations and, again, edit for mechanics.

It is difficult to imagine any proposal that could not be improved by the reporting of actual preliminary work. Whether it is to demonstrate the reliability of scores produced by the proposed instrumentation, the practicality of procedures, the availability of volunteers, the variability of observed events as a basis for power tests, subjects' capabilities, or the investigator's skills, the modest pilot study is the best possible basis for making wise decisions in designing research.

The pilot study, for example, is an excellent means by which to determine the sample size necessary to discover significant differences among experimental treatments.[1] Sample size estimation or "power analysis" recently has become commonplace in quantitative research. Although it always has been an important component of good research, it has become more frequently used because of the availability of easy-to-use books and computer programs. A particularly useful introduction to power analysis, replete with examples and tables, is *How Many Subjects?* (Kraemer & Thiemann, 1987). In addition, a number of computer programs are available—some of which only require the user to answer a few questions before calculating sample size. Because software is being updated quickly, we suggest inquiring at your

Table 4.3 Guidelines for Evaluating the Related Literature Section

After you have written the first draft of the related literature section (Steps 1-11 in Table 4.2) and then again as you prepare the penultimate draft (Step 14 in Table 4.2), answer the appropriate questions below. Mark the manuscript where the answers to each of these questions are located.

1. Is there a paragraph outlining the organization of the related literature section?
2. Does the order of the headings and subheadings represent the relative importance of the topics and subtopics? Is the order of headings logical?
3. Are there summary paragraphs for each of the two or three major sections and an overall summary at the end?
4. Is the relation of the proposed study to past and current research clearly shown in the summary paragraphs?
5. What new answers (extension of the body of knowledge) will the proposed research provide?
6. What is distinctive or different about the proposed research compared with previous research? Is this clearly stated? Is this introduced in just a few paragraphs?
7. Have the results from your own pilot studies, when appropriate, been interwoven into the synthesis of the related literature?
8. What are the most relevant articles (no more than five) that bear on this research? Underline these references. Are they listed under the first topical heading?
9. Are these articles presented in a way that denotes their importance? Are some cited so many times they lose their power through repetition?
10. Has the evaluation of these key articles, as well as all other articles, been presented succinctly in terms of both procedures and interpretation of results?

institution's statistical consulting office or computer center to find out what is available. In addition, you also should inquire about sample size calculators that are available on the Internet. These may serve your purpose, but you will want to make certain that the one you choose does the desired task using algorithms that are appropriate to your research.

Pilot data and a few decisions (primarily related to the error rates you want for the study) allow researchers to estimate the sample size needed to find significance, if in fact it exists in the data. It is possible that the estimated sample size required will be so large as to be prohibitive, in which case method and measurement tools should be reexamined. In the ideal case, a power analysis will inform the researcher of an appropriate sample size based on permissible error and the data—not just on arbitrary guessing. It is better to find the appropriate sample size in advance, rather than after the fact. Both a sample size that precludes a significant finding and the use of more subjects than needed are a waste of time and effort.

The use of even a few subjects in an informal trial can reveal a fatal flaw before it can destroy months of work. The same trial may even provide a fortunate opportunity to improve the precision of the investigation or to streamline cumbersome methods. For all these reasons, students and advisors should not insist on holding stringent, formal standards for exploratory studies. A pilot study is a pilot study; its target is the practicality of proposed operations, not the creation of empirical truth.

Examples of purposes that pilot studies might serve include the following:

1. To determine the reliability of measurement in your own laboratory, or under field conditions like those proposed for the study.
2. To ensure that differences that you expect to exist, do in fact exist—that is, if you are studying the different effects of gender on motivation, make sure the gender difference exists.
3. To "save" a sample that is difficult to obtain until the real research project is undertaken—that is, it is prudent to test available subjects until procedural bugs are worked out before testing world-class athletes.
4. To determine the best type of skills to use as an independent variable; for example—to study the effects of different ankle braces on knee mobility—test jumping vertically, horizontally, and while running, then select one.
5. To determine the feasibility of collecting audio recordings of participant talk in an environment where there is a great deal of background noise and high-speed verbal interactions.

The presentation of pilot study results sometimes does create a troublesome problem. Readers may be led inadvertently to expect more of pilot work than it can reasonably deliver. Their concerns with the limitations may distract from its limited use in the main line of the argument being advanced by the author. Accordingly, the best course is to make no more of the pilot study than it honestly is worth—most are no more than a report of limited experience under less than perfectly controlled conditions—and do so only when the report will best illuminate the choices made in the proposal.

Brief reference to pilot work may be made in supporting the broad research strategies selected consequent to the review of research. Some pilot studies may, in fact, be treated as one of the works worthy of review. More commonly, however, the results of exploratory studies are used in supporting specific procedures proposed in the section dealing with methodology.

When the pilot study represents a formal and relatively complete research effort, it is proper to cite the work in some detail, including actual data. When the preliminary work has been informal or limited, it may be introduced as a footnote to the main text. In the latter case, it may be desirable to provide a more detailed account of the work in a section of the appendix, leaving the reader the choice of pursuing the matter further if desired.

Murphy's Law: Anticipating the Unexpected

Murphy's Law dictates that, in the conduct of research, if anything can go wrong, it probably will. This is accepted by experienced researchers and research advisors but rarely is considered by the novice.

Within reasonable limits, the proposal is the place to provide for confrontation with the inexorable operation of Murphy's Law. Subject attrition cannot be prevented, but its effects can be circumscribed by careful planning. The potentially biasing effects created by nonreturns in questionnaire studies can be examined and, to some degree, mitigated by plans laid carefully in the proposal. The handling of subjects in the event of equipment failure is far better considered at leisure, in writing the proposal, than in the face of an unanticipated emergency. Field research in the public schools can provide a range of surprises, including indisposed teachers, fire drills, and inclement weather, all better managed by anticipation than by snap decisions forged in the heat of sudden necessity.

Equipment failure may interrupt carefully timed data collection sequences or interview protocols, or temporary computer breakdowns may delay data processing and analysis. At best, such accidents will do no more than alter the time schedule for the study. At worst, they may require substitutions or substantive changes in the procedures. Each step of the research process should be studied with regard to potential difficulty, and plans in the event of a problem should be stated in the appropriate place within the proposal. For instance, if unequal subject attrition occurs across groups, the type of analysis to be used with unequal *N*s should be stated in the analysis section of the proposal.

It is impossible to anticipate everything that can happen. A good proposal, however, provides contingency plans for the most important problems that may arise in the course of conducting the study.

Anticipating the Analysis: Do It Now

The proposal is the proper place to reveal the exact nature of the analysis, as well as anticipated plans in the event of emergency. For many students, especially master's candidates, the analysis, if statistical, may represent new knowledge recently acquired and not fully digested. In addition, the customary time limitation of 12 to 16 months, by which the master's candidate is bound, adds to the difficulty. The candidate may even be in the middle of a first formal course in techniques of data reduction and analysis during the same period of time used for constructing the proposal. Consequently, students find themselves in the awkward position of having to write lucidly about their analytic tools without yet knowing the entire armamentarium available. As untenable as this position is, and as much sympathy as may be generated by the student's advisor or friends, the omission of a full explanation of the analysis in the proposal may prove to be disastrous. Countless unfortunates have found themselves with files full of unanalyzable data, all because the analysis was supposed to take care of itself. A step-by-step anticipation of the analysis to be used is also a double check on the experimental design. Finally, as we suggest in Chapter 5, in the case of qualitative research forethought in planning the analysis can identify problems in data collection that might have a direct bearing on the persuasiveness of later conclusions.

Descriptive, survey, and normative studies require extensive data reduction to produce meaningful quantitative descriptions and summaries of the phenomena of interest. Techniques for determining sample characteristics may be different from those anticipated on the basis of pilot results, or the study sample may be skewed, resulting in the need to discuss techniques for normalizing the data.

Statistical techniques are founded on assumptions relating to sample characteristics and the relationship between that sample and its respective population. The methods one intends to use to determine whether the sample meets the assumptions implicit in the anticipated analysis should be clearly stated. For example, many statistical techniques must be used only when one or more of the following assumptions are met: (a) normal distribution of the sample, (b) random and independent selection of subjects, (c) linear relationships of variables, (d) homogeneity of variance among groups (in regression analysis, this is called homoscedasticity), (e) independence of sample means and variances, and (f) units of measure of the dependent variable on an interval or ratio scale.

Accordingly, the process of selecting a statistic brings with it a number of subsequent questions. What methods will be employed to determine that assumptions have been met? What analyses will be used in the event the assumptions are not met? Will the planned analyses be appropriate in the event subjects are lost so that there are unequal numbers of subjects or trials in the different conditions?

The analysis section of the proposal should be outlined to correspond with the objectives of the study so that each analysis will yield evidence relating to a corresponding hypothesis. In addition, the reader should be able to determine how all data collected are to be analyzed. An efficient way of accomplishing this is by using a model similar to the one shown in Table 4.4.

If data are to be presented in tabular or graphic form in the completed report, an example of one such table, including predicted figures, often will

Table 4.4 Example of a Table Showing How Each Hypothesis Will Be Tested

Hypothesis	*Variable*	*Analysis*
1. The self-concept of 7-year-olds is higher than that of 5-year-olds.	Sum of self-concept subscales	MANOVA with follow-up if age factor is significant
2. The self-concept of 9-year-olds is higher than that of 7-year-olds.	Sum of self-concept subscales	MANOVA with follow-up if age factor is significant
3. There is no difference between rural or suburban children's self-concept for 5- and 7-year-olds.	Sum of self-concept subscales	MANOVA with follow-up if location factor is significant
4. Nine-year-old suburban children will have higher self-concept than 9-year-old rural children.	Sum of self-concept subscales	MANOVA with follow-up if location factor is significant
5. The peer sociability of 7-year-olds is higher than that of 5-year-olds.	Peer Sociability Index	MANOVA with follow-up if age factor is significant
6. The peer sociability of 9-year-olds is higher than that of 7-year-olds.	Peer Sociability Index	MANOVA with follow-up if age factor is significant
7. Suburban children will have higher peer sociability than urban children.	Peer Sociability Index	MANOVA with follow-up if location factor is significant

NOTE: This is presented for illustrative purposes. An actual study may have additional hypotheses.

be helpful to the reader. The purpose of a table or figure in a research report is to summarize material and to supplement the text in making it clearly understandable. Tables and graphic presentation may serve the same purpose in a proposal.

Because of their display quality, the inclusion of tables in the proposal may expose errors of analysis or research design. For instance, some committee readers may not detect the use of an incorrect error term from a reading of the text, but one glance at the degrees of freedom column in an analysis of variance table will reveal the error. In analysis of variance comparisons, inclusion of several tables may expose the presence of nonindependent variables. Although most completed theses and dissertations no longer include analysis of variance tables, at the proposal stage they may be valuable for stimulating feedback from your committee.

If the analysis activity of the project is studied carefully in advance, many headaches as well as heartaches may be avoided. It may seem to take an inordinate amount of time to plan the analysis, but it is time that will not have to be spent again. As the analyses are completed, the results can immediately be inserted in the results section, and the researcher can complete the project with a feeling of fulfillment rather than a frantic scramble to make sense out of a puzzle for which some of the pieces may prove to be missing.

The Statistical Well: Drinking the Greatest Draught

Students usually can expect help from their advisors with the design of statistical analysis. At minimum, an experienced advisor will have some suggestions about the type of analysis that would be most appropriate for the proposed investigation. Many departments include specialists who have statistical consultation with graduate students as a primary part of their professorial responsibility. Other departments work closely with outside statistical and computer consultants who may be in departments of educational psychology, psychology, computer science, or business administration.

You should not, however, operate under the faulty impression that when the data are collected they can be turned over to a handy statistical expert who, having an intimate relationship with a computer, will magically transform raw data into a finished form of findings and conclusions. Just as you cannot expect the analysis of data to take care of itself, neither can you expect a statistical consultant to take care of it.

The assistance of a statistician or computer consultant, invaluable though it may be, ordinarily is limited to the technology of design and data analysis and help with using a packaged statistical program on a computer. The conceptual demands of the study and the particular form and characteristics of the data generated are the investigator's province—to be explained to the consultant, not vice versa. Likewise, the interpretation of results is a *logical,* not a technical, operation and thus is a responsibility for which only the investigator is properly prepared.

On Finding a Friendly Computer

A variety of interrelated decisions must be addressed as you contemplate the analysis phase of the study. First, what statistical tools are needed to analyze the data correctly? Then, after the statistical methods are selected, what is the most efficient way to do the analysis? (Note that it is not a good idea to reverse the order of these two questions.) At one time, students had few choices with regard to methods of analysis. Either they calculated the statistics themselves on hand or table calculators, or they used the university mainframe computer to complete the analysis. Today, a variety of statistical packages are available for computers. These packages are extremely versatile and, after a user learns the basics, are relatively easy to use. Part of research planning involves making choices among a growing number of attractive options for analysis.

The first step in selecting a software program to analyze your data is to find out what packages are available. Your advisor, other graduate students, and the university computer center can be helpful. Many universities have consultants to help students and faculty and may have agreements with software companies that greatly reduce the cost of the software. In addition, if a needed program is not available on a computer in your department, most universities have them available in a centralized site. It is important to note here that such careful search is just as important when selecting software for analysis of qualitative data as for the numerical data of quantitative designs.

The first step in deciding how to analyze your data is to eliminate all packages that will not allow you to do the desired analysis. This may involve more than assuring yourself that the proposed statistical operation is available. For instance, you should be interested in the number of subjects and variables the program or computer can handle, whether data can be converted to other forms if you need to conduct additional analyses at a later date, and, should data assume an unexpected form, whether other tests that might be needed are available with the same computer and package.

Fortunately, the most common and powerful statistical packages (Statistical Analysis System [SAS] and Statistical Package for the Social Sciences [SPSS]) now are available for personal computers. The packages can accommodate very large data sets for analysis and storage (the limits, of course, will depend on the configuration of the personal computer if one is being used for the analysis), can perform virtually any statistical test researchers require, and are available on most university campuses. Although some may find these packages slightly more difficult to master than most spreadsheets that have statistical options and those programs designed to perform a single analytic task, several benefits accrue from learning to use one or more of the large statistical packages. Among these advantages are:

1. You will be able to obtain help with your data analysis because many faculty members, graduate students, and consultants are experienced with the larger, well-known statistical packages.
2. If the data must be reanalyzed, it is very likely that the package can complete the analysis without the time-consuming problem of reformatting or retyping data. In addition, these programs can easily export data into formats that can be used by other more specific software programs.
3. Familiarity with a commonly used program will serve a student well when moving on to an academic appointment—the packages are available on most campuses and little start-up time will be wasted on learning a new program to analyze data.

No matter what type of computer or package you choose, it is absolutely essential to back up your files. This can be accomplished by simply saving computer files on both the hard disk and on a flash drive or CD. No matter how diligent you are about saving and backing up your files, you also should save a hard copy (a printout of your data) in more than one place. This applies equally to word processing text for the proposal document. As a matter of practice, have both a printout and an electronic copy of the data in at least two places. An extra word of caution—save frequently. You can do this manually or by using the autosave feature of your word processor, or, just to be safe, both. Remember, data are like eggs—they are most secure when stored in more than one basket.

Selecting the wrong statistical package rarely is fatal in the research process. The selection of a software package that does not meet all your needs may require additional time for entering data or may delay completion of the analysis. A little careful planning, however, may eliminate waste and reduce aggravation at a later date.

The Care and Nurture of Consultants

To obtain technical help from an advisor or consultant, you should be prepared to provide basic concepts about the content domain of the investigation, including a concise review of what is to be studied, a clear picture of the form data will take, and a preliminary estimate of alternative designs that might be appropriate to the demands of the proposed research. In addition, whether advice concerning design, statistics, or computer programming is sought from your project advisor, from a departmental specialist, or from an expert source external to the department, basic rules exist that must be considered if you are to glean the most information and help for the smallest cost in valuable consultation time.

Rule 1: Understand the consultant's frame of reference. As with any other situation involving extended communication, it is useful to know enough about the language, predilections, and knowledge base of the consultant to avoid serious misunderstandings and ease the process of initiating the transaction. Research consultants are professionals whose primary interest is in the process of research design, statistical analysis, and the application of computers in research. They use a system language unique to statistics and data management and appreciate those who understand at least the rudiments of this vocabulary. Correspondingly, your consultant will not necessarily understand the system language to be used in the proposal, nor the peculiar characteristics of the data. For example, it cannot be assumed that the consultant knows that some of your data consist of repeated measures. Similarly, it would be unlikely for a statistician to know whether these data are normally distributed across trials.

The consultant cannot be expected to make decisions that relate to the purpose of the study, such as those regarding the balance between internal and external validity. Some designs may maximize the validity of the differences that may be found, but correspondingly trade off external validity, and thus the generalizability of the findings. Decisions concerning the acceptability of such research designs must be made by the proposer of the study. The grounds for such a determination rest in the purpose of the study and thus in conceptual work completed long before the consultation interview.

Consultants can be expected to evaluate a proposed design, assist in selection from a group of alternative designs, suggest more efficient designs that have not been considered, and propose methods for efficiently completing the analysis. Often they can be most helpful, however, if preliminary models for design and statistical analysis have been proposed. This provides a starting place for discussion and may serve as a vehicle for considering characteristics of the data that will impose special demands. Consultants can provide information about computer programs, the appropriateness of a particular program for the proposed design, and the data entry techniques required of these programs. Again, some preliminary preparation by the student can make the consultant's advisory task easier and work to guarantee an optimal selection of procedures for processing raw data. This preparation might include talking with other students presently engaged in computer use, reviewing material on statistical programs, and visiting the computer center for an update on available services.

Normally, the statistics and computer specialists in a university setting are besieged by frantic graduate students and busy faculty colleagues, all in addition to the demands of their own students. Further, they may be responsible for the management of one or more functions in their own administrative unit or in the computer center. Finally, as active scholars they will be conducting their own research. Both the picture and the lesson should be equally clear to someone seeking assistance: Statisticians and consultants are busy people. They can provide effective assistance only when investigators come with accurate expectations for the kind of help a consultant can properly provide and come fully prepared to exercise their own responsibilities in the process.

Rule 2: Learn the language. The system languages of measurement, computers, experimental design, and both inferential and descriptive statistics are used in varying degrees in the process of technical consultation for many research proposals. No one, least of all an experienced consultant, expects fluent mastery in the novice. You must, however, have a working knowledge of fundamental concepts. These ordinarily include measures of central tendency and variability, distribution models, and the concept of statistical significance. Basic research designs, such as those described in introductory research method books, should be familiar to any novice.

It is, of course, preferable to complete at least one statistics course before attempting any study that will demand the analysis of quantitative data. If, as sometimes is the case, the student is learning basic statistics concurrently with the preparation of the proposal, special effort will have to be concentrated on preparing for consultations concerning design and analysis. The situation will be awkward at best, although many consultants will remain sympathetic and patient if students are honest about their limitations and willing to exert heroic effort once it becomes clear which tools and concepts must be mastered.

Beyond the problem of mastering enough of the language to participate in useful discussion is the more subtle problem of understanding the particular analysis and techniques selected for the study. You must not drift into the position of using a statistical tool or a measurement technique that you really do not understand—even one endorsed and urged by the most competent of advisors. Ultimately, you will have to make sense out of the results obtained through any analysis. At that point, shallow or incorrect interpretations will quickly betray a failure to understand the nature of the analysis. You also will have to answer questions about the findings long after the advisor is no longer around. Expert technical advice can be an invaluable asset in devising a strong proposal, but in the final analysis, such advice cannot substitute for the competence of the investigator.

Rule 3: Understand the proposed study. If the novice researcher does not understand the study sufficiently to identify and ask important and explicit questions, that lack is a major obstacle to a successful consultation. Only when the consultant understands the questions of central interest in the study is it possible to translate them into the steps of statistical analysis and selection of the appropriate computer program. Even if you employ a consultant only to help you with data preparation and analysis, if you cannot communicate exactly what you want, you may get back a printout from an analysis that is irrelevant to your needs. Further, a host of specific constraints associated with the nature of the study will condition the consultant's decision about which analysis to recommend.

You should be ready to provide answers to each of the following questions:

1. What are the independent variables of the study?
2. What are the dependent variables of the study?
3. What are the potential confounding variables of the study?
4. What is the measurement scale of each variable (nominal, ordinal, interval, or ratio)?
5. Which, if any, of the variables are repeated measures?
6. What, for each variable, are the reliability and validity of the scores produced by the instruments?
7. What are the population distribution characteristics for each of the variables?
8. What difference between dependent variables would be of *practical* significance?
9. What are the monetary, safety, ethical, or educational risks involved if a Type I error is made?
10. What is the nature of the loss if a Type II error is made?

In summary, before consulting with a technical specialist, you must be able to express exactly what the study will be designed to accomplish, identify the help needed in producing such a design, and provide all the explicit details the consultant will need in formulating advice.

The Scientific State of Mind: Proof, Truth, and Rationalized Choices

Scientific inquiry is not so much a matter of elaborate technology or even rigorous method as it is a particular state of mind. The processes of science rest, in the end, on how scientists regard the world and their work. Although some aspects of scientific thinking are subtle and elusive, others are not. These latter, the basic attitudinal prerequisites for the conduct of scientific inquiry, are reflected in the way a novice speaks and writes about proposed research. More directly, the proposal will reflect the degree to which the author has internalized critical attitudes toward such matters as proof, truth, and publicly rationalized choices.

What matters is not the observance of particular conventions concerning phrasing, but fundamental ways of thinking that are reflected in the selection of words. When, for example, students write, "The purpose of this study is to prove (or to demonstrate) that . . . ," there always is the dangerous possibility that the intent is to do just that—to prove what they have decided must be true.

Such phrasing cannot be dismissed simply as awkward or naive. Students capable of writing such a sentence without hearing at once its dangerous implications are students with a fundamental defect in preparation. They should be allowed to go no further until they apprehend both the nature of proof and the purpose of research in the scientific enterprise, for clearly neither is understood.

Proof, if it exists at all in any useful sense, is a probabilistic judgment based on an accumulation of observations. Ordinarily, only a series of careful replications can lead to the level of confidence implied by the word "proved." Research is not an attempt to prove or demonstrate, it is an attempt to ask a careful question and to allow the nature of things to dictate the answer. The difference between "attempting to prove" and "seeking proof" is subtle but critical, and a scientist must never confuse the two.

If scholars have no illusions about proof, it is wrong, nevertheless, to believe that they never care about the direction of results obtained from their research. As humans, they often are painfully aware of the distinction between results that will be fortunate or unfortunate for their developing line of thought. As scientists, however, they recognize the irrelevance (and even the danger) of allowing personal convenience or advantage to intrude in the business of seeking knowledge. In the end, researchers must sit down before their facts as students and allow themselves to be instructed. The task lies in arranging the context for instruction so that the answers to questions will be clear, but the content of the lesson must remain in the facts as revealed by the data.

A second critical sign of the student's ability to adopt the scientific viewpoint is the general way the matter of truth is treated in the proposal. When students write, "The purpose of this study is to discover the actual cause of . . . ," there is danger that they think it is possible to do just that—to discern the ultimate face of reality at a single glance. The most fundamental remediation will be required if such students ever are to understand, much less conduct, scientific inquiry.

Experienced researchers seek and revere veridical knowledge; they may even choose to think of research as the search for truth, but they also understand the elusive, fragile, and probabilistic nature of scientific truth. Knowledge is regarded as a tentative decision about the world, always held contingent on the content of the future.

The business of the researcher is striving to understand. Correspondingly, a high value is placed on hard-won knowledge. Truth is held gently, however, and the experienced investigator speaks and writes accordingly. It is not necessary to lard a proposal with reservations, provisos, and disclaimers such as "it seems." It is necessary to write with respect for the complexity of things and with modesty for what can be accomplished. The researcher's highest expectation for any study is a small but perceptible shift in the scale of evidence. Most scientific inquiry deals not in the heady stuff of truth, "establishing actual causes," but in hard-won increments of probability.

A third sign by which to estimate the student's scientific maturity is the ability (and willingness) to examine alternative interpretations of evidence, plausible rival hypotheses, facts that bid to disconfirm the theoretical framework, and considerations that reveal the limitations of the methodology. It is important not only to lay out the alternatives for the reader but also to explain the grounds for choice among them. The student who neither acknowledges alternatives nor rationalizes choices simply does not understand research well enough to bother with a proposal.

The mature researcher feels no compulsion to provide perfect interpretations or to make unassailably correct choices. One does the best one can within the limits of existing knowledge and the present situation. The author of a proposal is compelled, however, to make clearly rationalized choices from among carefully defined alternatives; this is one reason readers outside the scientific community find research reports tedious in their attention to detail and explanation. It is the public quality of the researcher's reasoning that makes a community of scientific enterprise possible, not the construction of a facade of uniform certainty and perfection.

Student-conducted research often contains choices that must be rationalized less by the shape of existing knowledge and the dictates of logic and more by the homely facts of logistics: time, costs, skills achieved, and available facilities. The habit of public clarity in describing and rationalizing choices must begin there, with the way things are. An honest accounting of hard and often imperfect choices is a firm step for the student toward achieving the habits of a good researcher—the scientific state of mind.

Note

1. Statistical significance, of course, is not synonymous with scientific significance in terms of the evolution of knowledge, or practical significance in terms of solving professional problems. Statistical significance largely depends on sample size and selection of an *alpha* level (the level of confidence necessary to reject the null hypothesis). It can be demonstrated between almost any two groups using almost any variable selected, if the sample size is large enough and the power of the test sufficiently high. Such differences between groups may be statistically significant but scientifically trivial and professionally worthless. The pilot study is an excellent device by which the probability of a Type I error may be estimated and an appropriate sample size selected. In this way, the investigator can increase the probability that a statistically significant result also will reflect a difference of scientific and practical significance.

5

Preparation of Proposals for Qualitative Research

Different Assumptions

The Only Constant Is Change

When this book first reached print (1976), the probability that any of our readers would elect a qualitative study for their dissertation or grant proposal was small. Only students in sociology or anthropology would have been likely to know that such an option even existed. In that year, with the exception of the small number working in history or philosophy most graduate students and young scholars would have begun their apprenticeship in research with studies cast in the familiar quantitative mode of natural science.

Those studies would have presumed views of the world and the process of inquiry that were then so pervasive in the disciplines of natural and social science (and applied professional fields such as education, nursing, and social work) as to be called simply "the scientific method." It was an orderly, understandable, and innocent time. There was only one way to do good research; one learned it, and then did it. Because science is not a static set of prescriptions, however, the natural evolution of the enterprise was to produce some dramatic alterations in that familiar landscape.

What changed was not the viability of the then-dominant natural science tradition. Experimental and quasi-experimental designs (with all their assumptions about the nature of truth and reality) remain the choice for many scientific purposes. What changed was our growing understanding that quantitative measurement, manipulative experiments, and the search for objective truth are not the only way to do research—and certainly not the only means of systematic investigation that deserve to be called scholarship.

A reconsideration of assumptions about such fundamental things as the nature of reality, what constitutes knowledge, and the role of human values in the process of research led scholars to challenge the adequacy of some of the established norms for inquiry. Such challenges led, in turn, to the development of new strategies for formal inquiry in the social sciences. Those alternatives created both the necessity of an expanded curriculum in research training programs and new options for research proposals.

As a convenient simplification, the alternative way of thinking about research questions (and the new forms of inquiry that it produced) is referred to in this text as "qualitative research." As an alternative paradigm, some forms of qualitative research have had long histories of use in particular areas of social science (for example, cultural anthropology) but until recently were not a significant part of mainstream scholarship or research training in other disciplines or applied professional fields.[1] In the last two decades, however, contributions from qualitative research have burgeoned in the literature of virtually every area of social science.

The same has been true of publications about qualitative research. At the end of the 1960s, only a handful of relatively obscure books and journal articles dealing with qualitative research existed. Suddenly, a mountain of print appeared containing discussions of theory, alternative designs for inquiry, and debates about technical applications and qualitative standards. Inevitably, then, qualitative research has been a "work in progress." Full of zesty academic disputation and exploratory studies pushing the envelope of acceptable science, qualitative research is slowly being defined by the uses of its practitioners.

That evolution has been reflected in successive editions of *Proposals That Work*. Certainly, our own vision of what constitutes a sound proposal for qualitative study has changed. For example, readers with access to earlier versions of this chapter would detect that we now have introduced attention to the particular problems of writing proposals for focus-group research, a format for qualitative inquiry that we had previously elected to ignore. Also, over time we have progressively altered our advice concerning the use of mixed (qualitative and quantitative) methods, the necessity for including a comprehensive review of the literature in the proposal document, and the appropriate way to address threats to validity. We make no apology for those (and other) changes in opinion and shifts in emphasis. To the contrary, we believe that the experiences of our own scholarship, as well as what we continue to learn from our students and colleagues, have given us a better understanding of the particular demands and problems involved in writing proposals for qualitative research.

Disagreements and Diversity

We recognize that some of you will come to this chapter unfamiliar with (or unclear about) qualitative research. For your use (as well as for the purpose of establishing a common point of reference for all readers), we will provide a brief introduction to the qualitative paradigm. First, however, we must establish several important caveats.

Because this type of inquiry is relatively new in some areas of scholarship and because it is everywhere undergoing a singular spurt of development and diversification, the field is anything but tidy. In consequence, we anticipate that some of the definitions employed in this chapter will be unsatisfactory (if not outright heresy) to some of our academic colleagues. This book, however, was not written for established researchers who already appreciate the subtle distinctions in the field—and who already know how to prepare proposals.

Further, there is no agreement on a universal label for this kind of research. In the literature of social science and applied professional fields, such terms as *interpretive, naturalistic, constructivist, ethnographic,* and *fieldwork* are variously employed to designate the broad collection of approaches that we call simply qualitative research. Some of those terms reflect important distinctions in the minds of the people who employ them. In contrast, we have selected the generic label "qualitative" as an arbitrary convenience.[2] It is intended to be a working term for writing about research proposals and should not be assigned any particular theoretical or ideological connotation.

Also, despite the deceptive simplicity of the single term *qualitative,* you will be confronted at the outset by the need to choose from among a number of possible formats for inquiry. Every research proposal must reflect the author's selection from among alternative approaches to doing research. If, for example, traditional quantitative inquiry seems most appropriate to a research question, there still remains the problem of determining which design offers the best fit. Will you employ an experiment, quasi-experiment, descriptive survey, case study, or mixed methodology?

Qualitative researchers must face a similar question. If the broad assumptions of the qualitative worldview seem appropriate (for reasons that typically mix the demands of the question with the investigator's practical intentions, the availability of resources, prior training, and personal dispositions), there still remains the problem of deciding which of the various qualitative traditions will best serve the needs of the study.[3] Do you think it appropriate to utilize the assumptions and methods that characterize ethnography, grounded theory, phenomenology, critical theory, or mixed methodology—or would a more generic qualitative format best serve your purposes?

Although true beginners may have the freedom to make such decisions more in theory than in practice (in graduate schools, for example, the availability of advisors often constrains the actual range of possibilities), they nevertheless should be aware of the alternatives. The choice is important, the different research traditions present very real advantages and limitations, and the decisions made in selecting from among them have profound consequences for the proposed study.

To survey some of the traditions under the generic umbrella of qualitative research, we suggest one or several of the following resources. They provide not only the economy of an overview (often with emphasis on contrasts across several perspectives and their characteristic strategies) but also a sense of the tensions that attend the process by which scholars begin to stake out territorial claims in a new enterprise.

In his textbook *Research Design: Qualitative, Quantitative, and Mixed Method Approaches* (2nd ed.), Creswell (2003) focuses on the relative advantages (and demands) of qualitative and quantitative designs for study. In a companion textbook (1998), however, he describes the conceptual and operational consequences of selecting a particular qualitative approach. To illustrate, he examines the respective traditions of biography, phenomenology, grounded theory, ethnography, and case study.

In addition, edited collections such as *The SAGE Handbook of Qualitative Research* (3rd ed.) (Denzin & Lincoln, 2005), and *The Handbook of Qualitative Research in Education* (LeCompte, Millroy, & Preissle, 1992), contain chapters that illuminate distinctions among various qualitative approaches. Several authors also have explored the major qualitative traditions for the purpose of developing useful taxonomies that clarify both differences and similarities. Among those are Jacob (1987, 1988, 1989), and Thornton (1987).[4]

Some of the textbooks that introduce qualitative research also give careful attention to helping novice researchers distinguish among the traditions within the paradigm. Among those are three that our students have found particularly helpful. Freebody (2003) devotes several chapters to sorting out major "categories" of qualitative inquiry, Morse and Richards (2002) use "selecting a method" as a central theme in organizing their text, and Rossman and Rallis (2003) treat alternative traditions as "genres," offering illustrative vignettes to suggest how graduate students might make selections from among them.

As part of a guidebook designed to assist people in the task of reading and understanding research reports, we developed our own list of "types" (Locke, Silverman, & Spirduso, 2004). Intended for a reading audience that includes both research consumers and entry-level researchers, the text offers brief descriptions of five distinctive approaches to qualitative inquiry. Each of those traditions is illustrated with the example of a published study.

As you can see, qualitative research is not a single, monolithic way of doing empirical research. We have made this a particular point of emphasis because obtaining an overview of the diversity within qualitative research will have the salutary effect of reminding you to hold your commitments lightly—until you have a complete map of the territory.

If you have the advantage here of having read the reports of some qualitative research it is virtually certain that you have encountered studies that could not be assigned to any particular tradition within that paradigm. Many published reports, perhaps even the majority, employ a generic approach that clearly is qualitative in its assumptions and methods, but that is not obviously the product of a single tradition. The concept of a "generic" qualitative approach was first forwarded by Merriam (2001, 2002). We have previously included it among the types of research described in our companion textbook (Locke et al., 2004), and will make use of it here as the "plain vanilla" version of qualitative inquiry.

Given such rich possibilities for inquiry, you might reasonably ask why we have opted for a generic model in the introduction that follows. Aside from the need to avoid overwhelming readers who are just starting to learn about research and the obvious need to keep this text to a reasonable size, the answer lies in our purpose.

All the forms of qualitative research with which we have experience have at least some characteristics in common. For example, a concern for maintaining flexibility in the execution of the research design is virtually universal among the different traditions. We also have found it true that all qualitative approaches make some demands that are different from those encountered in quantitative designs. A clear illustration is seen in the fact that the relatively straightforward recipes used to confront investigator bias in some quantitative studies (such as the double-blind experiment and analyses that identify investigator effect) are not possible in qualitative research. In consequence, no matter which of the qualitative research traditions is involved, the problem must be approached in a manner very different from, for example, what might be used in an experiment.

Observations like those have led us to conclude that for the purpose of helping novices make decisions about proposals, it is the similarities that bind distinct traditions into a broader family that really matter. Our simple, dichotomous division of paradigms into quantitative and qualitative, and our use of a rudimentary generic model to exemplify the latter, are practical, if inelegant, teaching strategies for encouraging you to focus first on the fundamentals. We leave to others the task of discerning and illustrating all the fine distinctions that lie both between and within the broad paradigms for inquiry in the social and behavioral sciences.

A Brief Description of Qualitative Research

What is qualitative research, and how is it done? On first hearing, the answer seems disarmingly simple. It is a systematic, empirical strategy for answering questions about people in a particular social context.[5] Given any person, group, or locus for interaction, it is a means for describing and attempting to understand the observed regularities in what people do, or in what they report as their experience.

For example, one of the most common purposes of qualitative research is served when investigators pose the basic question, "What's going on here?" Venues for their question might include a religious community, a hospital's administrative staff, a halfway house for paroled felons, a classroom, or a school district in which textbook selection has created controversy. Alternatively, the experience of being a first-year social worker, an older adult returning to a community college, a Little League coach, or a nurse in a hospital's intensive care unit might be the focus for study. In each instance, it is the total context that creates what it means to be present, to be a participant, to be a member, and to have a role to play. It is the participant's experience in that context that the researcher seeks to capture and understand in this kind of qualitative investigation.

The matter of definition is complicated, however, by the situation we described in the preceding section. The qualitative paradigm actually is a collection of research traditions, each with its own priorities, political agenda, preferred means of data collection and analysis, and—unhappily for the beginner—technical jargon. "What's going on here?" is just one of the many kinds of questions that can be raised about people. Further, such questions can be answered in very different ways when investigators start with different assumptions about what matters, and where and how to search. That is exactly the case for the different approaches to inquiry that are commonly grouped under the rubric of qualitative research.

The sometimes uneasy and frequently fractious scholarly bedfellows collected under the qualitative umbrella are bound together (albeit loosely) by several key assumptions about the social world, the nature of social realities, and the consequent nature of inquiry. It is not always easy, however, for the beginner to discern exactly what those shared assumptions may be. Part of the problem arises from the distinction between theoretical models and real life.

Because the development of a qualitative paradigm was stimulated by a rejection of the assumptions identified with quantitative research (the form of inquiry shaped by its roots in the philosophy of logical positivism), it has been common to describe qualitative research by noting how it differs from that older paradigm. Lists of assertions that purport to distinguish the belief systems of investigators are used to define quantitative and qualitative ways of understanding the world—and doing research.

We have some grave doubts, however, about the veracity of such dichotomous portrayals when applied to real people. Those reservations may prove to be useful as you negotiate your proposal through networks that may include advisors, reviewers, or co-investigators, all of whom will bring their own assumptions to your document. To be direct about our advice, you should not be disconcerted to discover that lives as lived rarely imitate science as performed.

Our careers have allowed us to become acquainted with a considerable number of active research workers from a variety of disciplines and applied fields. We can testify with confidence that it is difficult to find any investigator who will profess to all the beliefs that purportedly are required of an adherent to either qualitative or quantitative research. Put another way, when it comes to worldviews and personal philosophy, researchers are like most other people. Not only do they display the usual wide range of individual differences but they also seem perfectly comfortable with some remarkable inconsistencies in their thinking.

Accordingly, our advice is to take those lists of various assumptions about the world as pedagogical tools that can help you understand the qualitative paradigm, but not as portraits that accurately describe the beliefs of everyone you actually encounter on that (or any other) side of the paradigmatic street! With that caveat in mind, we offer you the following brief description of assumptions commonly attributed to (if not perfectly shared by) all who employ the qualitative paradigm.

> Qualitative researchers assume that there are aspects of reality that cannot be quantified. More particularly, they believe it is both possible and important to

> discover and understand how people make sense of what happens in their lives. That includes asking research questions about the meanings people assign to particular experiences, as well as discovering the processes through which they achieve their intentions in particular contexts. It also is assumed that all persons construct their individual accounts of each event in which they are participants. Those subjective constructions are accepted as the realities of the social world. Thus, what is real is regarded as invariably multiple and immutably relative to person and context.
>
> Given those assumptions, it is presumed appropriate and effective to inquire about specific social processes or particular persons' perspectives through direct contact with those involved—observing, interacting, and asking questions—in natural contexts where people function. In doing so, it is accepted that the investigator must be the primary instrument for data collection, and thus part of rather than separate from whatever is investigated. In turn, that requires the assumption that the researcher's own perspectives and values inevitably will become part of the research process and, ultimately, the findings and conclusions.

Over time, the assumptions laid out above have led qualitative researchers to generate a plethora of different designs for study, methods of data collection and analysis, and conventions for discourse about work within each tradition. Although the result sometimes resembles an academic Tower of Babel, those designs, methods, and conventions do reflect (with varying degrees of explicitness) the broad philosophic perspective assumed by the overarching paradigm.

To illustrate the conceptual difficulties that confront newcomers to the world of qualitative research, we can point to the incorrect (though commonplace) assumption that there are such entities as "qualitative research methods." We say "incorrect" because, absent the underlying assumptions, there is nothing in the long list of research procedures commonly employed by qualitative researchers that could not be employed in a quantitative study. Use of a method of data collection commonly employed in qualitative research, however, does not make a quantitative study one whit less quantitative in its scientific nature—unless that method is used (and the accumulated data subsequently interpreted) in accord with the assumptions of the qualitative paradigm. Of course, the reverse also is true. The further possibility of actually mixing research models (paradigms) within the same study is a topic we will address later in this chapter.

Having argued that no research methods are exclusively "qualitative" in nature, however, we have to admit that some research tools and conventions are closely identified with qualitative inquiry. In that sense, it is fair to say that some strategies for inquiry are characteristic of much, if not all, qualitative research. Those characteristics include many of the following (though in any one study it would be rare for all to be represented).

1. Qualitative researchers usually work inductively, trying to generate theories that help them understand their data. This is in contrast, for example, to the experimental tradition in quantitative research in which hypotheses usually are set a priori and then deductively tested with the collected data.

2. In most qualitative studies, the central problems are to identify how people interact with their world (what they do), and then to determine how they experience and understand that world: how they feel, what they believe, and how they explain structure and relationships within some segment of their existence.

3. Interviews and various forms of observation are the most common means of data collection, though they are sometimes supplemented by the collection of documents.

4. Data most commonly take the form of words (field notes, interview transcripts, diaries, etc.), although quantities, frequencies, and graphic representations also can be used.

5. It is common for reports of qualitative research to contain detailed descriptions of participants, as well as both the physical and social structures of the context within which the study takes place.

6. In many forms of qualitative research (though certainly not all), the investigator collects data in the field—the place where the behaviors of interest naturally occur.

7. Qualitative research designs frequently involve collection of data from different sources (sometimes by means of different methods) within a setting for the explicit purpose of cross-checking information, a procedure called *triangulation*. Inspection of such data sets and subsequent follow-up where discrepancies appear make this a primary means for establishing the truthfulness of sources.

8. It is rare for a qualitative researcher to introduce a deliberate intervention in the field of study. For the most part, investigators try to be non-intrusive, reducing the causes of participant reactivity to the smallest possible number. The exceptions here are participatory action research studies in which the investigator plays an active (though circumscribed) role.

9. It is common for qualitative researchers to have a primary interest in identifying and understanding the social processes by which particular end results are created, rather than simply describing the results themselves.

10. Although a researcher may make use of interview guides, systematic formats for recording observation data, and even material from responses to questionnaires, in the final analysis the researcher is the primary instrument for inquiry in qualitative research. With rare exceptions, he or she must interact directly with study participants, determining from moment to moment how to behave, what to notice and record, and how a particular line of inquiry does or does not offer promise for answering the research question at hand.

11. Qualitative researchers try to be conscious of the perspective they bring to a study. For that reason, they often explain their own background and particular interest in the research question as part of the research report. Researcher bias, however, in the sense of a vested personal interest in producing a particular finding, is regarded as a different matter. Bias must be controlled if the results of a study are to seem truthful. Accordingly, tactics for countering the inclination to see and hear what is desired often are central elements in qualitative research designs.

12. Irrespective of the paradigm, participant reactivity to the investigator or to the conditions of the study is a threat to the integrity of research. For a variety of reasons, however, this is a particularly sensitive problem in qualitative research. Accordingly, many studies include tactics intended to limit that source of data distortion.

13. Only rarely are samples of participants created by random procedures. Selection is more likely to be purposeful, with the intention of maximizing the utility of data for the research goals intended.

14. Designs for qualitative studies usually are carefully thought out during a period of planning and preparation. In some instances, the plan may be specified in considerable detail in the form of an extensive written proposal. Nevertheless, absolute fidelity in execution of a particular design does not offer the same benefit it yields in quantitative studies. Instead, it is common in qualitative research for plans to be regarded as tentative and contingent on the realities presented by data collection and analysis. At least in the case of experienced investigators, in-course adjustments are regarded as part of doing good research rather than fatal breaches of protocol.

15. Qualitative research reports often are written in the first person and may employ expressive language intended to make findings both accessible and powerfully persuasive.

Those are characteristics typical of studies that would belong under the qualitative umbrella. A particular scholar might add or delete one or several, or might modify some of our explanations, but we believe that when taken together, the 15 items collectively come close to a consensus model of what it means to do qualitative research.

As comforting as that assertion may be, however, you should be aware that there is disagreement about the exact location of boundaries for the paradigm. A type of inquiry called "critical theory research," for example, is considered by some scholars to differ so sharply in its basic assumptions as to constitute a separate and distinct paradigm. In other words, whether a critical approach to study is inside or outside the theoretical boundary of qualitative research depends on who is doing the looking.

Critical research is well established in both the social sciences and some areas of professional study (notably nursing, social work, and education). Although it shelters a sometimes bewildering number of its own permutations, there is a core of characteristics that does give definition to a critical view of both society and the tasks of inquiry. Because you are sure to encounter reports of critical studies, we will detour briefly here to suggest several sources that will make it easier to appreciate what you are reading.

For most readers, Thomas (1992) will serve to introduce this complex and sometimes controversial approach. For readings that offer more detail there is Madison's (2005) treatise on critical ethnography or any of several examinations of critical theory as applied in the field of education (Carr & Kemmis, 1986; Carspecken, 1996; Gitlin, 1994). None of those make heavy demands for background in the areas of social science and philosophy. There are source readings that go far deeper into critical theory, but most of our own students felt more than sufficiently introduced when they had finished the chapters on that subject in the handbook edited by Denzin and Lincoln (2005).

To really learn more about what constitutes the critical approach to research requires more than reading. It is our judgment that access to a mentor who has actually performed critical studies is a support for which books can never fully substitute. In accord with that opinion, except for the brief comment that follows, we will not attempt here to address the myriad implications of critical theory for the production of a research proposal.

As with its paradigmatic neighbors, critical theory includes a number of traditions that are only loosely (and not always comfortably) related. Participatory, empowering, action, materialist, and feminist research perspectives are among those. At the most fundamental level, however, what they share is an interest in and a concern for the ways that power is distributed and maintained in social settings—and how those arrangements can be challenged.

On the surface, that kind of interest appears not to require a new set of assumptions about inquiry. As the concerns of critical theory begin to shape the relationship between the researcher and the researched, however, they begin to have important implications for both method and the investigator's purposes in the study. In some forms of critical theory, for example, research becomes a vehicle for urging or facilitating the redistribution of power and improving the life circumstances of the participants. At that point, where the imagined line between politics and scholarship begins to blur, you can be sure that we have moved into new and controversial territory. Our own guide for reading and understanding research reports (Locke et al., 2004) describes several studies that serve to illustrate some of the dilemmas created by critical approaches to inquiry.

It is the very nature of those dilemmas and controversies, however, that will attract some individuals at the outset of their research careers. Personal values and deep commitments concerning justice and equity may present compelling reasons to explore critical theory as a mode of inquiry. Nevertheless, whatever you may discover about your own motives, it will be vital to remember that a sound critical study must begin with a sound proposal for becoming critical. Given that injunction, this brief introduction to critical theory will close with notice of one last variation on the theme.

Some investigators who do what is called "feminist" research also have made claims to a separate paradigmatic status for their perspective—a tradition that they regard as distinct from both critical theory and the larger collectivity of qualitative research. Whether or not that proves to be a useful perspective, we suggest that if you are interested in the way gender enters into research, you take a short excursion into the often lively literature on that topic. Good places to begin would be Lather (1991), or the collections edited by Eisner and Peshkin (1990), or Gitlin (1994). A somewhat more cautionary text edited by Ribbens and Edwards (1997) emphasizes the considerable difficulties that have attended efforts to pursue a consistently feminist viewpoint in qualitative research.

Preparing a Proposal for Qualitative Research

With regard to advice concerning proposals for qualitative research, we wish to make one point clear from the outset: Virtually all of what has been said about function, development, writing style, organization, and format for quantitative proposals will apply here. The qualitative proposal is not substantively a different kind of document. Our experience, however, has convinced us that it does present a particular set of problems that will demand your attention—either because they constitute common sources of difficulty for reviewers, or because they are matters better confronted in the proposal than at the later point of preparing a report.

In drawing up the following list of 12 key points, we have made three assumptions about your situation. First, we have presumed that your proposal will not be subject to review by someone who does not accept qualitative research as genuine scholarship. When that is the case, the proposal must undertake to defend the legitimacy of a paradigm—a task we think is better reserved for seasoned scholars. On the other hand, reviewers who simply are unfamiliar with qualitative research usually can be dealt with by patient explanation and by providing a judiciously short selection of introductory readings.

Second, we have assumed that at least one of the reviewers (for graduate students, most commonly the committee chair) will be familiar with the design and methods you propose, as well as the literature that explicates those choices. Without the support of such expertise, the burden of explaining, certifying, and persuading may be greater than can be born by a document of modest length.

Third, and finally, we have assumed that reviewers for your proposal will be looking for answers to a familiar set of questions. For example, do you appear to know the conceptual and methodological turf? Does your plan reflect careful thought? Do the parts of the proposal fit together? Is there evidence that you are fully aware of the problems to be overcome? Is the nature and scope of the study reasonably well matched to your skills and resources? If the five assumptions noted above are met, the following items can be used as a checklist and, when coupled with the generic advice about proposals offered in the other chapters of this book, should help you deal successfully with the problems that are particular to qualitative research.

In our experience, these are the proposal topics that most commonly attract the attention of reviewers.

1. *Why qualitative?* Make absolutely clear that a qualitative design is appropriate to both the study's general purpose (why you are doing the study) and the more specific research goals (such as formally stated research questions). Your training and personal values are not irrelevant to this argument. In the end, however, it is the match between the paradigm and the problem that must carry the day.

2. *Plan flexibility*. Present a plan describing what you will do from the outset to the finish of your study. If you are a novice, adhering closely to the general specifications of that plan is prudent policy. Qualitative research frequently involves some circumstances, however, for which a degree of anticipated flexibility also is wise. Some procedures must be responsive to what actually happens during data collection, as well as to the nature of the data that begin to accumulate. If there are such points within your proposed plan, showing that you have anticipated the necessity of selecting (or devising) alternative courses of action always is reassuring to reviewers.

A delicate balance has to be maintained in this aspect of a proposal for qualitative research. Present a careful plan and stick to it unless there are compelling reasons not to do so (dependence on "emergent design" is for experienced investigators). Give clear indication, however, that you have given careful thought to alternatives—should they be required.

3. *Build a framework.* Present a conceptual framework that helps to explain and clarify your proposed design. Define the main constructs and show their relationship to one another, to the research questions, to the methodology, and to the related literature. Absent a contrary regulation, this should not take the form of an extended general review of the literature. Previous scholarship is best limited to assistance in defining the precise conceptual territory of the proposed study. The literature can provide construct definitions, theoretical frameworks, examples of successful research strategies used in parallel circumstances, and a display of where your study would fit into the ongoing conversation among scholars. The primary emphasis, however, should be on the concepts and relationships assembled for your own study. A graphic format often is useful for achieving clarity here (see the excellent advice and examples provided by Maxwell, 2005).

4. *Articulate the parts.* Take special care at each step to write brief but explicit explanations of how the parts fit together—purpose with question, question with framework, framework with methods, and collected data with means of analysis. In the absence of the structure provided by standardized designs, it is easy for authors of qualitative proposals to lose the sense of cohesive unity among the several parts.

5. *Plan for validity.* Deal directly with the issue of validity.[6] If you complete the proposed study, everyone who reads the report will have a perfect right to ask "Why should I believe you?" If you want to be prepared with a persuasive answer, the proposal is the place to search out the threats to validity inherent in your plans. Morse and Richards (2002) have an excellent chapter on convincing readers about the rigor and trustworthiness of your study, and Maxwell (2005) goes even further to suggest that qualitative proposals should have a separate section devoted to answering the question, "How could I be wrong?" That is a tough question, but a healthy one.

   At the least, you must deal with the three threats to validity that most commonly attend the procedures used in qualitative research. (a) How will you ensure that descriptions of participants and context are accurate and complete? (b) Are your personal biases a threat? If not, why not, or if so, what do you plan to do about them? (c) In what ways and to what degree will participant reactions to you (and to the procedures used in the study) impede acquisition of valid data, and what are your plans for dealing with that problem? Again, experiences (and data) cited from pilot work are powerful ways of showing that you are prepared to deal realistically with threats to validity.

6. *Plan for records.* Explain exactly how you will maintain a paper trail. For example, if you propose to use some form of category system for analysis of transcripts, how and when will you record the exact source of each category? Where will you document revisions as they become necessary? Likewise, how will you record your speculations about the data, the participants, the study, or yourself when such thoughts cannot be handled as bracketed insertions into field notes or interview transcripts? We can confirm that such records will be essential when writing the report—many weeks or months later. The length and complexity of most qualitative studies virtually guarantee that you will lose important information if there is no planned regimen for recording it promptly and in adequate detail. We also urge that all records be maintained in duplicate—at separate locations.

7. *Demonstrate procedures.* Avoid the sin of nominalism. Because qualitative research traditions usually are rich with specialized nomenclature, it is easy to slip into the habit of using the names of complex operations as though they were magic incantations. To say that you will produce "grounded theory," or employ "analysis through constant comparison," "triangulation of data sources," or a "peer debriefer," tells the reader little more than that you know how to spell the words. Explaining why you will employ the operation, showing exactly how you will use it in the context of your design, and giving citations for the literature sources you have consulted constitute a far more persuasive presentation.

8. *Don't anticipate findings.* Be careful about using language that might appear to build your personal expectations about findings into the study procedures. For example, a research question such as, "How do student interns deal with feelings of hostility toward supervisory staff?" presumes that such affective states will be experienced by the participants. Whether that assumption is correct or incorrect, the question too easily translates into interview questions that can cue participants as to how they "ought" to feel.

   Of course, you will have anticipated at least some aspects of what is going on for your participants. In most studies, the conceptual framework itself reflects, directly or indirectly, what the investigator suspects is going on—or at least what he or she believes is worthy of attention. Such expectations, however, become bias (threats to validity) when they go unrecognized, unmonitored, and unchecked. The proposal should be written in a manner that is sensitive to such dangers.

9. *Be explicit about relationships.* Your proposal should demonstrate that the nature of your relationship with participants has been thoughtfully planned, and will be carefully monitored during the course of the study. What people say to you and how they behave in your company is conditioned in large part by the nature of your relationship. Thus, what happens between you and your participants will reflect how you present yourself and how the mutual perception of roles is progressively defined by subsequent interactions.

Presenting yourself (and then acting) as an interested and respectful visitor, a professional colleague, a genuine friend and companion, a needy supplicant (common among doctoral students), a potential political ally, an omniscient scholar (occasional with professorial types), a totally dispassionate and objective observer, or a warm and sympathetic listener, should be a decision that is made consciously and for deliberate purpose. Make no mistake; how you structure relationships with participants will have an effect on what will be collected as data. It is inevitable that those who read your proposal will ask whether those social interactions will serve the purpose of completing a sound study.

10. *Plan entry and exit.* Think through procedures for entry to and exit from your research context (both the site and the human relationships) and make those plans explicit in the proposal. Negotiating conditions for your presence and departure can be delicate matters that have both ethical and practical consequences. This aspect of your study is likely to contain problems that would not be encountered in a typical quantitative study—not least of which are close personal relationships with some or all of the participants.

11. *Treat transfer cautiously.* Be careful to write about the potential generalizability (application to populations outside your study) of your conclusions in ways that match the proposed procedures for selecting participants. Absent random sampling, claims to acquisition of valid knowledge about other groups (either within the study context or external to it) almost always are inappropriate. Careful and thorough description of the context and participants can make it possible to challenge readers of your report with the later question, "Why would these conclusions not apply in another context?" That, however, is not the same as attempting to generalize your findings to another setting.

12. *Name your own perspectives.* Either in the main body or in an appendix to the proposal, include a brief statement highlighting those aspects of your personal biography (work experiences, education, mentors, salient events) that have shaped your perspective on the proposed study—its questions, participants, venue, and general purpose. Reviewers at this stage, as well as readers who will later consult your report, have important reason to know what baggage you bring to the proposed study in the form of relevant beliefs, values, concerns, commitments, and intentions. You will be the primary research instrument, and publicly naming the ways you relate to your study is a vital part of preparing for qualitative research.

New Territory: Proposals for Focus Group Research

We have three reasons for inserting a special note here about proposing the use of focus groups for data collection in qualitative or mixed method studies. First, this is a research strategy that has been applied to social and behavioral research with increasing frequency in recent years. There is a strong possibility that you will consider using focus groups either as an adjunct to other data collection techniques, or as the primary vehicle for inquiry.

Second, this is a particularly attractive format for gathering information that allows insight into participants' feelings, attitudes, and perceptions about a selected topic. Unlike individual interviews, for example, a focus group presents a more natural environment because participants appear to be influencing and influenced by others—just as they are in life. In turn, that sense of authenticity can lend a useful degree of authority to the data.

Third, and finally, for the novice investigator there is a kind of obviousness about the method that allows brief acquaintance to obscure the truly difficult aspects of its application. If you are ready to use focus group technique with skill it can be a powerful tool. If you are not ready it also can be a trap that lures the unwary and unpracticed into wasting time, or, worse, into erroneous conclusions masquerading as profound insights.

Our advice on this topic can be didactic and explicit:

- Begin by reading some references that go beyond the brief descriptions found in standard research texts. We have found that Krueger and Casey (2000), Morgan (1997) and Puchta and Potter (2004) are ideal for that purpose. In addition, to begin sensitizing yourself to the mysteries of focus group moderation, you can do no better than to visit the world of marketing research in two books by Greenbaum (1998, 2000). You also should include inspection of some critiques and cautions about focus group methodology such as Kidd and Parshall (2000) and Webb and Kevern (2001). And, finally, be sure to read several reports from your own discipline in which focus groups are employed in a variety of different research designs.
- Find a mentor, whether a colleague or an academic advisor, who has had experience in leading focus groups as part of a research study. Talk with him or her to explore the advantages and limitations of the method, giving particular attention to the demands of skillful group leadership and the complexities of analyzing transcriptions of group interaction.
- If the use of focus groups continues to be an attractive option, at the earliest opportunity try your hand at both leading group sessions and working with actual transcriptions of data. Some pilot study experience will reveal rather quickly whether preparation and practice will allow you to become comfortable with the tricky nuances of focus group dynamics. Likewise, when confronted by the enormously messy realities of what people actually say in conversations, you will learn whether you really have the patience, penchant for rigorous use of an analytic system, and the necessary eye for subtle regularities within convoluted text that are required to extract useful meaning out of focus group recordings.

- By the time you have taken those three prior steps, the preparation of an effective proposal will present no mysteries. Aside from the requisites for any sound plan for research, the keys to making your proposal persuasive will be: (a) explaining exactly why focus group methodology is appropriate to your purpose, and (b) demonstrating that you actually have "been there and done that" with a credible record of familiarity and facility with the technique.

New Territory: Proposals for Mixed Method Research

If there is a growth area in qualitative research right now it has to be the swirl of interest in mixed method designs. And, although quantitative researchers have long made use of such tactics as interviewing subjects to obtain insights that supplement their primary analysis of numeric data, and qualitative researchers have for generations been noting frequency counts of all sorts of objects and events as part of their "thick descriptions," the rules for such casual and mostly opportunistic blending of methods have changed.

In fact, the major change is that there now actually are such rules! Conferences, workshops, college courses, monographs, textbooks, and special journal issues, all devoted to the theory and practice of using mixed method designs, make it impossible to be anything other than conscious, cautious, and deliberate in making such a choice. Being nicely sensitive to the changing winds of investigatory fashion, our own graduate students have become increasingly eager to suggest that they might do a mixed method dissertation. It is almost certain that you will at least consider such a possibility for your own proposal.

It is not difficult to trace some of the roots for the new status of mixed methods. The tensions that attended the growth of interest in qualitative research produced a lively debate about its legitimacy and, ultimately, about the compatibility of quantitative and qualitative paradigms for inquiry. Inevitably, that led to some uneasiness about studies that employed any method of data collection which ordinarily was associated with the procedures of another paradigm. From that concern scholars moved on to disputes over the question of whether it was possible for one investigator to simultaneously hold conflicting world-views about the nature of research. For a time, the whole topic appeared to be spiraling into chaos.

In the natural course of events in the research community, however, some level-headed and industrious people decided to define terms, parse the possibilities into orderly categories, decide which debates were not worthy of continued effort, and, in general, set the stage for researchers to get on with their work. That effort achieved much more than just lowering the volume of disputation. Once the possibilities for mixing methods were examined more closely, it was immediately clear that casual use had not allowed scholars to discover significant advantages that could be achieved by deliberately selecting particular formats for combining methods.

At last count we could identify no fewer than 23 distinctive designs for making use of mixed methods. Not one of those plans, however, is defined by the particular methods being mixed. Instead, the designs are defined by how the mixtures of methods are distributed, ordered, executed, and utilized in the subsequent data analysis. Thus, although some members of the 23 designs are no more than close cousins within a cluster of related strategies, you can be sure that in preparing your proposal there are clearly defined mixed-method choices to be made, and that making the right choice can have positive consequences for a study.

It is generally accepted that all methods have their limitations as well as their strengths. Thus, the idea is to mix them so that the strengths are complementary. Such blends can produce a convergence of evidence that reinforces findings, can eliminate or at least minimize otherwise plausible alternatives to your conclusions, or can enrich your conclusions by revealing divergent aspects that would otherwise be invisible.

We will not undertake even a survey of the possibilities for mixed method designs. What we can offer is to identify what we think are the best sources (most understandable and most economical of time) from which to extract a sound introduction. Ignoring journal articles and limiting the list to books and monographs, these are commonly available and generally reliable sources.

- Creswell, J. W., & Plano-Clark, V. L. (2006). *Designing and conducting mixed methods research*. Thousand Oaks, CA: Sage. (Creswell has a well-deserved reputation for constructing step-by-step guides for the decisions that must be made in designing studies. The book is oriented to the needs of graduate students, and focuses on producing a proposal that envisions a manageable mixed method project for the beginning researcher. If you are entirely new to the idea of mixing methodologies, this book offers three chapters that will be particularly helpful: Chapter 1, "Understanding Mixed Methods Research: Purpose and Organization"; Chapter 3, "Locating and Reviewing Mixed Methods Studies"; and Chapter 9, "Questions Often Raised About Mixed Methods Research.")
- Tashakkori, A., & Teddlie, C. (Eds.). (2003). *Handbook of mixed methods in social and behavioral research*. Thousand Oaks, CA: Sage. (There are 26 chapters here with varying levels of transparency and usefulness. We suggest starting with those by Teddlie and Tashakkori [Chapter 1], Green and Caracelli [Chapter 3], and Morse [Chapter 7]. As your interests and needs may dictate, there are more specialized chapters that offer discussions of topics ranging from computerized analysis of mixed methods research, through methods for teaching about mixed methods, to writing reports of mixed method studies.)

- Thomas, R. M. (2003). *Blending qualitative and quantitative research methods in theses and dissertations.* Thousand Oaks, CA: Corwin Press. (Written in transparent prose, and without any scholarly pretensions, this book deals directly with what the title proposes. Rich with illustrative examples from actual proposals that employ mixed methods, and built around a simple model of five kinds of research purposes, this book will get you started. If nothing else, the first chapter will take you further into the subject in 13 pages than many other sources can manage in 100.)
- Reichardt, C. S., & Rallis, S. F. (Eds.). (1994). *The qualitative-quantitative debate.* San Francisco: Jossey-Bass. (To make sense of where the mixed method movement began, you will need at least a little history. Here, in a small paperback from the publisher's series on *New Directions for Program Evaluation,* the contributors have captured the essence of the controversy. Remember, those who do not read history are destined to repeat it!)
- Green, J. C., & Caracelli, V. J. (Eds.). (1997). *Advances in mixed method evaluation: The challenges and benefits of integrating diverse paradigms.* San Francisco: Jossey-Bass. (From the same series as the item above, this collection offers one of the first clear descriptions of the benefits to be derived from careful construction of mixed method designs. The fact that the authors are primarily interested in research as a tool for evaluation is not an impediment to the utility of this source.)

In closing this brief section on mixed methods we want to make clear that we are not unalloyed fans of combining approaches to research. For example, folding in a frequency count of something is hardly likely to produce severe stress in a qualitative study. But assuming that the investigator can easily swing back and forth between the world views that are characteristic of positivistic and naturalistic science (as in so-called "sequential mixed-model" designs) is a far more troublesome requirement. As Datta (1994) has succinctly pointed out, "mixed-up" methods and models (those that are used incorrectly, inconsistently, or without reference to any kind of consistent philosophical orientation) are not to be confused with mixed methodology!

While the mixing of paradigmatic models (as distinct from methods) clearly is feasible, we still think it a less than prudent option for the average graduate student. He or she must have sufficient flexibility of mind to move back and forth between two vantage points when contemplating the data. That is asking a great deal of anyone who is not yet perfectly confident of his or her own grasp of research technique.

Resources for Qualitative Research

Our recommendations for resources that will assist in preparing a qualitative proposal have been divided into 12 topical areas. We caution you not to treat these as an exhaustive set designed to include all the resources that are relevant to qualitative research. Our purpose here is more limited. We have selected books and journals that attend directly to topics that novice investigators (principally our own students) typically have found relevant to the two steps of creating a study design and writing the proposal. The order of topics begins with the needs of a novice at the earliest stage, and then adds in serial order those investments in reading that would accompany preparation of a qualitative proposal.

1. *User-Friendly Introductions*—The first step in considering use of a qualitative approach is to gain some sense of what it can and cannot accomplish as a research strategy, and what sort of skills are required to design studies, collect and analyze data, and write reports. Our personal favorites for this purpose are Bogdan and Biklen (2003), Glesne (2006), Lofland, Snow, Anderson, & Lofland (2005), and Merriam (2001), all four of which have the virtue of careful revisions after long periods of extensive use. Delamont (2001) and Rossman and Rallis (2003) are equally competent as introductions, however, and as they have distinctive styles they may better fit your taste in textbooks. Finally, Creswell (2003) is widely used in introductory research courses because its juxtaposition of quantitative, qualitative, and mixed models for research helps students begin to grasp the range of formats that are available for inquiry.

For proposals involving smaller scale qualitative studies, Denscombe's *The Good Research Guide for Small Scale Research Projects* (2003) is aptly named and thoroughly practical. For case study research, Stake (1995) offers an eminently readable survey of its many permutations, and the several texts from Yin (2002a, 2002b) are widely used by students who are contemplating use of the case format.

Finally, even at the introductory level you will encounter a good deal of unfamiliar vocabulary, much of which will not be found in a standard dictionary. Fortunately, a dictionary devoted exclusively to the language conventions (a kindly label for jargon) of qualitative inquiry is available (Schwandt, 2001). With suggested readings and cross-references given in most entries, this is an instance for which cover-to-cover reading of a dictionary might represent an eminently practical strategy.

2. *Specimens*—Almost any modern research journal in the social sciences or applied professional fields will contain qualitative studies. It will be wise to locate journals that publish such reports in your own area of interest, and we urge you to do so early in the process of exploring the qualitative option. Only concrete examples can make theoretical discussions come alive. If you need a convenient starting place, both *Qualitative*

Sociology and *Qualitative Health Research* offer many reports that are relatively brief, intrinsically interesting for most readers, and quite undemanding of technical background.

Many areas of study also have collections of qualitative research reports in book format. Excellent examples include Merriam's *Qualitative Research in Practice* (2002), Milinki's *Cases in Qualitative Research* (1999), Yin's *Case Study Anthology* (2004), and Riessman's *Qualitative Studies in Social Work Research* (1994). If you wish to read reports that are further out on the cutting edge of evolving methodology for qualitative study, the collection edited by Denzin and Lincoln (2002) will take you there.

3. *Theory*—Books dealing with the theoretical foundations of qualitative research are notoriously difficult—and for the uninitiated, ponderously dull. Sooner or later, however, you will have to make a start. Many beginners find it best temporarily to bypass texts that deal with the epistemological roots of the paradigm, and start instead by reading material that deals with how to do research that is faithful to those philosophical origins.

The standard in the field for that purpose is Lincoln and Guba (1985), a book that is consulted, cited, quoted, and, in most cases, owned by virtually all who are active in qualitative research. A treatment of foundational theory that will be even more accessible for many readers is offered by Patton (2001). This is a textbook on theory and methods for research and evaluation that is justifiably famous for the author's light touch. Another gentle way to ease yourself into the literature dealing with theoretical foundations is to consult the book by Creswell (1998), in which the author illustrates how each of five different qualitative research traditions shapes the nature of study design. If you become serious about doing a qualitative study, there will be many more challenging theoretical mountains to climb. For the present, however, this is enough.

4. *About Qualitative Proposals*—If you have gone this far, you probably are going to write a proposal for qualitative research. Presently two textbooks are substantially devoted to that topic, and in this case our suggestion is that you purchase and use both of them. Marshall and Rossman (2006) and Maxwell (2005) are both ideal for the novice in qualitative research. With contrasting styles and emphases, they form a perfect complementary pair.

If you are preparing for a dissertation, the next step would be to read Piantanida and Garman (1999), who offer a four-chapter treatment of the proposal process. Richly illustrated with real-life examples of how graduate students struggle with that task, this is a tour through all of the notorious tough spots in gaining approval for a qualitative study. Finally, available on its website (http://www.ssrc.org), the Social Science Research Council provides a copy of its guide for authors, *The Art of Writing Proposals* (Przeworski & Salomon, 1995). Prepared by veteran reviewers, the advice offered is eminently practical and applicable to proposals for either qualitative or quantitative research.

5. *Ethics in Qualitative Research*—We placed ethics here in the topical order before the categories dealing with methodology because we think this is where it belongs. Thinking through the design for a study should be framed in terms of the ethical consequences that might attend each decision. To do so requires acquainting yourself with such problems early in the proposal process. As we noted in Chapter 2, the topic too often receives short shrift in research training at all levels and, sad to say, that continues to be true of qualitative research. Aside from a passing encounter with a human subjects review protocol, most beginners give scant thought to the question of how they wish to treat their participants—until they walk headlong into one of the nasty dilemmas that abound in qualitative research.

That it may be necessary to consider the topics of participant anonymity and confidentiality is fairly obvious. But how many novices would anticipate the need to deal with situations in which it is the participant who makes anonymity impossible? Likewise, it is one thing to plan for development of attentive and sympathetic listening skills when interviewing. It is quite another thing, however, to anticipate the need to handle interview situations in which the participant discloses sensitive and potentially dangerous information. Even the seemingly simple question of determining when a participant is free to withdraw from a study can be more complicated than it might seem. Decisions made in the relative calm of preparing a proposal are almost always better than those made in the field when the right and wrong of things is so easily obscured by panic.

You certainly can begin your preparation for designing an ethically responsible study by reviewing our introduction to that topic. From there, however, you will need to expand to resources that deal more particularly with the wide range of ethical dilemmas encountered in applied social research (Kimmel, 1988) and the processes by which your institution will enforce ethical standards when reviewing your proposal (Sieber, 1992).

Ultimately, of course, you will have to consult readings that are more directly focused on qualitative inquiry. One way to begin that process would be by simply surveying the full range of ethical problems that can arise in the conduct of qualitative studies. For that purpose, a recent text by Mauthner, Birch, Jessop, and Miller (2002) will serve admirably. As a more traditional alternative, however, you might consult the chapters devoted to ethical questions in foundational sources such as Eisner and Peshkin (1990), Denzin and Lincoln (2005), or LeCompte et al. (1992).

If they provide a closer fit with your interests, more focused treatments of ethics are available for qualitative studies in the field of education (Simons & Usher, 2000) and for designs that fall under the broad rubric of critical ethnography (Madison, 2005). Finally, in a small paperback prepared explicitly for undergraduates and beginning researchers, Carol Bailey has elegantly underscored the ubiquitous nature of ethical concerns when investigators intrude into people's lives. Each successive chapter in her *Guide to Field Research* (1995) contains a discussion of the sticky ethical problems that can entrap the unwary. If an honest appraisal of your background in research suggests that a book for novices would be appropriate, this one might be a sound investment.

6. *Methods*—Taken in the generic sense, methods are the tools for doing research. They include the procedures and instruments used by the investigator to generate data, as well as the techniques used to analyze data. Although we ordinarily think of interviewing (listening and conversing), observing (watching people), and document analysis (reading) as the primary means for collecting data in qualitative studies, a quick survey of published reports will reveal that there are many others. Questionnaires, surveys, systematic observation instruments, unobtrusive measures, videotapes, and photographs also serve as data sources. Likewise, dozens of methods are available when organizing data for the purpose of analysis. No single source can cover all methods, so it is necessary to narrow any search to what can be found in more specialized texts. We have provided suggestions below for such sources in the broad topic areas of interviewing, field notes, computer management of data, and analysis.

A useful first step in retrieving sources that explain particular research methods is to use a dictionary or glossary of qualitative terminology to look up the synonyms, definitions, and standard references associated with the method about which you want information. Again, we suggest Schwandt (2001) as particularly helpful for that purpose. A second step would be to consult the index to any of the introductory-level textbooks recommended above in order to track down citations for method-related articles and books.

If you think it would be helpful to browse articles that treat different aspects of qualitative methodology, the collection edited by Michael Huberman and the late Matthew Miles (2002) is a sound and generally accessible place to begin. Finally, we suggest that you take a few minutes to scan recent catalogue listings from publishers that offer books and monographs dealing with social research in various social science disciplines and applied fields (for example, Corwin Press, Falmer Press, Jossey-Bass, Longman, Pine Forge Press, Routledge, Sage Publications, and Teachers College Press). Methodology for qualitative research is a particularly active area of publication and many new resources appear each year.

7. *Interviewing*—Because this is a particularly common form of data collection, interviewing is accorded a chapter in nearly every qualitative research textbook. In that regard, we think that Patton (2001) and Merriam (2001) constitute good places to begin. Seidman (2006) and Rubin and Rubin (2005) are more specialized, yet quite accessible. Finally, Kvale (1994) has catalogued the most common objections to the use of interview data—and some appropriate responses.

8. *Field Notes*—Again, virtually every basic textbook covering qualitative research offers a chapter on the art of recording observations in the field—the ubiquitous "field notes." Lofland et al. (2005) provides a thoughtful treatment that has guided several generations of students in the social sciences. More extended instruction in the techniques of writing ethnographic field notes is available in Emerson, Fretz, and Shaw (1995).

9. *Computer Management of Data*—Computers make it possible to perform the complex tasks of data management, coding, retrieval, and manipulation with a speed, economy, and accuracy never before available to qualitative researchers. They also make it possible to waste time and resources, make egregious errors, and create the illusion of substance where there is none—more swiftly than ever before. Read, consult, reflect, and plan before you decide whether (or how) a computer might serve your proposed study.

Although many of the older textbooks can provide an adequate introduction to the potential uses and abuses of computer software for manipulating qualitative data, the frequent appearance of new software quickly serves to date most of their commentary on particular systems. At the time of this writing, Morse and Richards (2002) and Bazeley and Richards (2000) have published recent guide/workbooks for the use of NVivo, the most widely used software package for qualitative data analysis. By the time the present text is in your hands, however, there surely will be other such resources available, and, quite possibly, new software systems as well.

10. *Data Analysis*—Whether you use a sophisticated computer program or something as simple as a large accounting ledger, analysis requires a plan, and the place to sketch out the initial shape of that element of your methodology is in the proposal. The strategy you employ for making sense out of

your data will be determined in the first instance by the particular qualitative tradition you have adopted for the study. Some traditions offer little flexibility, whereas others make no prescription about the particular form of analysis to be employed. The purpose of your study, the nature of the data set, and even how much time you have available will influence the decision, and thus the resource texts, that will be relevant.

By far the most comprehensive collection of analytic strategies can be found in the second edition of Miles and Huberman's sourcebook *Qualitative Data Analysis* (1994). On a scale that more closely resembles standard textbooks, however, several other sources offer excellent overviews of what is available. Creswell (1998) compares techniques of analysis and data representation across five qualitative traditions, and Coffey and Atkinson (1996) use a single data set to demonstrate how different techniques of analysis can be employed in a complementary fashion.

You will find that one of the most carefully defined and fully explicated forms of analysis is employed in the social science tradition of "grounded theory." Although the term sometimes is used in nonspecific ways to refer to any approach for developing theoretical ideas that somehow begins with data, when employed to indicate a specific qualitative research tradition something very different is denoted. Grounded theory involves a specific set of highly developed, rigorous, and intellectually demanding analytic techniques for generating substantive theories of social phenomena.

Among various steps and techniques, the grounded theory model of analysis employs the operation called "constant comparison." That term, unfortunately, has been considerably misused by graduate students to give an air of scientific respectability to proposals for analysis by means of unsystematic data-snooping. If you wish to avoid the stigma of such amateur usage, we urge you to read *Basics of Qualitative Research* by Strauss and Corbin (1998). Although demanding, with reasonable diligence it is an accessible source, particularly when coupled with the extensive collection of reports and readings found in *Grounded Theory in Practice* (Strauss & Corbin, 1997).

11. *Writing and Publication*—There now is no scarcity of resources intended to help you write the report of your study. Although that task might be considered beyond the purview of the proposal, that is not at all the case. There are aspects of the report that are not encountered in the task of writing the proposal, but the work of crafting clear and precise prose about a study, whether proposed or completed, is the same. Just a few hours expended on a survey of what will be demanded in a sound report can have a powerful influence not only on how you write, but also on what you propose to do during the study.

We suggest that you begin with Becker's *Writing for Social Scientists* (1986), an elegant and highly personal exposition by a master of the craft, and continue with Golden-Biddle and Locke (1997), a lively introduction to the qualitative report as a form of "storytelling." In a more traditional textbook format, Holliday (2002) recounts the practical problems which writers face when they attempt to transform rich data from real-life research into a formal document. The book is richly diagrammatic, transparent in style, and appropriate for students in any discipline or applied field.

From there, you can turn to texts that have been honored by virtually universal use. Although specific to a single tradition, Van Maanen (1988) has long been a standard reference for the preparation of ethnographic reports. And, more generic in its purview, Wolcott's (2001) small monograph on "writing up" qualitative research (volume 20 in Sage Publications' *Qualitative Research Methods* series) is widely regarded as a classic and, in itself, constitutes a model of good writing.

Finally, if you can allow yourself to dream beyond the labors of writing, there are, of course, far more attractive activities. Those may include presenting papers, designing poster sessions, locating an appropriate outlet for publication, perhaps the somewhat less salubrious task of dealing with your first rejection notice, and, ultimately, the heady thrill of discovering that your efforts have influenced the thinking or practical decisions of other people.

To sample some of those delights, we suggest the unusual collection edited by Morse (1997), prepared by 24 active researchers in the field of health care. The focus of *Completing a Qualitative Project* is on what remains to be done after your data analysis is finished, as you move through the final stages of a long journey for which your proposal constitutes the first step.

12. *Standards for Qualitative Research*—If there were clear, explicit, and reasonably parsimonious standards for quality at the other end of the research pipeline (criteria for evaluating the adequacy of completed qualitative studies), those could be used for critiquing and strengthening the designs presented in proposals. We are sure you will not be surprised to learn that this is not possible, at least not in any simple and straightforward way. One obvious impediment rests in the diversity of qualitative research traditions, many of them still evolving, and all of them infused with distinctive intentions and commitments that would have to shape judgments about the adequacy of each study. Less apparent to the beginner, however, will be the fact that there has been a distinct reluctance on the part of members of the research community to engage with the question of qualitative criteria for qualitative studies.

The most common explanation for this aversion has been a desire not to impose rigid guidelines for process within a paradigm that puts the creative and individual journey toward understanding at the center of inquiry. Whether that explanation is entirely true or not, there recently has been some tentative movement toward elucidation of standards to be held when reviewing a report of qualitative research.

If you wish to make use of the literature on standards as part of your preparation for writing a proposal, we suggest beginning with an examination of what constitutes evidence in qualitative research (Morse, Swanson, & Kugel, 2001) and then opening the topic of standards with the brief but illuminating introduction offered by Rossman and Rallis (2003). If that seems helpful, more extensive discussions of the art and science of critiquing qualitative research may be found in Lincoln (1995) or in the triad of chapters authored, respectively, by Sally Thorne, Phyllis Stern, and Judith Hupcey in the collection edited by Morse (1997).

Some of the best writing about reasonable expectations for quality has appeared in discussions about the nature of validity in qualitative research. A useful introduction to that topic can be found in either Kvale (1995) or Johnson (1997). If that brief overview seems fruitful for thinking about your proposed study, however, you will wish to consult much more thorough discussions such as those offered by Eisenhart and Howe (1992), Mays and Pope (2000), or Maxwell (1992).

In closing this section dealing with resources, we draw your attention to several more that, while overlapping many of the 12 topical categories above, constitute by themselves an important repository for information about qualitative research. Whatever you are unable to find in the places we have suggested, you surely can locate in one of the following.

In 1986, Sage Publications initiated a series in *Qualitative Research Methods*. Through 2002 the series had accumulated 48 volumes, making the collection one of the most comprehensive efforts ever undertaken to provide support for qualitative research. Packaged in the form of paperbound monographs of 50–100 pages in length, the series provides beginners and veteran researchers with a resource that is both inexpensive and carefully targeted at the most troublesome aspects in the planning and execution of qualitative studies. Though not limited to qualitative designs, the parallel *Applied Social Research Methods* series from Sage Publications contains a number of monographs that may be equally useful in preparing a proposal. Those dealing with case study research (Yin, 2002a, 2002b), ethnography (Fetterman, 1998), participant observation (Jorgensen, 1989), and qualitative research design (Maxwell, 2005) represent excellent (and inexpensive) introductory sources.

The handbook format has become a standard means for periodically collecting and reviewing research in various disciplines, as well as presenting discussions of technical matters and broad issues attending the research enterprise itself. The Sage *Handbook of Qualitative Research* (Denzin & Lincoln, 2005) provides a generic overview of the paradigm, including its many component traditions, strategies for inquiry, methods of data collection and analysis, practical applications, and future development. Although less ambitious in scale and more specific to the field of education, *The Handbook of Qualitative Research in Education* (LeCompte et al., 1992) offers coverage of many broad topics that relate to qualitative research in any applied field of professional service.

A number of journals in the social sciences and applied fields have long included qualitative research reports in their coverage, but five have been devoted exclusively to that form of inquiry. The *International Journal of Qualitative Studies in Education* began publication in 1988 and now represents an invaluable resource for locating original reports, research reviews, theoretical and technical articles, and book reviews. Now in its third decade of issue, *Qualitative Sociology* deals with the qualitative interpretation of social life. Including both research reports and articles on theoretical and technical topics, the journal is made particularly useful by its frequent thematic issues and the inclusion of book reviews in an extended review-essay format. In the broad field of health care, the journal *Qualitative Health Research* offers an interdisciplinary forum for studies that employ qualitative methods. It has been particularly effective as a resource for encouraging alternative approaches to inquiry in the applied fields of medicine. Finally, relative newcomers to this specialized group, *Qualitative Inquiry* and *Qualitative Research* focus primarily on theoretical issues related to interdisciplinary and cross-paradigm research. Accordingly, research reports appear only when they present points of special methodological or theoretical interest.

The Decision to Go Qualitative

The decision to undertake a qualitative study brings two types of problems: those that are external and mostly antecedent to the proposal, and those that are internal and associated with devising an appropriate design for the study (and then writing the proposal document). To this point, this chapter has been concerned with the latter—the design and its written presentation. Here, however, we want to turn briefly to the precursors and surrounding circumstances that will exert their influence on that process.

The problems that precede the preparation of a qualitative proposal begin with the author as a person. Everyone who is tempted to employ a qualitative design should confront and honestly answer one question: "Why do I want to do a qualitative study?" Some novice researchers, traumatized by a fourth-grade encounter with fractions, see qualitative research as a way of avoiding numbers in general and statistics in particular. So long as question and paradigm truly are well matched, however, a choice made on such personal grounds, taken by itself, is neither improper nor inevitably dysfunctional.

It also is true, however, that finding certain kinds of questions appealing is not the same as having the personal capacities and intellectual interests demanded in the conduct of a qualitative study. If avoiding statistics is not so much a bad reason for electing to go qualitative as it is an *irrelevant reason,* then personal values that are compatible with a qualitative worldview are not so much an irrelevant rationale as they are *insufficient.* Determining a mode of inquiry that matches both your research goals and your research capabilities requires a more elaborate calculus.

We are not suggesting here that qualitative research must presume unusual capabilities or exceptional intellect. To the contrary, our experience has been that a large proportion of properly trained individuals can do perfectly competent qualitative research—if they are strongly motivated to do so. The demands, however, are very real, and your capacity to meet them should be considered with honest care. At the least, factors such as interactive social skills (people skills), a sensitive ear for nuanced language, and analytic capability (pattern recognition) are far more relevant to success than either an interest in social dynamics or a bad case of math anxiety!

In the same vein, a graduate student who elects qualitative research because it appears either to be relatively "quick" in terms of time commitment or "easy" in terms of intellectual demands has, in the first instance, simply never talked with anyone who has completed such a study and, in the second instance, not read published reports of field research with much care. Qualitative studies are never quick and rarely are completed within the projected time lines. The analysis of qualitative data demands a sustained level of creative thought rarely required of the investigator once data are collected in a quantitative study. Qualitative research may be enormously valuable for many purposes, as well as immensely satisfying to the investigator, but quick and easy it is not.

The vital codicil to this discussion of personal factors in selection of a research paradigm is that whatever one's abilities or interests, it still is necessary to match the procedures of inquiry with the purposes of inquiry. Put another way, a close and comfortable match between your abilities and predilections on the one hand, and the nature of qualitative research on the other hand, as genuinely desirable as that may be, still will not make it possible to fit round methods into square questions. If you start out committed to doing qualitative research, then you have limited yourself to only those questions that best yield to that scientific paradigm. Selection of the means for inquiry prior to identification of the question always bears that restriction.

A second preproposal problem centers on the investigator's ability to move out of the quantitative mode of thought. For many graduate students, the process of adopting assumptions that are consistent with the qualitative perspective means breaking well-worn habits of thought. For most of us, the assumptions of quantitative research have been presented and learned as "science" through many years of school and university education. The unspoken premise of the perspective used in the physical and natural sciences is that if something truly exists, it must exist in some quantity, and exist "out there" in some finite form. Although that certainly represents the most extreme formulation of positivist thought, in functional terms it marks the distinction between what is "real" and "not real" in much of our everyday life. To adopt a different and unfamiliar perspective is sometimes more difficult than one might expect.

To think and write in a consistent fashion with the assumption that people construct reality, allowing truth to reside as much in our heads as "out there," requires a sharp alteration in the habits of intellect. Even imperfectly accomplished, this is difficult for most, and, as experience warns us, impossible for some. It is important to confront this problem during the early apprenticeship stage of research training, when patient and sympathetic mentors can assist with the difficult transition between familiar ways of perceiving and conceptualizing, and a different vantage point—the qualitative worldview.

Before making the final decision to go qualitative, there also are logistic questions to consider. Preparation for qualitative research is most effective when it takes the form of apprenticeship, with intensive field experiences and closely supervised opportunities to practice the analysis of actual data. Where those opportunities are not available, some hard questions must be addressed. Will it be cost-effective to acquire training and experience through some alternative source? Is there sufficient time to invest in both a reasonable level of preparation and a lengthy study? Will the best solution be to transfer to another department or institution? These are hard questions, but better raised early than too late.

Having fulfilled our obligation to sound reasonable cautions, however, the last thing we would wish is to leave the impression that the opportunity to do qualitative research is anything less than an exciting, thoroughly fascinating, and deeply fulfilling option. Here is a form of research that invites questions that deal with how real people think and feel. Here is a way to produce findings that are thoroughly grounded in the stuff of a recognizable reality—the world as it is experienced. Finally, here is an opportunity to join with other scholars in an enterprise characterized by fresh ideas, energetic expansion, and as-yet-unexplored possibilities. If all of that sounds appealing to you, then yes, you should consider going qualitative.

Notes

1. The word *paradigm*, as used here, denotes a conceptual framework that provides a particular way of thinking about meaning in the context of formal inquiry. Thus, taken collectively, the beliefs, values, perspectives, commitments, and consequent methods of inquiry shared by a group of investigators constitute a scientific paradigm. Just as each person has a cognitive schema for making sense out of his or her daily experience in the world, scholars who share the assumptions of a paradigm have a particular way of making sense out of their scientific world. Social scientists who perform experiments, for example, share a general perspective on their work that is distinctly different from that shared by investigators who do studies in the feminist tradition of qualitative research. Accordingly, it would be asserted, for example, that experimental (quantitative) and feminist (qualitative) research have their roots in different scientific paradigms. Any modern textbook on qualitative methodology will provide a starting place for defining the qualitative paradigm, but nearly all of them will lead you back to Lincoln and Guba (1985), and from there to the modern origins of the term *paradigm* in Kuhn (1996).

2. Our use of the term *quantitative* also is arbitrary. The simple presence of numbers in a study, per se, does not serve to distinguish one paradigm from another. Both quantitative and qualitative research can employ quantification. It is the underlying assumptions about those numbers that provide the distinctive differences.

3. The term *tradition* in this context represents a convention for collectively designating the various distinctive forms of qualitative research. It does not carry the general, common-use sense of something passed down, generation to generation, over a long period of time. Instead, it is employed in the narrower sense of designating a coherent body of precedents intended to govern some set of actions—in this case, a mode of thought and a related set of research procedures. Hence, phenomenological research is a tradition within the qualitative paradigm, as are ethnography, life history, symbolic interactionism, grounded theory, and case studies.

4. Taxonomies for qualitative approaches to research may be extended beyond what has been touched on in the present chapter by including artistic as well as scientific modes of inquiry. For an introduction to that topic, see Eisner (1981) and Thornton (1987).

5. Our use of the term *empirical* here is not employed as it sometimes is in the discipline of philosophy, to designate the family of theories called *empiricism*.

We use it only in the common-use sense of designating inquiry based on the data of experience, things that the investigator saw or heard that can then be employed as the warrant for a claim. In that sense, all qualitative research is empirical.

6. Readers should be aware that the word *validity* is not commonly used in the rhetoric of qualitative research. In fact, there is considerable discomfort with and even some rejection of the construct. Our decision, nevertheless, to employ it throughout this chapter is based on two simple facts. First, most of our readers will be familiar with the general use of the term to denote a datum that accurately represents the phenomenon to which it refers (it is true), or a research finding for which persuasive evidence has been presented (it is certain). Second, the formal language that has been invented by qualitative researchers to replace the construct of validity is extensive, complex, and far from sufficiently universal to constitute a reliable system language. If you decide to write a proposal for qualitative research, it will be necessary to assimilate this new language (we suggest that you begin with Lincoln and Guba, 1985; Guba and Lincoln, 1989; Morse et al., 2001, and Kvale, 1995), but for our present purpose the single term *validity* will suffice.

6

Style and Form in Writing the Proposal

The writing style of the thesis or grant proposal may be the most important factor in conveying your ideas to graduate advisors or funding agencies. Even experienced researchers must critically evaluate their writing to ensure that the best laid plans are presented in a clear, straightforward fashion. The sections that follow represent primary concerns for proposal writers.

Praising, Exhorting, and Polemicizing: Don't

For a variety of motives arising principally from the reward system governing other writing tasks, many students use their proposal as an opportunity to praise the importance of their discipline or professional field. Some use exhortative language to urge such particular points of view as the supposed importance of empirical research in designing professional practice. Others use the proposed research as the basis for espousing the virtues of particular social or political positions.

There is no need or proper place in a research proposal for such subjective side excursions. The purpose of a proposal is to set forth for a reader the exact nature of the matter to be investigated and a detailed account of the methods to be employed. Anything else distracts and serves as an impediment to clear communication.

As a general rule, it is best to stick to the topic and resist the temptation to sound "properly positive and enthusiastic." Do not attempt to manipulate the opinions of the reader in areas other than those essential to the investigation. The simple test is to ask yourself this question: "Does the reader really need to consider this point in order to judge the adequacy of my thinking?" If the answer is "no," then the decision to delete is clear, if not always easy, for the author.

Quotations: How to Pick Fruit From the Knowledge Tree

Too often, inexperienced writers are inclined to equate the number of citations in a paper with the weight of the argument being presented. This is an error. The proper purposes served by the system of scholarly citation are limited to a few specific tasks. When a document has all the citations needed to meet the demands of those few tasks, it has enough. When it contains more citations, it has too many and is defective in that regard. Reviewers deem the use of nonselective references as an indication of poor scholarship, an inability to discriminate the central from the peripheral and the important from the trivial in research.

The proper uses of direct quotation are even more stringently limited than the use of general citations for paraphrased material. The practice of liberally sprinkling the proposal with quoted material—particularly lengthy quotations—is more than pointless; it is self-defeating. The first truth is that no one will read them. The second truth is that most readers find the presence of unessential quotations irritating and a distraction from the line of thought being presented for examination. When quotations are introduced at points for which even general citations are unnecessary, the writer has displayed clear disregard for the reader.

There are two legitimate motives for direct use of another scholar's words: (a) the weight of authoritative judgment, in which "who said it" is of critical importance, and (b) the nature of expression, in which "how it was said" is the important element. In the former instance, when unexpected, unusual, or genuinely pivotal points are to be presented, it is reasonable to show the reader that another competent craftsperson has reached exactly the desired conclusion, or observed exactly the event at issue. In the latter instance, when another writer has hit on the precise, perfect phrasing to express a difficult point, it is proper to employ that talent on behalf of your own argument. The rule to follow is simple. If the substance of a quotation can be conveyed by a careful paraphrase, followed, of course, by the appropriate credit of a citation, with all the clarity and persuasive impact of the original, *then don't quote*. In almost all instances, it is best for the proposer to speak directly to the reader. The intervention of words from a third party should be reserved, like heavy cannon in battle, for those rare instances when the targets are specific and truly critical to the outcome of the contest.

A beneficial technique for students who recognize their own propensity toward excessive quotation is to use the critical summary form of note-taking. In this format, after carefully recording a full citation, each article is critically examined and then paraphrased on reference cards in the student's own words. During note-taking, a decision is made on whether the aesthetics of phrasing or the author's importance in terms of authority justify the use of direct quotation. Except in rare instances, quoted material is not transferred to the note cards. Thus, direct quoting becomes less tempting during the subsequent writing phase when the student has recourse to notes. This technique also prevents unintentional plagiarism.

If using a computer for note storage and retrieval, similar precautions should be taken. When retrieving information from the computer, you should make certain that each item can clearly be identified either as your paraphrase or as a direct quotation. It is possible to lose this information as you switch back and forth between notes, computer, and proposal document. One way to ensure identification, which can be used both on the computer and on handwritten note cards, is to use quotation marks for all direct quotations, listing the page number on which the text was found in parentheses immediately after the closing quotation mark. As you work between notes, computer, and writing of the proposal, transfer all of this to your draft.

Clarity and Precision: Speaking in System Language

The language we use in the commerce of our everyday lives is common language. We acquired our common language vocabulary and grammar by a process that was gradual, unsystematic, and mostly unconscious. Our everyday language serves us well, at least as long as the inevitable differences in word meanings assigned by different people do not produce serious failures of communication.

The language of science, specifically the language of research, is uncommon. The ongoing conversation of science, for which a research proposal is a plan of entry, is carried on in system languages in which each word must mean one thing to both writer and reader. Where small differences may matter a great deal, as in research, there must be a minimum of slippage between the referent object, the word used to stand for the object, and the images called forth by the word in the minds of listeners and readers.

The rules of invariant word usage give system languages a high order of precision. Minute or subtle distinctions can be made with relative ease. Evaluative language can be eliminated or clearly segregated from empirical descriptive language. More important, however, the language of research affords the reliability of communication that permits scientists to create a powerful interdependent research enterprise rather than limited independent investigations. When a chemist uses the system language of chemistry to communicate with another chemist, the word "element" has one and only one reference, is assigned to that referent on all occasions, is used for no other purpose within the language system, and consistently evokes the same image in the minds of everyone, everywhere, who has mastered the language.

Various domains of knowledge and various research enterprises are characterized by differing levels of language development. Some disciplines, such as anatomy or entomology, have highly developed and completely regularized language systems, whereas others, particularly the behavioral sciences, employ languages still in the process of development. Irrespective of the area of investigation, however, the language of any research proposal must, as a minimum requirement, be systematic within itself. The words used in the proposal must have referents that are clear to the reader, and each must consistently designate only one referent. When the investigation lies within a subject area with an existing language system, then, of course, the author is bound to the conventions of that system.

Obviously, the researcher should be familiar with the system languages that function in the area of proposed investigation. Reading and writing both the specific language of the subject matter area and the more general languages common to the proposed methodology (statistics, experimental design, psychometrics, qualitative traditions, etc.) are clear requirements for any study. Less obvious, however, is the fact that research proposals, by their exploratory nature, often demand the extension of existing language into new territory. Operations, observations, concepts, and relationships not previously specified within a language system must be assigned invariant word symbols by the investigator. More important, the reader must carefully be drawn into the agreement to make these same assignments.

Advisors and reviewers misunderstand student proposals far more often than they disagree with what is proposed. The failure of communication often occurs precisely at the point where the proposal moves beyond the use of the existing system language. This problem involves a failure of careful invention rather than a failure of mastering technique or subject matter. The following rules may be of some help as you attempt to translate a personal vision of the unknown into the form of a carefully specified public record.

1. Never invent new words when the existing system language is adequate. If the referent in established use has a label that excludes what you do not want and includes all that you do want, then it needs no new name.
2. If there is reasonable doubt as to whether the word is in the system or the common domain, provide early in the proposal the definition that will be used throughout. Readers may give unnecessary time and attention to deciphering the intended meaning unless you put their minds at ease.
3. Words that have been assigned system meaning should not be used in their common language form. For example, the word *significant* should not be used to denote its common language meaning of "important" in a proposal involving the use of statistical analysis. The system language of inferential statistics assigns invariant meaning to the word *significant*; any other use invites confusion.
4. Where a system language word is to be used in either a more limited or a more expanded sense, make this clear when the word first is introduced in the proposal. If local style requirements permit, this is one of the legitimate uses of footnotes to the text.
5. Where it is necessary to assign invariant meaning to a common language word to communicate about something not already accommodated within the system language, the author should choose with great care. Words with strong evaluative overtones, words with a long history of ambiguity, and words that have well entrenched usage in common language make poor candidates for elevation to system status. No matter how carefully the author operationalizes the new definition, it is always difficult for the reader to make new responses to familiar stimuli.
6. A specific definition is the best way to assign invariant meaning to a word. When only one or two words require such treatment, this can be accomplished in the text. A larger number of words may be set aside in a section of the proposal devoted to definitions. The best definition is one that describes the operations that are required to produce or observe the event or object. For example, note how the following words are assigned special meaning for the purpose of a proposal.
    a. A common language word is assigned invariant use:
    Exclusion will be deemed to have occurred when both of the following happen: The student no longer is eligible to participate in extracurricular activities under any provision of school district policy, and the student's name is stricken from the list of students eligible for extracurricular activities.

b. A system language word is employed with limitations not ordinarily assigned:
The curriculum will be limited to those after-school activities that the current *School District Manual* lists as approved for secondary school students.
c. A system language word is operationalized by describing a criterion:
Increased motivation will be presumed when, subsequent to any treatment condition, the time spent in any extracurricular activity rises more than 10% of the previous weekly total.
d. A common language word is operationalized by describing a criterion:
Dropouts are defined as all participants who fail to attend three consecutive activity meetings.
e. A system language word is operationalized by describing procedure:
Reinforcement will refer to the procedure of listing all club members in the school newspaper, providing special hall passes for members, and listing club memberships on school transcripts.
f. A common language word is operationalized by describing procedure:
Instruction will consist of five 10-minute sessions in which the club sponsor may employ any method of teaching so long as it includes no fewer than five attempts for each student to complete the activity.

Editing: The Care and Nurture of a Document

A proposal is a working document. As a primary vehicle for communication with advisors and funding agencies, as a plan for action, and as a contract, the proposal performs functions that are immediate and practical, not symbolic or aesthetic. Precisely because of these important functions, the proposal, in all its public appearances at least, should be free from distracting mechanical errors and the irritating confusion of shoddy format.

At the privacy of your own desk, it is entirely appropriate to cross out passages, add new ones, and rearrange the order of paragraphs. The series of rough drafts is part of the process through which is proposal evolves toward final form. When, however, the proposal is given to an advisor, sent to a funding agency, or presented to a seminar, the occasion is public and calls for an edited, formally prepared document. The document should be easy to read—for which a good printer and high-quality duplication are the first essentials.

Every sentence must be examined and reexamined in terms of its clarity, grammar, and relationship with surrounding sentences. A mark of the neophyte writer is the tendency to resist changing a sentence once it is written, and even more so when it has been typed. A sentence may be grammatically correct and still be awkward within its surroundings. The tough test is the best test here. If, in reading any sentence, a colleague or reviewer hesitates, stumbles, or has to reread the sentence to understand the content, then the sentence must be examined for possible revision—no matter how elegant, obvious, and precise it seems to the author.

Aside from meticulous care in writing and rewriting, the most helpful procedure in editorial revision is to obtain the assistance of colleagues to read the proposal for mechanical errors, lack of clarity, and inadequacies of content. An author can read the same error over and over without recognizing it, and the probability of discovery declines with each review. The same error may leap at once to the attention of even the most casual reader who is reading the proposal for the first time. One useful trick that may improve the author's ability to spot mechanical errors is to read the sentences in reverse order, thus destroying the strong perceptual set created by the normal sequence of ideas.

Although format will be a matter of individual taste or departmental or agency regulation, several general rules may be used in designing the layout of the document:

1. Use double spacing, substantial margins, and ample separation for major subsections. Crowding makes reading both difficult and unpleasant. Always number pages so that readers can quickly refer to a specific location.

2. Make ample use of graphic illustration. A chart or simple diagram can improve clarity and ease the difficult task of critical appraisal and advisement.

3. Make careful and systematic use of headings. The system of headings recommended in the *Publication Manual of the American Psychological Association* (2001) is particularly useful for the design of proposals.

4. Place in an appendix everything that is not immediately essential to the main tasks of the proposal. Allowing readers to decide whether they will read supplementary material is both a courtesy and good strategy.

In Search of a Title: First Impressions and the Route to Retrieval

The title of the proposal is the first contact a reader has with the proposed research. First impressions, be they about people, music, food, or potential research topics, generate powerful anticipations about what is to follow. Shocking the reader by implying one content domain in the title and following with a different one in the body of the proposal is certain to evoke a strong negative response. The first rule in composing a title is to achieve reasonable parity between the images evoked by the title and the opening pages of the proposal.

For the graduate student, the proposal title may well become the thesis or dissertation title and therefore calls for careful consideration of all the functions it must serve and the standards by which it will be judged. The first function of the title is to identify content for the purpose of retrieval. Theses and dissertations are much more retrievable than was once the case. In fact, they have become a part of the public domain of the scholar. The increasing use of the Internet has made the circulation of unpublished documents many times faster and far broader in geographic scope. Titling research has become, thereby, an important factor in sharing research.

In less sophisticated times, titles could be carelessly constructed and the documents would still be discovered by diligent researchers who could take the time to investigate items that appeared only remotely related to their interests. Today, scholars stagger under the burden of sifting through enormous and constantly increasing quantities of material apparently pertinent to their domain. There is no recourse other than to be increasingly selective in choosing which documents to actually retrieve and inspect. Hence, each title the researcher scans must present at least a moderate probability of being pertinent on the basis of the title alone, or it will not be included on the reading list for review. In short, the degree to which the title communicates a concise, thorough, and unambiguous picture of the content is the first factor governing whether a given report will enter the ongoing dialogue of the academic community.

Word selection should be governed more by universality of usage than by personal aesthetic judgment or peculiarly local considerations. Some computer retrieval systems classify titles according to a limited set of keywords. As we discussed in Chapter 4, researchers construct search plans that will identify all studies categorized by keywords known to be associated with their area of interest. Thus, both readers and writers of research reports must describe the research in similar terms or, in too many instances, they will not reach each other.

The title should describe as accurately as possible the exact nature of the main elements in the study. Although such accuracy demands the use of specific language, the title should be free of obscure technical terms or jargon that will be recognized only by small groups of researchers who happen to pursue similar questions within a narrow band of the knowledge domain.

Components Appropriate for Inclusion in the Title

What is included in a sound title will depend first of all on the type of research involved. The elements most commonly considered for inclusion in the title of an experimental study, for example, are the dependent and independent variables, the performance component represented by the criterion task or tasks, the treatment or treatments to be administered, the model underlying the study, the purpose of the study (predicting, establishing relationships, determining differences, or describing a setting), a specialized environment in which the research was conducted (e.g., health care facility), and any unusual contribution of the study. In contrast, titles for qualitative field studies often give prominence to words that describe participants, physical and social context, and the particular research tradition that frames the analysis of data.

A clever author can, by careful selection of words, provide information in the title that a theory is being tested by using a word that often is associated with the theory. For instance, the title "Generalizability of Contingency Management and Reinforcement in Second-Grade Special Education Classes" implies that the investigator is testing the applicability of behavioral theory to a specific population. Much has been communicated by including the single word "generalizability" in the title.

The ultimate purpose of the study in terms of predicting, establishing relationships, determining differences, or describing a setting can be expressed without providing an explicit statement. For example, when variables are expressed in a series, such as "Anthropometrics, Swimming Speed, and Shoulder-Girdle Strength," a relationship generally is implied. If the same study were titled "Anthropometrics and Shoulder Girdle Strength of Fast and Slow Swimmers," the reader would anticipate a study in which differences were determined.

Any aspect of the study that is particularly unusual in terms of methodology, or that represents a unique contribution to the literature, should be included in the title. A treatment that is unusually long or of great magnitude (e.g., "Longitudinal Analysis of Human Short-Term Memory From Age 20 to Age 80"), a method of observation that is creative or unusually accurate (e.g., "Hand Preference in Telephone Use as a Measure of Limb Dominance and Laterality"), a sampling technique that is unique (e.g., "Intelligence of Children Whose Parents Communicate with Hand-Held Devices"), and a particular site for measurement that sets the study apart from others (e.g., "Perceptual Judgment in a Weightless Environment: Report From the Space Shuttle") are examples of such aspects.

Components Inappropriate for Inclusion in the Title

Such factors as population, research design, and instrumentation should not be included in the title unless they represent a substantial departure from similar studies. The population, for instance, should not be noted unless it is a population never sampled before, or is in some way an unusual target group. In the title "Imbedded Figures Acuity in World-Class Chess Masters," the population of the subjects is critical to the rationale for the study. The population in "Running Speed, Leg Strength, and Long Jump Performance of High School Boys" is not important enough to occupy space in the title.

Similarly, research design and instrumentation are not appropriate for inclusion in the title unless they represent an unusual approach to measurement or analysis. The type of research method expressed in "Physiological Analysis of Precompetitive Stress" is common in studies dealing with stress, and surely some other aspect of the study would make a more informative contribution to the title. The approach in "Phenomenological Analysis of Precompetitive Stress," however, is unique and signals the reader that the report contains information of an unusual kind.

Mechanics of Titling

Mechanically, the title should be concise and should provide comfortable reading, free from elaborate or jarring constructions. Excessive length should be avoided because it dilutes the impact of the key elements presented; two lines generally should be adequate. Some retrieval systems place a word limitation on titles, thus enforcing brevity. Redundancies such as "Aspects of," "Comments on," "Study of," "Investigation of," "Inquiry Into," and "An Analysis of" are expendable. It is obvious that a careful investigation of a topic will include "aspects of" the topic, whereas the research report has as its entire purpose the communication of "comments on" the findings of a study. It is pointless to state the obvious in a title.

Attempts to include all subtopics of a study in the title sometimes result in elephantine rubrics. The decision to include or exclude mention of a subtopic should be made less in terms of an abstraction, such as complete coverage, and more in terms of whether inclusion actually will facilitate appropriate retrieval. One useful way to construct a title is to list all the elements that seem appropriate for inclusion, and then to weave them into various permutations until a title appears that satisfies both technical and aesthetic standards.

7

The Oral Presentation

The processes for consideration of your proposal may not require an oral presentation, and the written document may be the only opportunity to explain (and sell) the study. In many universities, however, formal presentation of your ideas for a thesis or dissertation (particularly at the point of an initial prospectus) is a standard expectation, often performed in the context of a graduate seminar. It is less common to have an opportunity to make an oral presentation of a grant proposal, but interviews with site visitors and even command performances before review panels do occur.

Our advice is that if an opportunity occurs to stand up before an audience and talk about what you propose to do—you should take it. Even if your oral presentation is delivered only to peers in a study or support group, do it! The feedback and exchange of ideas may prove to be invaluable as you shape the final form of the written document. Further, such exercises are excellent preparation for a later thesis or dissertation defense—the final hurdle before graduation at many institutions.

Of equal importance, however, is the valuable training in research presentation skills that will be used when you have later opportunities to share your completed work. Whether with teachers at a local school, with students and faculty at a nearby college, with professional colleagues at a state conference, or in the rarified atmosphere of an international research symposium, talking about your study and findings is an important extension of the research process. Learning to do it with clarity, economy, and confidence will serve both you and your audiences well—and explaining your proposal is the perfect way to develop and refine those skills.

Self-help guidebooks for graduate students who must write theses or dissertations have always included attention to preparing the proposal document. Likewise, for use at the other end of the study process, they usually have dispensed advice about successfully navigating the rigors of the so-called "oral defense." Only in the most recent of such texts, however, has space been allocated to assistance in preparing for the "proposal presentation." Among those with sections or chapters that you might find useful as introductions to the presentation task are *The Portable Dissertation Advisor* (Bryant, 2004) *Writing the Winning Dissertation* (Glatthorn & Joyner, 2005) and *The Dissertation Journey* (Roberts, 2004).

In contrast to the specifics of guidebook how-to-do-it instructions, we will focus here on the more general problem of how to think about the problems presented by explaining your proposal. Certainly we can't offer a tutorial on public speaking. What we will do, however, is note those points at which the task of explaining research to a live audience creates unique demands that require thoughtful preparation. The simple fact that in many presentations some members of your audience will already have read your proposal while others will be hearing about it for the first time adds a considerable complication. That is one of many reasons for concluding that careful planning is as much needed for presenting a proposed study as for reporting its results.

A slick presentation will not finesse a weak design, but a clear explanation of dilemmas encountered and hard choices that had to be made will improve the quality of assistance you receive. Further, a brisk and economical presentation recruits support because it shows that you understand exactly what is salient for a reviewer's consideration. Best of all, at the other end of the research process, a graceful and lucid discussion of your findings will allow others to share the fruits of your labor—and may even produce invitations to present your work to fellow investigators in exciting and distant venues.

Before launching into a consideration of oral presentations, it will be helpful to review briefly how talking about research is different from writing about it. Each of these differences bears consequences for how the task of speaking is planned and executed.

1. *Time*—The temporal limits for explaining your study to a live audience invariably are more restrictive than the time requirements for accomplishing a written presentation of the same study. Put another way, 20 minutes of a listener's time allow you to cover far less than 20 minutes of a reader's time. The reasons for this are complex and not all of them are

obvious, but the end result is a demand for selectivity in content and economy in delivery. Finally, because the listener is not free to go back and hear a presentation over again, as a reader is free to restudy a text, the speaker must achieve absolute clarity on the first attempt.

2. *Interaction*—Oral presentations of research almost always include interchange with the audience. Indeed, in the case of presenting a proposal to a committee or seminar, that exchange is the purpose of the occasion, some or all of the audience having already read your prospectus or full proposal. Questions, comments, advice, and discussion often follow the talk and sometimes are interspersed within the flow of presentation. This fact not only calls for the use of interactive skills (at the least, the need to think on your feet) but has consequences for planning and delivery as well.

3. *Adjustment*—The speaker can monitor the ongoing reactions of a live audience, responding to the array of nonverbal signals by adjusting pace and content. Boredom, puzzlement, surprise, distraction, and close interest are not visible on the reader's face while you are writing. While you speak, however, all of them can be observed from moment to moment, and such reactions can be useful information. Responding in appropriate ways not only increases the effectiveness of communication but also ties speaker and audience together in a mutual enterprise that is impossible to achieve through the written page.

4. *Anxiety*—If the range of skills demanded for speaking is not wider than the collective span of those needed for writing, at the least, many of them are different. For most students, the skills for making formal oral presentations are much less used, and thereby are less well honed. The need to display unpracticed skills in a public setting, when added to the normal concern about winning approval or obtaining needed assistance, produces an occasion for anxiety. For some, the spontaneous variables at play in a live setting add zest and a sense of rich possibility. For others, a live listening audience simply adds a discomfort for which there is no parallel in the work of writing. For those people, the consequence is a need to prepare in such a way that anxiety (or outright terror) is held within bounds and becomes facilitating arousal rather than debilitating inhibition.

Those are some of the most distinctive differences between writing and speaking. Their existence means that preparing and delivering an oral presentation involves all the problems of using a new medium to communicate familiar content. Accordingly, we have some advice that may provide users of this book with a helpful set of guidelines for getting started. Our suggestions cluster into seven areas that occur in natural sequence: (a) preparing content, (b) preparing materials, (c) practice and revision, (d) preparing the environment, (e) oral presentation, (f) managing interaction, and (g) the aftermath. What we have to say about each of these topics will seem different, in some respects, from what you may have seen in books or heard in public speaking courses. There are several reasons for that, and knowing them may help you digest what is to follow.

First, our advice is tailored to the task of talking about research to other researchers, or, at the least, to people who have working familiarity with scholarship. Customs of language, format, and demeanor allow (or demand) a form of communication that is different from what would be appropriate with another kind of audience. Although much may be shared among all kinds of public speaking, we are not dealing in this text with a generic form of public discourse. Second, the source here is our own experience as people who have frequent occasion to do our own oral presentations dealing with research. Although that background certainly is helpful in sorting out ideas about what distinguishes effective from ineffective public speaking, it does not make us specialists in oral communication. Accordingly, this chapter is about understanding the demands of explaining a plan for research to a small audience in a live forum. By extension, then, it also deals with designing contextually appropriate strategies for accomplishing that limited but vital task.

Preparing Content

In determining what to cover, the particulars of circumstance will dictate most of the decisions. How long do you have? Who will be present, and what will they expect you to do? What will the audience already know? What will your listeners need to learn from you to play their part in the occasion? For your purpose, what is absolutely essential to explain, and what is not? What content requires emphasis, and what may be given only a passing note? To what extent must you contend with uneven prior knowledge on the part of members of the audience? Is the presentation a follow-up of a previously distributed document (a review leading into discussion), or is it the only access the audience will have to your work? If you can specify the answers to those questions, you will have a rough map of what to include in your presentation. In the process of using that map, the following guidelines will assist in shaping the final product.

1. Don't confuse what you already know with what the members of the audience already know. If there was not something they needed to learn from you, why would you be doing the presentation? The right question to start with is *not* "What do I want to talk about?" or "What is in my proposal?" or "What were my results?" The right questions are "What don't they know that they need or want to know?" and "In what order do they need to learn things to make sense of what I have to say?" You are going to lead the listeners through a sequence of logic, or the events of a study, step by step. In that sense, a research presentation is teaching, not ritual display. You begin with, stay with, and end with a concern for what you want your audience to learn. That requires a highly selected version of what you know.

2. In deciding what to cover in the time you have available, less often is more. People will understand more, remember more, and make better use (in subsequent discussion) of what you explain if you explain less. Locate the essentials and make them the center of your presentation. It often will serve you well to note for the audience particular sections of detail you are leaving out (or simply designating by name) and, perhaps, why you are doing so. You even may make clear that you have the material and would be pleased to provide it at another time. That strategy prevents people from being distracted by thoughts about what you are not saying. Having less to talk about allows an easier pace; more opportunity for emphasis, illustration, and review; and less tension in the whole process.

If it is genuinely impossible to communicate what is essential to the purpose of the presentation in the time provided, don't complain about it in your speech. Negotiate an extension of your time, ask for a change in the goals to be accomplished by the presentation, or, as a last resort, make brief public note of what it has been necessary to exclude. Trying to crowd everything in is a lose-lose proposition. It won't work, it will irritate your audience, and it will leave you feeling inadequate.

3. Try to anticipate the needs and reactions of your audience, and build this in as part of your presentation. As with the point made above about noting omissions, there are a number of predictable concerns or needs among audience members that can be dealt with as you proceed. If there is a point about which you are unsure, acknowledge it and invite comment or advice in the following discussion. If there is a point on which you know some of the audience will disagree with what you say, make a polite note of the fact—and perhaps invite subsequent discussion of alternatives. If some people already know something that must be explained for others, recognize the situation and ask those who don't need your tutorial to bear with you.

If there has been discussion about a point on a previous occasion, or if there have been revisions in or additions to your work, note them, even if you elect not to discuss them.

All of this serves to assure the audience that you are thinking specifically about them and that you are aware of their unique relationships with the content of your presentation. In addition to helping to maintain their close attention, this strategy allows you to maintain a degree of control (or, at least, influence) over what will be given attention in any discussion that ensues.

Preparing Materials

Unless you are blessed with the happy facility of perfect memory, the primary task here is to prepare something to guide your presentation. The most obvious purpose of the guide (or script) is to ensure that you say the right things, about the right topics, in the right order. Less obvious are the reasons for trying so hard to get it right—the fact that you have just one chance to achieve transparent clarity, and the fact that making things easy for the audience is the best way to get them to be helpful (and friendly).

It does not matter what form your guide takes, so long as it serves the purpose of getting it right and is constructed to allow effective delivery of your message. The first rule of delivery is *never read!* The second rule is *if you have to read, never appear to be doing so!* By disconnecting speaker and listeners, reading defeats the purpose of doing an oral presentation. It is impossible for the audience to feel other than insulted. Accordingly, your guide has to take a form that allows you to spend most of your time looking at and reacting to your audience, and only brief instants being reminded of what comes next.

The possibilities for accomplishing this are endless. They range from small note cards with key words listed in bold type, through pages of text in traditional outline format, to verbatim scripts with key words or phrases highlighted in fluorescent colors. In designing your own guide, however, remember that there are other needs to be serviced beyond having a memory aid. For example, whatever you use should be visible in a way that requires only a quick glance to find your location and pick up a cue to trigger recall of the next sequence. If you want to move about while speaking, then your guide should be comfortably portable. Pages should be boldly numbered in sequence to avoid the unfortunate accident of missing or jumbled notes, and reminders for the use of each visual aid should be inserted exactly where it is to be introduced.

As an example, this is what one of the present authors does for each presentation. A rough script is typed with double spacing. Some sentences are laid out in complete form, whereas others are indicated just by the opening phrase. All, however, begin at the left margin (essentially, the format is a sequential list of individual sentences). Only the top one third of each sheet is used, the lower two thirds being left entirely blank. Then the typed pages are darkened and enlarged by 10% on a copy machine. (On a computer, all of this could be accomplished by using a larger font and the top half of the page.) Finally, a key word or phrase in each sentence is marked with a highlighter pen.

Placed on a lectern, the script, which occupies only the top portion of the page, now rests at a level directly on the speaker's line of sight to the audience. All the highlighted words are read from peripheral vision, and eye contact with the audience is never broken by having to look down to read the lower portion of the page. Sheets are transferred silently from right to left as used and never become visible distractions for the listeners. The effect given is that of articulate, completely extemporaneous speech, delivered with close attention to the audience.

Some of what the audience sees and hears is illusion, made possible by the idiosyncratic speaking guide described above—a guide that, in truth, serves to disguise a poor memory and poor eyesight. In a like fashion, you must devise a guide that will best serve your unique strengths and limitations as a speaker. The purpose is not to enhance your reputation for oral presentation beyond what you deserve. The purpose is to serve each audience—by making it easy for its members to understand what you have to say.

An alternative strategy is to make key words or phrases concurrently available to the audience by projecting them onto a front screen. This can be done effectively by means that range from the antique, such a series of simple transparencies displayed with an overhead projector, to such modern technologies as PowerPoint slides controlled from a laptop computer placed unobtrusively behind the lectern.

Sharing your talking notes in that manner can be very effective. Not only does it provide an agency for maintaining close attention by the audience, but it also frees you from the restrictions imposed by a lectern. In addition, there is reason to believe that visual display of key words will help some members of your audience to note and later recall main points in the presentation.

This strategy, however, has two potential drawbacks. One is irksome and the other is disastrous. Having the words on the screen tempts the speaker to look at the screen instead of the audience. That has the effect of making the projected image the primary object of attention rather than a supplement to the oral presentation. Such breaks in the psychological connection between speaker and listener impose the irritations and inefficiencies of having to switch back and forth between the two sources of communication. The second problem always lurks in the potential of electronic equipment to fail, leaving you with no guide at all. A paper copy of your PowerPoint slides or other projection sources should prevent any serious disaster, but it is prudent to add some practice with your life-preserver if you want to be completely safe.

In addition to some form of speaking guide, you may wish to prepare aids that will support your oral presentation. There are several reasons for electing to do so. Not the least of these is the fact that any audience of more than a few individuals will contain some people who learn best (or only) when they can inspect a concrete example. The more complex the object or relationship, the more they are disadvantaged by purely verbal imagery. Photographs, models, tables, figures, graphs, outlines, data specimens, diagrams, and demonstrations are just a few of the supporting illustrations that can give your words concrete representation.

There are other important reasons to consider the use of graphic aids. Economy of communication is one. Although not all pictures can replace a thousand words, some certainly can. Other graphics, such as diagrams or key word outlines, help the audience keep track of sequence and relationship in otherwise convoluted explanations. A good example is a flowchart that helps you take your audience through your methodology. Finally, some illustrations provide focus and heighten impact, as in an audiotape segment or transcript excerpt from an interview.

In deciding whether to make use of such aids, subject each possibility to this basic question: "Will the prop really help people (or at least some people) to understand what I am saying?" If you can't honestly conclude that introducing the supplement to your presentation will cause better or quicker learning, don't use it. Each use of an aid consumes valuable time. Each requires the transfer of audience attention from one object to another, with consequent risk of leaving someone behind. Each, therefore, involves a trade-off to be weighed with great care. Be sure your audience gets back more than it has to give.

One simple aid is almost always appropriate at a proposal presentation. A printed outline of what you plan to talk about (and do), laid out in sequence, is appreciated by most listeners. Keep it to a single sheet, include no details, and indicate beginning and ending times. Knowing what is to come (and being able to tick off major segments as they are completed) is a reassuring convenience. It also is the first sign of careful planning and concern for the listeners.

All visual aids such as PowerPoint slides, transparencies, newsprint diagrams, and chalkboard illustrations should be *used*—not simply displayed. Present them, explain their significance, allow the audience a brief moment to digest or ask questions, and then remove the aid from sight, firmly directing attention to the next element of the presentation. Leaving aids visible invites continued inspection or simple distraction (as with the case of projectors running without a slide or transparency in place).

There is a hard rule about quality for visual aids, one never to be violated. *If you use it, everyone should be able to see it!* Large print, high magnification, sharp contrast, and reduction of content to bare essentials are basic qualifications for any illustration. Further, every aid has to be judged relative to the limitations of the environment in which you will be speaking. If it can't be seen by everyone, in every part of the room, don't use it. Under most circumstances, this disqualifies newsprint, chalkboards, and even overhead projectors for any but small group presentations.

We would be remiss if we did not acknowledge the availability of technologies for the production and display of graphic material that now have relegated chalkboards, overhead projectors, and even the 35 mm carousel slide projector to the status of horse and buggy artifacts. Among those are computer-driven video projectors that permit computer images to be projected directly on the screen, Internet-linked projection, and multi-media presentations that integrate a variety of audio and visual sources.

Perhaps even more directly relevant to preparation for a typical student oral presentation are equipment and software that allow the creation of both an original graphic and the necessary slide with a personal computer. Some software programs allow you to print out miniaturized copies of your slides for use as an audience handout. Lined spaces adjacent to each slide allow viewers to conveniently take notes or write questions as they listen. To all that magic we say, "Use it, if you have it!" But do profit from the experience of many students (and professors) who have entered the wonderland of graphic technology before you. You might wish to copy these simple rules and pin them over your computer.

1. No matter what it says in the publisher's promotional blurb, you can bet that effective use of software will take more time and more skill than you expect. Learn how to use any new method for production of graphics well in advance of need—and practice it at least to the level of basic mastery.

2. Remember that with the best equipment and software in the world, if there is any step in the process that depends on someone else (e.g., film developing, slide duplicating, tape editing, scanning, and transferring audio files),

they may not be able to meet your demand for last-minute service. Give everyone involved, including yourself, plenty of time.

3. If a computer is involved in any step in the process, save early, save often, and save again!

4. Some of the same familiar rules that applied to now antique equipment still apply to high-tech displays. For example, don't clutter anything you show the audience—"white space" is a virtue when it serves to frame and enhance what you really want people to see. That applies as much to an LCD projector as it did to magic lantern glass plates or filmstrips.

5. When you want people to listen to you, as in transitions between slides or tape segments, don't distract them with too much action of any kind. Even changes in color background or framing in a display can engage attention at the wrong location.

6. Until you become a practiced expert, pay attention to the suggestions for color schemes that accompany production software. What looks really cool (to you) on your monitor may turn into a disastrous chromatic soup when projected onto a large screen. Even worse, some colors dim or even disappear when projected on some surfaces. If you do not leave enough time to preview and correct such defects, you will be in the unenviable position of asking your audience to "imagine" a line running from point "a" to point "b."

7. Technologies, particularly those involving computer projectors, become doubly prone to catastrophic failure when you try to employ them as a visitor in an unfamiliar venue. Connecting cables, port specifications, type of projector resolution, and other compatibility problems must be resolved in advance—a difficult accomplishment over the telephone, and often an impossible one at short notice. Under those conditions, our advice is to be old fashioned and at least bring a backup. Trial runs and provisions for backup make for a good sleep the night before.

8. Make certain you know how to use all the equipment required for your presentation—including what to do should any item malfunction. The time you take to learn and practice will be repaid by a smooth and satisfying performance. Using part of your presentation time to figure out what just went wrong—and what to do next—not only will distract your audience but also is likely to leave you flustered and prone to make other blunders. Electronic aids can be a wonderful enhancement for virtually any presentation. They also are a prime target for Murphy's Law—if anything can come unplugged, it probably will!

Some presentation formats appropriately utilize a continuous stream of graphic aids, as in large-scale lectures. Those special instances aside, the rule again is that less usually will be more. Even with high-quality graphics, there is the risk that overuse of aids will dilute the focus of what you are saying and begin to intrude between you and the audience. Select only the aids that will advance your cause, use them deliberately for that purpose, then never let them remain as distractors. As a general rule, with you to point out what is salient, a well-constructed aid should yield its main message in a few seconds and should do so for everyone in the room. This means that for your audience, illustrations should feel like a natural continuation of the presentation rather than an interruption.

A Special Note About the Use of PowerPoint

The use of PowerPoint (included in the Microsoft Office software package) has become ubiquitous in research presentations—given in every imaginable venue and for every conceivable purpose. Most graduate students will not require assistance in creating PowerPoint adjuncts to a proposal presentation. If, however, you do happen to be a genuine novice there probably will be ample sources of help at your own institution. If that is not true, the place to start gathering some self-help is at the Microsoft Office home page at http://office.microsoft.com where you can click on "PowerPoint" and then select the assistance package for the year of issue of your software.

As for advice concerning how to actually use PowerPoint as an effective component of a presentation, the Internet is replete with such sources, the only problem being selection of those that best serve your particular needs. There also are books, magazine articles, and a profusion of research reports that purport to identify what works best for particular presentational purposes. We will not attempt to add anything substantial to that cornucopia of instruction. What we can do is simply to identify a handful of problems that seem to persist among graduate students for whom the proposal presentation represents their first foray into talking about research before a "public" audience (one not composed of friends, family, colleagues, or casually assembled listeners). Committee members and research advisors always try to focus their attention on the content of the research plan rather than the ancillary aspects of presentational form. Nevertheless, some distracting elements can make that difficult to do—and a few can make it near impossible.

The audience will be small, with 4 to 12 as the usual number. The familiarity of individual members with the proposal will range from intimate and detailed to virtually none whatever. The room will be of small or modest size with a screen at the front and the distance to the projection device not more than 20 feet. Seating, table surfaces, and lecterns usually are adjustable. Beyond those points, however, the details become particular to individual locations. Here are a small set of things that we have seen undercut the effectiveness of PowerPoint in such settings. We will list first those that relate to construction of the graphic displays, followed by those that attach to how the displays are used in the presentation proper, and then close with our own version of general rules.

Construction

- One idea per slide. All subordinate points must relate only to that idea.
- Put nothing on a slide that will not be covered in your oral presentation.
- Use bullet points with short phrases rather than complete sentences.
- Keep content simple and don't crowd text. As a general rule, use a limit of 6 words per line and 6 lines per slide.
- Use a mix of upper and lower case letters. All capitals are harder to read.
- Even for use in intimate settings, the smallest font size should be 18 point and 28–30 point (or more) should be the norm.
- Use reveals and transitions sparingly, if at all. Keep the focus on content.
- There are no acceptable excuses for spelling errors. And please remember, *data* is a plural and *datum* is the singular form.

Proper Use

- Face the audience and talk directly to them, maintaining steady eye contact.
- Never talk to the screen.
- Never read from the screen.
- Introduce each slide either prior to or immediately with its appearance. Don't wait to use the slide as the cue for deciding what you are going to say.
- In discussion, touch on every point made in the slide.
- Stand next to the screen, not the projector. If you do not have remote control, have someone else run projection.
- If you do use a pointer (rarely needed in such presentations) be sure to practice its use, particularly with regard to switching, tracking, and holding it steady.
- It takes the average reader about 8 seconds to silently read 30–40 words. If you have properly introduced the slide, then touched on each sub-point in your discussion of the central idea, it should be rare to display a slide for longer than one minute. For slides containing limited content or a simple graphic, less time should be sufficient.
- Never leave a slide up when you move on to another topic in your presentation. If necessary, blank slides can be introduced into the projection sequence for that purpose.

The General Rules

- Don't use a slide just to use a slide. If the graphic, photograph, or clip art doesn't further the audience's comprehension of your presentation, then don't use it.
- Go easy on the bells and whistles. Presentations largely stand or fall on the quality, relevance, and integrity of their content. Audience boredom or inattention is a content problem and not a decorative failure.
- Slide displays are not compulsory. Visual aids can help make a presentation more effective, and PowerPoint is a flexible and potent aid. It remains true, however, that some of the most memorable proposal presentations we have witnessed employed nothing more than terse, clear explanations, judicious use of nonverbal support (gestures, facial expression, voice tone and volume, and changes in speaking pace), and the steady attentiveness of the speaker to the audience.

Practice and Revision

Explaining any research study or its results is a difficult task. To the extent that you genuinely care about communicating with an audience, it will be natural and proper to do the job well. If the quality of presentation enhances or restricts the possibility of achieving your goal, practice makes sense. To do so also signals respect for a significant form of craft knowledge in the world of scholarship, a craft that will serve your work—even if you are just explaining a proposal to the members of a small committee.

There are three main varieties of practice: silent reading, solitary speaking, and trial runs before a live audience. Silent reading gives familiarity to the guide and gradually allows memory to free attention from the bondage of literal reading. It does not, however, give a correct estimate of time required (silent reading always is faster than reading aloud) or give any clue as to how what you say will sound in the ears of an audience.

In contrast, solitary speaking gives both a better estimate of timing and an opportunity to listen to the sounds and sequences of your presentation. You will discover that the syntaxes and rhythms of written text are not those of normal speech, and adjustments will be required if you are to avoid the stilted and unnatural sound of reading aloud. The most common adjustments are simplification of content (often by painful excisions), longer pauses, and more frequent use of words that provide transition between topics.

The use of a mirror, a lectern, and a tape recorder can markedly increase the value of solitary practice, particularly if you now include actual use of graphic aids. This kind of practice accelerates both the fine tuning needed to give confidence and acquisition of automaticity in performing mechanical tasks—such as managing your notes. The latter is vital because it allows you to spend more and more time looking into the mirror that is standing in for a live audience.

Finally, practice before a real audience allows you to move toward personalization, the ability to speak with the relaxed informality of serious conversations in everyday social interaction. We say "move toward" because there are some conventions that govern formal presentations that are absent in ordinary conversation. Just the matter of how long one person speaks, who controls the topic, and who is free to interrupt mark the rules of formal presentation as different. The intent to instruct (and influence) also is different, not in kind, but in the degree of conscious purpose. Those distinctions (and others) notwithstanding, the audience always is made more comfortable and attentive by a speaker who adopts the tone and cadence of "talking with" rather than "talking at." To achieve that kind of grace, nothing equals live practice.

Friends, relatives, peers, and professors can be drafted for the purpose of live practice. The only absolute qualification is a desire to help you prepare. Of course, the presence of anyone with sufficient background to raise questions about content can elevate the trial to a new level. Questions and comments allow you to revise defective explanations or, at the least, to anticipate reactions and to determine whether the questions raised reveal problems in the proposed study that you must address or whether the questions reflect only an inadequate explanation. Further, now you can practice using aids in ways that truly support rather than interrupt, and you will discover how much longer everything is going to take than you had first imagined.

Even though not everyone in the audience may be able to help you with the details of the study, all are competent to give feedback about your delivery. Once they understand that critique of practice is helpful (albeit sometimes disconcerting) and not destructive, anyone can notice and report distractors such as the ubiquitous "OK?" or "aaah" that so often intrude without the speaker's awareness. Gestures or postures that erode audience attention such as rubbing the nose, gripping the lectern with white knuckles, staring fixedly above the heads of listeners, swaying from side to side, making eye contact only with individuals on the left, and fiddling with a pointer are so common as to be almost predictable in novice speakers. Good friends will tell you all about them.

When live practice and feedback yield substantial revisions, second and third trials may be needed. A good practice sequence to follow is solitary speaking followed by three presentations to (a) two or three good friends, (b) your advisor, and finally (c) a small group of people who have some competence in research—all prior to the official presentation of your proposal. The limit often is the availability (and goodwill) of your colleagues. Although there may be some theoretical risk of over-practice, it is unlikely to be a serious concern. As with any other task, work at it until you get it right, and then get on with the job.

Preparing the Environment

The rule about the environment, the space where you will actually present, is simple: *"Know it, and manipulate it!"* This means that you should inspect the room and arrange it (so far as possible) to suit your purpose. Where particular people sit and the general configuration of seating (as around a table or in a room with freely movable chairs) are first-order concerns. The presence of a lectern, location of light switches, the availability of doors to shut out distracting noise, and the location of power outlets (if in doubt, always bring an extension cord) are all essential information.

If there is a wall-mounted projection screen you intend to use, always pull it down for inspection. More than one unwary novice has discovered an obscene message scrawled there for all to contemplate as the introduction to the first slide! In larger venues, sound systems, projector stands, and lectern lights are never to be taken for granted.

If you are planning to employ one or more assistants to help with demonstrations or the operation of equipment, have them come with you. Where they will sit, how well they can see you, and their familiarity with the environment are matters for your concern.

If you plan to use handouts, having a packet already in place as people enter the room avoids the confusion of later distribution and signals a businesslike atmosphere. If you include a program outline, it can be placed on top with names marked so that advisors or special visitors can have precedence in seating. It always is good to have the most important listeners located where you can keep the closest possible contact with them during your presentation.

The more you can reduce unexpected intrusions and the more you can make the environment comfortable for the audience, the more closely they will attend what you have to say. Take the time to inspect the field of action. If there is not much you can do to adjust it in your favor, at the least let it hold no surprises.

Oral Presentation

There are two things most novice speakers do not know (or find difficult to remember). First, virtually everyone is nervous before a public presentation.

That is normal and, once you learn how to turn that excitement to your advantage, healthy. A second point is that under almost all imaginable circumstances, everyone in the audience wants you to do well. Even more, most of them are quite prepared to overlook all sorts of minor slips as part of giving you a full and fair hearing. How many times have you sat in the audience and wished the speaker would give a terrible presentation? The answer should afford some comfort for your next effort.

Having something at stake beyond just the challenge of performing well, as in a proposal review or dissertation defense, does add greater pressure to a presentation. In that regard, you will note that we have not said reassuringly, "Don't worry about it." Anyone in that situation will have at least some honest cause for concern. What we do want to point out is that with solid preparation, the possibility of a truly bad presentation is vanishingly small. Formal evaluation of what you have to say is out of our hands—and at the moment you begin, out of yours as well! There is nothing left to do but concentrate on a good clear presentation and let the rest take care of itself.

Take a deep breath, acknowledge the audience (large or small) with a friendly word or gesture of greeting, and then start at a slow, even pace. In the opening sentences, make a special point of trying to match your speaking voice to your normal conversational tone. Many speakers report that the sound of their own voice provides recurrent feedback. If they are able to sound relaxed and confident, they begin gradually to really feel that way. Likewise, if your voice, posture, and movements signal exceptional tension, your audience will be concerned and begin to reflect the tension back. Start easy, breathe, look at people, and just talk to them.

We did not say "smile" as part of our advice because nothing is worse than the frozen rictus of a false smile. Forget your face and concentrate on how your listeners are responding to your words. If you can do that, smiles will come when they are a natural part of the process.

Particularly in the opening minutes, remember to pause when you have made an important point. Do then the kinds of things we all do in conversation: Nod to stress a point and raise your eyebrows to ask your audience if they understand. Punctuate with a simple gesture, or just wait for a moment as we do when encouraging a listener to contemplate the significance of what we have said.

Later on in your presentation, you will want to consider the need for stimulus variation. Changing the display is one of the key factors in sustaining attention, and the need for it grows as the presentation lengthens. Throughout the presentation, you might change your speaking location, briefly introduce a graphic display, or simply alter your tone and speed of delivery. Any small change can break the dull grip of too much of one thing.

If possible, you should have someone tape record your presentation. It is unlikely to be the last you give, and although being perfect is not a reasonable expectation, getting better is. Be sure, of course, to use long-playing tape or digital recorder to avoid the distraction of having to flip a cassette in midcourse. In addition to making sure you can profit by listening to your own performance, purchase some further insurance by asking several friends to watch for and record the things you will not be able to hear on a tape. Nonverbal signals, for example, are a powerful part of every communication. Knowing about what your body language is saying, good and bad, is part of learning to do effective presentations.

As you close, three things are almost universally appropriate. First, revisit with your audience the essential points that you wish them to retain. Second, provide a graceful transition by reminding the audience about what will happen next, whether that be a break, a period of questions and discussion, return of control to a moderator, or termination of the meeting. Third, and finally, any audience deserves the courtesy of your thanks. While you are struggling to master the difficult craft of presenting research, a patient and forgiving audience will have earned a good deal more. Remember that, when next you are on the other side of the lectern.

Managing Interaction

Whether you serve as your own moderator or have someone else to preside over a period of questions and discussion, a little advance planning will make it go more smoothly. If you can tape the discussion that follows your presentation, be sure to do so. Valuable points of comment or advice that seem vivid in the excitement of interaction with the audience may be lost to recall. Few people find it possible to take notes and talk at the same time.

If the audience contains both people charged with some official function (such as reviewing your proposal) and others who are visitors or supporters, either you or your committee chair should make a point of setting the ground rules at the outset. It is good to let everyone who has attended have an opportunity to comment or question, but give priority of first opportunity to those who require it. If you are managing the flow of discussion, be sure to spread interaction around. Some people are not skilled in taking turns, and others need encouragement if their comments are to be heard.

In fielding questions, the first rule is *Get it straight!* Be sure you understand what is being asked before attempting to answer. Repeating or rephrasing complex or unclear questions serves the dual purpose of detecting miscommunication and reassuring the questioner that you do understand what he or she wants to know. Repeating every inquiry, of course, is tedious and to be avoided.

When you just do not know the answer, say so. When you just can't understand the question, say so—but also ask if it can be taken up later when there is more time for clarification. If you know something that may be helpful but you suspect it is not exactly what has been requested, ask permission to alter the question as a way of advancing the discussion. If the question is clear enough but you think it is inappropriate to the occasion, acknowledge it politely and suggest that you would be glad to take it up after the meeting is concluded.

If you are pressed for an agreement that, in good conscience, you cannot give, try acknowledging the existence of alternative or competing views. If that does not close the point, ask to defer the matter until there is sufficient time for deeper examination of the argument. Normally, by this time someone will have interceded to redirect the discussion. If help is not forthcoming, there is little to do except to hold your ground, perhaps asking for comment by others in the audience.

Finally, keep an eye on the clock. Just as the presentation proper should end on time, any follow-up discussion should be limited by prior announcement, mutual agreement of those present, or just plain common sense.

The Aftermath

Within several days, make a point of obtaining and processing all the available feedback from your presentation. Listen to the tape, review notes taken by friends, and solicit feedback on both content and delivery. Some people will do this with greater skill than others, and some will feel more comfortable in providing an honest critique. Predictably, there will be conflicting opinions on some points. What one finds helpful, another may not. Rather than attempt to respond by planning some change to accommodate every comment, listen for patterns. Where concerns overlap lie the points at which effort for redesign or further practice will yield improvement.

Finally, encourage people to indicate what they liked and found helpful in your presentation. This is not so that you can revel in their praise but because you may not be aware of what was most effective. If you can learn what worked, strength can be built on strength.

PART II

Money for Research

The two chapters that follow are all about money—money to pay for research. The first, Chapter 8, deals with finding where the money is and how to go about the task of working with the people who hold the purse strings. The second, Chapter 9, deals with all the details that go into writing the document with which you will present yourself and your study as worthy of their favor—a grant proposal.

The chapter titles and text headings should make the organization of Part II self-explanatory. If you are already committed to the idea of seeking research funds, Chapters 8 and 9 should be read in sequence. If you are a student for whom the idea of funding student research is new (or even a bit unimaginable), then Chapter 8 may contain a genuine revelation. In fact, if you still are at the browsing stage of inspecting this text, we think it might be to your advantage to locate the several places where funding for student research is discussed in Chapter 8 and begin your journey toward writing a proposal right there rather than doing the traditional thing and beginning with Chapter 1!

Whether you are a graduate student facing a thesis or dissertation requirement, or a young faculty member embarked on the road to tenure and promotion, the time to start thinking about funds to support your research is right now. If you are not entirely clear as to why money is such an issue, you can be forgiven because in the undergraduate years neither professors nor textbooks offer much explicit advice about that topic. Accordingly, in Part II we will offer some of the arguments for treating grant seeking as an essential part of your education and career.

Let us start here with a baseline proposition. *Money often makes for better research and happier researchers.* First there are the technical reasons for that to be true. Money can pay for larger sample sizes or participants with characteristics that are more nearly ideal, for instrumentation that is more reliable and precise, travel to acquire training in special skills, budget for expendable items and materials, needed extensions of time in the field to gather a richer pool of data, and several semesters of freedom from other obligations such as teaching and student advisement. Then there are the social and spiritual reasons to think that there is virtue in finding funds for your research. Some of the money doesn't have to come out of your pocket, and you might just be able to have a life while moving toward graduation—or tenure and an academic career.

Whenever you think you may need more information than we have provided in Part II, please consult the sources we have recommended in each chapter (full citations for books are in the reference list, and many website addresses are organized in handy tabular form). Finally, some of the key sources have been abstracted for your consideration in the Appendix.

8

Money for Research

How to Ask for Help

Every research study has a price tag. The cost may be indirect but simply calculated, as in the value of time expended by a solitary investigator engaged in an anthropological field study. Or the cost may be direct but cal culable only from a complex audit trail, as when all the equipment and personnel required for a biomedical experiment are assembled and put to work. In a like manner, answering the question "Who pays the tab?" may be simple and straightforward or surprisingly complex and even opaque.

The expenses of a study simply may be borne out-of-pocket by the graduate student author of a doctoral dissertation. Financing for a young faculty member's research may be drawn from an institutional research fund. Or a major inter-institutional project may be supported through a research contract with a federal agency. And, of course, what is written on the price tag for inquiry can range from a few hundred dollars to many millions. Thus it is that the arrangements for financing and the sums involved will vary widely, but at the bottom line one fact remains for every study. There always is a price to be paid.

This chapter will be about the inevitable price tag that dangles from your proposal, and how to locate and obtain the funds to pay for the study you want to undertake. Before launching into that discussion, however, there are several caveats we want to underscore. First, the ability to obtain financial support is a skill that may be critical to the success of a study. Sometimes it even plays a central part in how individuals establish successful careers in research. *Nevertheless, money, by itself, does not make good research.* Good researchers do that. All money can do is allow designs that take advantage of opportunities for producing strong results.

A second caution is not to underestimate the ubiquitous power of funding to exert a restriction on research proposals of all kinds and at all levels of sophistication. In preparing your first research proposal, you will design a study for which you either believe you already have sufficient resources in hand, or for which you can reasonably believe funding will be available. In virtually every case, that will involve devising acceptable compromises between the ideal design and what constitutes a supportable design.

Implacably, as the number and magnitude of necessary compromises increase, the potential power of the study will decrease. And, if absolute power corrupts the political process absolutely, you can be assured that inadequate funding corrupts the research process absolutely! The rule applies just as much to the modest and relatively inexpensive doctoral dissertation as to a research program involving a series of elaborate and costly investigations. Sound research can be done on tight budgets (and in the academic world that is the usual condition), but there is a limit to how much belt-tightening can be applied without damaging the product.

Money matters and it can matter mightily in the delicate process of designing and executing research. In case you missed our point here, what all of this means is that finding money for research probably is going to matter *to you!* Recruitment of participants with ideal characteristics, purchase of software for data analysis, investments in equipment, employment and training of assistants required for reliability checks, hiring out tasks that demand special technical competence, support for needed extensions of time in the field, travel to obtain training in special skills, and the budget for expendables and overhead are only a sample of what may have to be paid for—by somebody.

Learning how to locate and entice that "somebody" is part of your preparation for doing research. So read on with close attention to how the grant-garnering game is played. Think hard, think realistically, and think early about what it will cost to execute the study you are proposing, and begin right now to consider how you can answer the question, "Who will pay the tab?"

We will begin by suggesting how to search for funding sources that are likely to be interested in your study, and then how to work effectively with the various and distinctively different kinds of organizations you may encounter. Next, the core of this chapter describes the proposal document that is submitted, how reviewers judge such appeals for support, and how they arrive at final decisions about which ones to fund.

We also offer some advice concerning how to deal with the consequences produced by a successful application, as well as some instruction for making lemonade out of the bitter fruit of a rejected proposal. The chapter closes with a discussion of the special case of student researchers: why they should search for funding; what types of support are earmarked for students; and how the process of preparing a grant proposal can add substantial value to a graduate program.

Having asserted above that there may be a connection between the adequacy of the funding you obtain and the quality of the research you produce, we want to close by turning to a very different but equally consequential benefit of money. In case personal experience has not yet drawn this point to your attention, researchers are people. They have careers to nurture, friends and family to love and support, usually a number of other work-related roles to play, and personal lives to live.

Notwithstanding that indisputable fact, there often is a strong tendency for researchers to risk erosion of their status as persons by taking the costs of doing their scholarship entirely out of their own financial (and social) hides. Given the attractions of inquiry, that is a perfectly understandable and, perhaps, even laudable disposition. Nevertheless, paying your own way can have corrosive consequences that extend beyond the quality of the studies you perform to the quality of the life you live. Scholarship need not involve a vow of genteel poverty or a monastic disconnection from the responsibilities and rewards of a rich social life. That sort of price need not be a requisite cost of doing research. Indeed, we argue in this chapter that it neither is nor ought to be so.

Money Is Available to Support Research

Good researchers at all levels of career development can find money to support their research. Those funds are not evenly distributed across disciplines, do not always come without restrictions, often involve competitive review of proposals, and certainly are not just lying about waiting to be picked up. Nevertheless, the money is there, and it is there in truly astonishing amounts. Thousands of foundations and agencies (both public and private) have the support of research as one of their missions.

Apply early and apply often. Unlike the rule on submission of manuscripts to multiple publication outlets, there is absolutely no prohibition against applying to several sources of financial support for the same study. In fact, once you have the main elements of a strong research plan (and the necessary written boilerplate for description), multiple applications should be the rule and not the exception. Further, rejection of one or several requests should not end your search for support. You should learn from each failure, and subsequent applications (and reapplications to the same source) will be more effective as well as easier to complete.

If (surprise!) you receive favorable responses from more than one source, you often need not decline one in favor of another. Most agencies do want to know (and will ask you to report) if your proposal will be funded by another source. "Double-dipping" (the practice of taking research funds from two agencies without informing both) is unethical and could end your career before it starts. Nevertheless, it is not uncommon in the research funding community for agencies to work together in supporting different aspects of the same investigation. For example, one source might be willing to fund a strong pilot study or necessary instrument validation work, while another supported the main investigation. Always (of course) be completely honest about your circumstances in any negotiation, and work for the arrangement that best fits your study.

Tapping into financial sources, however, should not be done haphazardly or hastily. Preparation of a successful grant application requires knowledge and skills that are not the same as the requisites for performing an investigation. In most cases, the process also involves preparation of a research proposal that differs in some important ways from those written for theses and dissertations. Sophistication about all of this has acquired the name "grantsmanship." This chapter and Chapter 9 in Part II are designed to introduce you to the art of grantsmanship (with its amazing diversity, local customs, and jargon), the process of preparing a grant proposal for support of a study, and the idea that students (undergraduate, graduate, and postgraduate) can be successful in shaking the money tree.

If you are already committed to the idea of seeking research funds, you will find Chapters 8 and 9 to be very helpful and interesting. If you are a student for whom the idea of funding student research is new (or even unimaginable), then we hope that before you decide to skip this chapter, you will look ahead to the section, "Resources Earmarked for Student Research." We think you will find our discussion to be reassuring, and for some readers, outright motivating. Wherever you encounter a need for more detailed information than we have provided, please consult the sources we have recommended in each chapter (full citations are in the reference list), and, particularly, those books we have abstracted for your use in the Appendix.

Finally, we draw your attention to the fact that funding grants may be sought for two distinctly different kinds of undertaking: (a) projects in which the objective is to provide a needed service or (b) research studies in which the objective is to produce knowledge, with any service provided being purely incidental. Specialized guides are available for proposal writers who seek funds for the development of programmatic services. Some of these, such as the texts by Coley and Scheinberg (2000) and Bauer (2003), will be of interest to all novices in the grantsmanship game. As you already have detected, however, this book is directed specifically to proposals involving research, rather than service to a client population.

Locating the Money Tree

The first step in preparing a successful grant proposal is to know where to apply. Finding the funding source that provides the highest probability of successful application often is a time-consuming process. Given the time that will be expended in preparation of the proposal, however, it is a relatively small investment that can return significant dividends.

General Strategies

An effective strategy is to find an individual, university, or service group that maintains a grant library or monitors information sources concerning agencies and foundations. With their assistance, you can identify a preliminary list of grantors with possible interest in your study. Not only can such sources offer current information on available research support, many of their services are computerized to allow highly efficient search procedures.

Most research universities also provide postings of grant opportunities that are available for anyone who accesses their website (see Table 8.1). Many of these sites also have a section on regional, state, or local resources. Some universities also have specialized lists of resources, such as those shown in Table 8.1, that are specially targeted for graduate students. In addition, all governmental agencies and large foundations have a website, and much information about availability of grants can be obtained from those locations. Thus, the absence of information resources at smaller institutions is no longer the obstacle it was before the advent of the World Wide Web.

Many senior professors who have large research programs not only maintain current information about sources of funding in their particular area but also know and maintain regular communication with key individuals in granting agencies. Such contacts are invaluable for what they can reveal about priorities (and changes in priorities) within those organizations.

It often is beneficial for a novice to collaborate with a senior researcher on one or two applications for support before attempting a grant proposal on his or her own. Not only does this provide a valuable opportunity to acquire

Table 8.1 Examples of University Websites Providing Information About Research Funding

University	*Web Address*	*Type of Funding*
University of California, Berkeley	http://research.chance.berkeley.edu/	General
	http://www.grad.berkeley.edu/financial/deadlines.shtml	Graduate
	http://research.berkeley.edu/	Undergraduate
Columbia University	http://www.cumc.columbia.edu/research/website/3/3/1/index.html	Information on governmental, non-governmental, foundation, and student funding
Illinois Research Information Service	http://www.library.uiuc.edu/iris	Computerized searches for research funding sources
University of Massachusetts at Amherst	http://www.umass.edu/research/ogca/funding/	Information on governmental, non-governmental, and foundation funding
The University of Texas at Austin	http://www.utexas.edu/research/grants/funding.html http://www.utexas.edu/ogs/funding/	General graduate support

insight into the process, it is likely to be a successful and therefore reinforcing experience as well. The presence of a well-known researcher's name on a proposal can be very helpful because reviewers have confidence in known quantities and often confer that same confidence on collaborators. Of course, not every newcomer to grantsmanship is in the enviable position of association with a senior researcher or research staff. Many young scholars must undertake the unfamiliar task of locating the money tree on their own.

In this process, the first rule is to throw a wide net and keep an open mind about what might constitute a possible source. Although it sometimes does happen serendipitously, do not expect to find an agency with interests exactly matched to your particular research idea. With some creative adaptation, however, it often is possible to adjust the focus of a study sufficiently to interest a funding agency without distorting the primary intent of the investigation. For instance, a proposal involving the role of motor or play behavior in child development might be of interest to sources as widely disparate as the Center for Research for Mothers and Children in the National Institute of Child Health and Human Development, or an early childhood project within the Institute of Education Sciences, all depending on how the research questions are formulated. For this reason, a flexible point of view is helpful in assembling the initial list of potential funding sources.

Another source of information is the identification of funding agencies that are supporting research published or presented at national or regional conferences. The granting agency or foundation always is listed at the bottom of the first page of a research report published in a scientific journal (positioning of information about funding sources is more variable in unpublished papers, but the information generally will be present). Similarly, abstracts printed in conference programs and research session posters usually will note the research grant number and source for any funding. These can tell an observant reader not only what agencies are presently supporting research in particular topic areas but also something of the specific interests that may be present among those who review proposals in those organizations. For information of that type, conference abstracts and poster session notes often are the most current resource.

The most efficient way to find sources, however, is to use the wide variety of available retrieval systems. Most are computerized, although some continue to employ printed versions or CD-ROM technology. Many systems also provide both computerized retrieval and notification services. A sampling of these sources along with their Web addresses are shown in Tables 8.1 through 8.3, representing institutional, governmental, and nongovernmental retrieval resources.

As soon as the general topic of your proposed research has been defined in your mind, begin collecting and reviewing application forms from prospective sources. Because a growing number of granting agencies will not respond to either telephone or mail requests, that restriction will serve to force you into the medium of communication they prefer—electronic mail on the Internet. One of the things you will discover in reviewing application instructions is that many agencies require electronic application. Whatever the format used for information, instructions, and applications, having all that material at hand is a strong encouragement to consideration of research plans that will have at least some capacity to attract money—while serving your scholarly interests.

Table 8.2 Examples of Governmental Websites From Which Research Funding Information May Be Obtained

Institution/Agency	*Web Address*
Army Research Institute for Behavioral and Social Sciences	http://www.hqda.army.mil/ari/research/index.shtml
Army Research Office	http://www.aro.army.mil/Forms/form2.htm
Office of Naval Research	http://www.onr.navy.mil/education/
Air Force Office of Scientific Research	http://www.afosr.af.mil/
Catalogue of Federal Domestic Assistance	http://www.gsa.gov/fdac/
Department of Energy	http://www.sc.doe.gov/grants/grants.html
Department of Health and Human Services	http://forms.psc.dhhs.gov
Federal Register	http://www.gpoaccess.gov/fr/index.html
General Services Administration	http://www.gsa.gov/forms
National Endowment for the Humanities	http://www.neh.gov/grants/index.html
National Institutes of Health	http://www.grants.gov/ http://www.nih.gov/grants/guide http://grants.nih.gov/grants/guide/index.html
National Science Foundation	http://www.nsf.gov/funding/ http://www.nsf.gov/home/ebulletin/
National Technical Information Service	http://www.ntis.gov/index.html
Small Business Funding Opportunities	http://grants.nih.gov/grants/funding/sbir.htm

Using Institutional Retrieval Systems

A good example of institutional retrieval systems is the Illinois Research Information Service (IRIS), which is one of the most comprehensive and widely utilized computer retrieval services. If your institution subscribes to IRIS, you will quickly realize the benefits of a computerized search. IRIS

Table 8.3 Examples of Non-Governmental Websites From Which Research Funding Information May Be Obtained

Institution/Agency	*Web Address*
The Foundation Directory	http://fconline.fdncenter.org/
The Foundation Center	http://fdncenter.org/
	http://fdncenter.org/funders/
Grantsmanship Center Magazine	http://tgci.com/
Foundation News and Commentary	http://www.cof.org/
GrantSelect	http://www.grantselect.com/

permits those seeking grants to search on a wide variety of keywords (e.g., *clinical psychology*) and situations (e.g., *graduate study or research training programs*). In addition, you can create a personal profile of your interests and have periodic notices e-mailed to you to alert you to new funding possibilities.

Many institutional libraries also provide access to more generalized retrieval systems that include not only funding sources maintained by governmental agencies but those sponsored by private foundations, industries, and professional organizations. Learning to use such multi-purpose indexes often requires an artful combination of assistance from library staff and discovery through personal exploration.

Finding Federal and State Dollars

The second category of resources, and perhaps the largest, is governmental resources. Information relating to the research programs and procedures of the two largest sources of federal funding, the National Institutes of Health and the National Science Foundation, are easy to find on IRIS, discussed above, or by googling their names. State agencies are also listed in these systems, but states generally have few dollars for unspecified (noncontract) research.

The U.S. *Federal Register*, which is provided by the National Archives and Records Administration (NARA), can be accessed online and provides notices of requests for applications for all of the federal agencies and organizations (Table 8.2). It is also available in hard-bound format. In addition to the government's Web-based retrieval systems, many library reference departments offer standard works in hard copy or CD-ROM format that may be useful for beginning or extending your search for federal dollars. For example, a comprehensive listing of assistance available from the federal government is contained in the *Catalog of Federal Domestic Assistance (CFDA)* (United States General Services Administration, semi-yearly). The *CFDA* also publishes material that describes their research support programs and application procedures. Examples are the National Institutes of Health's *Guide for Grants and Contracts* and the National Science Foundation's *E-Bulletin.* Updated regularly, and now also available in a searchable format on the Internet, *CFDA* is an invaluable tool both for locating agencies with appropriate interests and for obtaining general information about support programs within those agencies.

The database of the National Technical Information Service also provides information on federal funding sources. Not only does this service provide information about the availability of money for research and procedures for obtaining it, but in many instances it also offers downloadable forms for electronic submission of grant proposals.

Tapping Into Non-Governmental Sources

The third category of resources includes non-governmental sources such as private, community, operating, and corporate foundations, and grant-making public charities. An excellent resource is the Foundation Center, which is a leading authority on philanthropy in the United States. The Center provides free access to grant information in print, CD-ROM, and online resources. In addition, it also provides extensive coverage of non-governmental sources through its publications (The Foundation Center, 2005a, 2005b, 2005c, 2005d, 2005e) and its regional centers. The Center is particularly useful for identifying smaller, local organizations with interest in the support of research.

The Center's website provides information, profiles, and deadlines for 47,000 granting foundations and agencies. The Center, supported by grants from foundations and corporations, also includes educational materials dealing with grant-seeking, news about research and philanthropy of interest to those seeking funding, and research on philanthropy. The Foundation Center publishes the *Foundation Directory,* which contains records provided by a variety of the Center's other publications. These yield information about more than 41,000 grantmakers. *The Foundation Directory* (The Foundation Center, 2005a, 2005b) is also available as an online retrieval service to which individuals must subscribe, but that option is expensive.

If your institution has a grants and contracts office, staff there can inform you as to whether the search capabilities are available for your use. Most large universities have the *Foundation Directory* available in several places on campus, and anyone with the proper identification and password can access it. If you do not have access to university or other institutional resources but are fortunate enough to live near one of the Foundation Center libraries (in Atlanta, Cleveland, New York City, San Francisco, and Washington, D.C.) you can perform a search there without charge.

Directories that list grant opportunities for other areas include the *Directory of Grants in the Humanities* (Oryx Press, 2005b), *Directory of Biomedical and Health Care Grants* (Oryx Press, 2005a), and *Grants, Fellowships, and Prizes of Interest to Historians* (American Historical Association, 1998). A researcher is likely to use these large and expensive compendia in the library or at a research center.

Considering Commercial Search Services

A number of commercial publications and services can meet the need for help in searching out sources for research support, and samples of their websites are shown in Table 8.3. Some of these provide valuable information about approaching foundations, new agency priorities, formulating more competitive proposals, and the legislative processes that precede new governmental programs. Used as a supplement to direct contact with funding agencies, information from such publications as *Grantsmanship Center Magazine* (Grantsmanship Center) and *Foundation News and Commentary* (Council on Foundations, Inc.) can help both in correctly targeting proposals and in improving their quality.

Other computerized grant retrieval and notification systems, such as *GrantSelect*, available through Oryx Press, provide subscription services to individuals. Some of these, such as the *Annual Register of Grant Support* (Information Today, Inc., 2005), the *Directory of Research Grants* (Oryx Press, 2005c), and *The Grants Register* (Palgrave Macmillan, 2005), are shelved in large library volumes. Although the *Directory of Research Grants* and *The Grants Register* may contain some overlapping information, they also have unique features that justify your attention.

The Bottom Line

Whether the retrieval system is institutional, as with IRIS, government-produced, as with the *Federal Register*, or individual, as with *GrantSelect*, the inclusion of notification capability offers a powerful advantage in the grantsmanship game. By obtaining prompt electronic access to the grant announcement as well as to other vital information concerning application procedures, the valuable time you save can be used to begin working on a proposal.

Business Possibilities

Finally, a potential source of long-term funding is the development of contractual partnerships with business and industry. If you want to pursue this possibility, we recommend reading *Get Funded!* (Schumacher, 1992) before you start. Even though this book is more than a decade old, it continues to stand alone as a practical guide for making contacts with corporations, understanding corporate politics, and working with business contacts to organize and maintain a successful research program.

The commercial sector needs research findings from a wide array of academic disciplines. Businesses of all kinds can profit from information that improves their operation in such diverse areas as "planning, finance, facilities, accounting, operations, marketing, sales, customer service, human resource management, training and education, retention of personnel, public and community relations, and so on" (Schumacher, 1992, p. 43). Faculty scholars in almost any discipline may be able to convince corporate executives that their research project merits support if they bring creative ideas and are willing to persevere in the effort to obtain a hearing.

Most large businesses have research and development units assigned primarily to the tasks of applied investigations, although some devote a portion of their resources to basic research. In either case, some of the research work may be subcontracted to sources outside the organization. This, of course, is a primary point of entry for the development of business/university partnerships. Thus, in realistic terms, you are likely to find that funding for specific research studies primarily is available to university faculty members who hold permanent appointments. Graduate students, postdoctoral scholars, and untenured junior faculty are rarely considered unless they are included in a team of university researchers.

It is not surprising that businesses are unlikely to be interested in funding short-term projects. Given the highly specialized needs of businesses, when research is done for them, it often involves expensive services, equipment, and facilities. Contracting to support such costly activity makes sense only when a partnership between the corporation and the researcher can be sustained over a substantial period of time. Businesses must look to the bottom line of profitability, and that goal is best served when funding can purchase a reliable source of continuing research capability.

Research support from commercial enterprises is a double-edged sword. Not only does it hold potential for support that is substantial in quantity, but additional funding also is much easier to obtain once the relationship is established. The disadvantage, however, is that support must go up or down with company fortunes. When profits go down, management changes, or priorities shift, the company can lose interest, and funds may quickly dry up.

Whether you locate potential funding sources in the government, foundations, agencies, or businesses, each has a processing culture that is unique. Just as important as finding the right source of support is to move through the system by "following the rules" as conscientiously as you can.

Resources Earmarked for Student Research

Money is available for student research, in some areas it even is plentiful, and there are hundreds of sources for small grants that require no more than a few pages of application information. For starters, although not all the funding available for support of graduate study can be used for support of a particular research project, university fellowships are a promising source. Once used exclusively to designate support for tuition, books, and living expenses while pursuing advanced study, the word *fellowship* now includes funding that may, or may not, include research. The words *grant* and *award* now are used so variously that only inspection of funding agency specifications can reveal whether money is available for support of student research.

Many other particularly fruitful locations exist to search for funds to support theses, dissertations, and other forms of student research projects. These locations are in the form of large volumes that catalogue sources of research support (in some cases these have associated sites on the Internet). It is unlikely that any one of them will identify every possible fellowship, grant, or award that might be of interest to you, but if you will work your way through each of those we recommend below, we think it will be equally unlikely that you will come up empty. We will list these resources in the order of priority we would assign to them, but your particular needs may make it sensible to use a different sequence.

Assuming that you already have consulted the reference librarians and the offices dealing with external and internal research funding at your institution, we think the best place to begin your search will be within a series of volumes produced by the Reference Service Press (RSP) that we have listed below. They allow you to select a retrieval source based on your primary field of research interest.

- Schlachter, G. A., & Weber, R. D. (2006a). *Money for graduate students in the social and behavioral sciences, 2005–2007*. El Dorado Hills, CA: Reference Service Press.

- Schlachter, G. A., & Weber, R. D. (2006b). *Money for graduate students in the biological & health sciences, 2005–2007*. El Dorado Hills, CA: Reference Service Press.
- Schlachter, G. A., & Weber, R. D. (2006c). *Money for graduate students in physical & earth sciences, 2005–2007*. El Dorado Hills, CA: Reference Service Press.
- Schlachter, G. A., & Weber, R. D. (2006d). *Money for graduate students in arts & humanities, 2005–2007*. El Dorado Hills, CA: Reference Service Press.
- Schlachter, G. A., & Weber, R. D. (2006e). *How to pay for your degree in education & related fields, 2006–2008*. El Dorado Hills, CA: Reference Service Press.

With the exception of the last volume, each of the Schlachter & Weber books contains a listing of more than 1000 fellowships, awards, grants, traineeships, and other funding programs set aside to support graduate study—of which 250–300 are specifically directed to support of research and creative activities. The latter will be displayed in a section devoted to "Money for Graduate Study or Research in . . . (academic field identified in the title of the volume)."

The listing of grant sources is alphabetical and thus requires some browsing time to digest. Each entry includes contact information for the grantor, a brief summary, eligibility, financial data, duration, number awarded in each application cycle, and deadlines. Specifically, funds are available to support theses, dissertations, and small scale studies such as pilot investigations. Some have restrictions to a particular location, but others do not.

Within the present RSP series, the last named reference above is devoted to study in the field of education and has no analogue for students in other academic or professional disciplines. If your area of graduate study does happen to be in education, however, this expansive volume offers more than 40 research grant sources that are specific to a variety of teaching specializations ranging from early childhood education to the teaching of organic chemistry.

Given our advice about building a track record, it is significant that within this group of funding sources are several research awards that are "mini-grants" providing modest funding for small projects. Finally, if you will inspect the grant eligibility information with care you may be surprised by the number of funding programs open to undergraduates. Several in this RSP volume even specify that the applicant must be a pre-service teacher trainee who proposes to do a research project mentored by a senior faculty member. We can't imagine a better way to start your track record!

The publisher of this series maintains a useful website at http://www.rspfunding.com and serious student grant-seekers will want to spend some time there reviewing all of the publisher's products related to research funding.

In particular, there are a number of even more specialized source guides in the RSP series including those listing funds available for particular racial and ethnic groups, and a volume devoted to the extensive collection of resources available for study and research outside the USA. These are:

- *Directory of Research Grants 2005*. (2005c). Westport, CT: Oryx Press.
  We have previously directed your attention to this volume (widely known by its acronym as the *DRG*) as a general resource in the search for research support. One of the 38 categories for organizing the several thousand grant opportunities within the massive *DRG*'s covers, however, is a section devoted exclusively to "Dissertation and Thesis support." It lists nearly 200 grants, fellowships, and other awards for support of graduate student research. The title of the section notwithstanding, if you read closely you will discover that a number of the listed sources are available for small-scale student inquiry projects other than theses and dissertations.
  Each listing includes information about requirements, restrictions, amount, duration, deadline dates, and contact information. The publisher's website at http://www.greenwood.com contains a full description of the *DRG* as well as other publications related to research support.
- *The Grants Register 2006*. (2005). New York: Palgrave Macmillan.
  The Grants Register (TGR) also has been previously introduced as a general resource for locating research support. What is not generally recognized, however, is that despite its subtitle, "The complete guide to postgraduate funding worldwide," this vast collection of grant sources does include funds for thesis and doctoral research, and even for support at the pre-proposal stage of work.

Your entry into *TGR* must be through your area of study (nursing, education and teacher training, arts, humanities, etc.) wherein everything is listed in alphabetical order (there is no general index for the full volume). This requires some investment of time to browse through what may be a lengthy list of potential sources. Depending somewhat on your subject field, however, the hunt can be worthwhile and almost always turns up some surprises.

TGR does specialize in grants for postdoctoral research, and the listed sources are weighted accordingly. Nevertheless, if you will inspect the information concerning "eligibility" and "level of study" you will discover that in the convoluted language used in the world of research grants words such as "post-graduate" may signal the availability of money for "pre-doctoral" investigations. It will take some study, but the *TGR* is a treasure map that can be deciphered and put to use in your quest.

Finally, examine the websites of the professional associations of which you are a member. Many of these national and state associations provide scholarships and/or awards for promising students who are interested in making research an active part of their career.

Working With the Money Source

The ratio of proposals presently being funded to those being rejected is very low. Fewer than 25% of first-time applicants to the National Institutes of Health and large foundations are successful, and standards almost as rigorous are in place in many other funding sources. It is obvious that the task of writing and submitting a proposal cannot be approached in a cavalier manner. The process is very competitive. Because of this, the preparation of grant applications and the entire process of interacting with the funding agencies has become a skill that often is as important as inquiry skills themselves.

Each government agency and independent foundation has a specific form to be completed and a unique process through which the proposal must be evaluated. Thus, only if the potential investigator of a research project is skilled, diligent, persistent, patient, and, above all, meticulous in following the sometimes unique regulations and procedures of the funding agency can support be acquired. Below are some suggestions for working with four of the sources you are most likely to tap for research funds: (a) internal resources from your own institution, (b) federal and state agencies, (c) foundations, and (d) business.

Home Institutional Resources

Your own college or university is the best place to start. Relatively speaking, you are "family," you are already familiar with the language and the electronic culture of the institution, and the institution wants you to succeed. It is a good place to practice your skills. What is available? Your university website's home page, with links such as "For Faculty," "Faculty Resources," "Student Research," or "Research Funding," will probably take you to funds that the university makes available for research. These sometimes take the form of summer salaries which must be used to conduct a small pilot research study, special equipment that may be requested (accompanied by a small proposal), computer upgrades available only for research activities, travel budgets for research-related activities, and so forth.

Internal sources ordinarily require only a modest proposal document (see Specimen Proposal 4 for an example) that is not as lengthy or complex as those required by external agencies. That advantage notwithstanding, you should make use of the seminars and workshops on grant-writing that may be available on your campus. Those experiences not only allow you to learn more about grant writing in general, but to become acquainted with specific procedures that are required by your institution. In addition, it is wise to locate and use the research consultants (usually located in the computer center or the graduate school), who can confer with you and answer questions specific to your proposal.

The most effective strategy for making successful use of internal funding is to land one or two small ($1,000–$15,000) institutional grants, conduct the investigations, submit the results for publication, and then use those reports as your track record to support a proposal for more extensive external funding. This process provides you with start-up funds to do the first studies and also indicates to outside agency reviewers that your work already has been critically and positively reviewed, not only by an editorial board but also by a university funding review committee.

The use of local grant sources also underscores how important it is to have envisioned a systematic research program rather than a single study. Each research project can build on the previous one, providing more and more evidence that your ideas are worthy of major funding from sources beyond the walls of your home base. In programmatic research, it is logical sequence, not the accumulated number of studies, that will impress reviewers. Clearly, a fundable proposal could never be supported by a host of small-scale studies aimlessly conducted on a hodgepodge of tenuously related topics.

Federal and State Agencies: Progressing Through the Maze

Federal sources are primarily the National Institutes of Health (NIH) and the National Science Foundation (NSF). They both also are the primary funders of theoretical research and theory-driven applied research. The NIH, which is composed of 27 Institutes and Centers through which more than 27 billion research dollars are distributed, is the primary federal agency that conducts and supports research to improve people's health (e.g., the health of children, men, women, minorities, and seniors, as well as topics related to wellness and lifestyle).

The NSF funds research in science, mathematics, engineering, technology, and studies of education pertaining to those fields. The NSF provides about 25% of federal support to academic institutions for basic research. In academic circles, achieving federal funding from either of these federal entities is highly prestigious, partly because these are the best sources for obtaining large grants, but also because the reviews of proposals are extremely rigorous.

The NIH review process serves as an example. Reviewers for the NIH cover all of the major sections of a proposal that we have already discussed in this text. They (a) evaluate the practical and theoretical importance of the question in the study, (b) determine the technical adequacy of the proposed study, (c) judge the applicant's ability to execute the research, (d) identify the extent of congruence between the research topic and the organization's mission, and (e) fit those evaluations into a calculus determined by the amount of funding available—to identify which proposal(s) will be supported. Because NIH must apportion limited funds in a careful and equitable manner within a particularly important field of research, its reviews are complex, thorough, rigorous, and not a little daunting to even a veteran investigator.

The NIH website provides guidelines, instruction pages, application forms, samples of completed form pages, a discussion of the peer review process, and suggestions for maximizing your chances to receive a positive review. The guidelines and instruction sheets must be followed explicitly and all applications must be submitted electronically.

Even before an NIH application can be sent on its way, there will be a host of institutional hurdles to navigate. Although the process differs across institutions and in many cases is all electronic, acquisition of required signatures alone may require at least a month. Included in a large-scale proposal might be some combination of officers in a variety of units and at several different levels: department chairs, program chairs, deans, provost, director of technical facilities (the computer center, for example), a human subjects Institutional Review Board, business managers, and, in some cases, even the chief executive officer of the institution or state system. Few of these signatures will come without delay, a fact that must be coordinated with the NIH deadline with the greatest care. Because the electronic submission process requires extreme precision and approval for submission will not be granted until each and every requirement is fulfilled, most investigators plan on starting the submission process at least one month before the actual NIH deadline for receiving the application.

Once the proposal is ready for submission, it is transmitted to the NIH Center for Scientific Review (CSR), usually from the institution's office of sponsored projects. At this point the proposal travels through an amazing complex of decisions, branch points, evaluations, and judgments, the details of which are well beyond the scope of this text. If you are interested in details of the process, they may be obtained at the NIH websites shown in Table 8.2.

In summary, Study Section members, all of whom are active researchers, will evaluate the scientific and technical merit of the proposal. They will reject the bottom 50% of the applications, and assign a percentile rank for scientific merit to the remaining 50%. These ranks, along with summary statements from the Primary Reviewers (and the Scientific Review Administrator) are forwarded to the appropriate Institutes (e.g., the Institute of Aging). The Advisory Council of the Institute, leaning heavily on the scientific merit percentile ranks that have come from the Study Section, balances this information with other factors, such as the implications of the research topic for the Institute's goals, the similarity of the proposed research to other proposals or to existing funded research programs, and the total level of funding available. In some cases, members of the council will recommend that particular proposals be given special consideration for funding, or they may determine that an application is of questionable relevance to the institute's mission and recommend that it be assigned elsewhere. A reassignment, of course, would require several more weeks to process. That point emphasizes how important it is for the applicant to do everything possible to write the proposal with a specific orientation and to direct it to an Institute that would find it attractive.

The National Science Foundation is the other large federal resource for grants. As is true for the NIH, competition for grant money is intense. NSF receives approximately 40,000 proposals each year for research, education and training projects, of which approximately 11,000 are funded. In addition, NSF processes several thousand applications for graduate support (see section on student funding at the end of this chapter). The agency operates no laboratories itself but does support National Research Centers, user facilities, certain oceanographic vessels, and Antarctic research stations. The Foundation also supports cooperative research between universities and industry, U.S. participation in international scientific and engineering efforts, and educational activities.

Another source of federal research funding is the United States Department of Defense, which funds many programs. The Army, Navy, and Air Force have active research programs and a process by which they fund investigators external to the military. Their home website addresses are shown in Table 8.2. Some of these military research programs even have opportunities for new researchers and pre-doctoral students. These sources, however, are generally very specific to the needs of particular branches of the military. Thus, the funding opportunities are more likely to be known to those in the military or to those who have contacts with a specific branch of the military.

Some of the larger states have research funds available, but generally state research funding is for state-specific agendas, such as increasing particular types of technology, or enhancing certain types of state resources in areas such as agriculture, fishing, or business. Most state education and health agencies support applied research, and these may be located by exploring the state website. Working with state agencies is similar to working with federal agencies. Find out who the contact people are, communicate with them briefly but frequently, and follow the instructions and regulations.

Foundations

Because every foundation has different procedures for working with grant proposals we cannot provide a single description, as was possible with the NIH example above. We can, however, provide advice about working with foundation staff and offer some suggestions that will improve your chances for a successful application.

The charter of each foundation establishes a mission that is particular to that organization. Some are constrained only to fund programs providing benefit in the community (for example, grants to support development of legal services for the poor) and do not fund research. Other foundations have been created specifically for the purpose of supporting research in particular areas (for example, mental health or education). Still others fund both public service programs and research. Foundations in this latter category often are more likely to support studies that deal with problems closely associated with the service aspect of their mission. Even though they are willing to support research projects, their first priority remains the impact of findings on their community and their particular constituents.

Accordingly, when investigating a foundation as a possible source of funding, try first to inform yourself about its mission. If the interests of the foundation appear to have some relationship with your research agenda, the next steps are to obtain a sense of the priorities articulated by its leaders and information about the kinds of studies that have been approved for funding in recent years. If proposals have been funded recently in areas that are related to your research interests, it will be particularly helpful if they can be obtained for review. If the proposals are not available, the final report most surely will be in a public document that the foundation can provide.

The initial contacts with foundation personnel are particularly important. Courtesy to everyone is a rule, whether it is an executive director, board member, or receptionist, and remembering that their time is valuable is the first way to show respect. If it becomes apparent that the agenda of the foundation does not provide a sufficiently close fit with your present ideas, make a point of thanking staff for their time and withdraw politely. You may wish to return at a future date.

In the highly competitive grant marketplace, trying to force a square proposal into a round foundation is likely to be a poor use of time and effort. If, however, preliminary contacts with staff do suggest that there is some potential for mutual interest, you may wish to extend the discussion with staff before developing and submitting a proposal.

Most foundations welcome the opportunity to have staff discuss possible proposals with prospective applicants. Exploring areas of shared interest and providing some preliminary screening of potential applicants are functions that are in the best interest of the organization. Foundation staff can alert you to deadlines, describe application procedures, indicate areas of investigation for which the foundation currently has particular concern, explain how the review process works, and identify the individuals within the organization who hold final authority for grant decisions.

As with government agencies, some foundations have grant application packets that provide comprehensive information about the required format for all proposals. Others will accept proposals in a variety of formats. Likewise, some foundations make funding decisions only at a specified point (or points) during the year, whereas others make decisions on applications as they are received. It is your responsibility to obtain the existing regulations for the proposal process—and follow them. For example, if the instructions specify that the pages of your proposal be hole-punched, do it (if you really want the money)!

Assuming your proposal reaches the foundation office on time and in the proper form, it will be reviewed. This process may be structured in any of several ways.

1. A single staff member reviews the application and makes a decision based on the technical merits of the proposal, his or her perception of your ability to carry out the research, and how the study relates to the priorities of the foundation. This is the least common form of review, although it still may be encountered in a few organizations.

2. One or more staff members review the application and provide comments on some (or all) of the points noted above. The foundation's board of directors (or a special committee of the board) then uses that information as the basis for making final funding decisions. This process may be as formal as the one described for NIH reviews—complete with rankings—or may be much less formal, with nothing more elaborate than a committee discussion about which proposal will produce the most benefit for the foundation's money.

3. Comments produced by staff review procedures are supplemented by recruiting the critiques of outside specialists with expertise in the area of the proposal. All comments are then forwarded to the next level of the review and decision process.

In the event that a proposal is not funded, it is to your advantage to obtain feedback from the review committee or foundation staff. Not only can you learn how to improve your proposals, but you also may wish to approach the foundation with another application in the future. The more you learn about their review process, and which aspects of the proposal are pivotal in determining acceptance or rejection, the more likely you are to succeed on a subsequent try. In any case, it is important to be businesslike and respectful even in the face of a rejection. As one veteran foundation executive told us, "No one likes whiners." Positive interactions with foundation staff are the basis for future receptivity.

If your proposal is funded, anticipate that the expectation for continued accountability may be higher than that exercised by governmental sources, especially if the foundation is local and the project is nearby. Local foundations expect to receive as much publicity about their investments as possible, because it makes them credible to their board of directors and also helps them raise more money. They may expect several interim reports, and they may expect you to provide oral presentations to their board, television audiences, or participants in community programs sponsored by the foundation.

Partnerships With Business

The process of shaking the corporate money tree may be very different from that required for government agencies or foundations. Access here often is achieved through contacts with personnel in the organization—frequently with researchers, technicians, or research managers. Usually it is necessary to have someone inside the organization to advocate for your proposal and assist in moving your ideas for a partnership through the channels of corporate management.

In her informative book *Get Funded!*, which is reviewed in the Appendix, Schumacher (1992, pp. 95–108) suggests a variety of ways you can identify and cultivate company employees who may be interested in sponsoring your proposal for a research partnership. Some of these are paraphrased in the brief list below.

1. Keep in touch with former students and colleagues who have found employment in business or industry.
2. Enlist the help of alumni who may have contacts in the corporate world.
3. Make use of university staff who have corporate relations as a responsibility.
4. Attend on-campus presentations by corporate representatives.
5. Talk to company recruiters when they come to your campus.
6. Make contact with representatives from the business sector who serve on any of your institution's various advisory committees.

7. Make contact with business organizations through your state's economic development agencies.
8. Use cooperative education programs and field placements to locate some of your students in companies with which you may have shared research interests.
9. Search out possible avenues of contact through the School of Business or Management at your institution.
10. Make contact with corporate researchers who publish in your field.

Business and Corporate Foundations

Most business foundations differ sharply from philanthropic foundations. The availability of research funding from such institutions often is not completely insulated from the profitability of the corporate parent and rises and falls accordingly. Frequently, the nature and purpose of the foundation is closely tied to the interests of its business sponsor, although it is common for the mission statement to be written in such general terms that it is difficult to discern the true priorities (and biases) of the organization.

Grants from business foundations often are made preferentially to particular institutions of higher education. This may be on the basis of some historical relationship, such as the undergraduate education of a corporate founder or the existence of familial ties between company and institution managers. For that reason, applications or letters of inquiry sent without the advantage of sponsorship may provoke little interest from the corporate foundation's management. As in the case of direct contacts with business organizations, it is best to locate an inside advocate who can sponsor your proposal for serious attention.

The Grant Application—The Big Picture

We have talked about how to find prospective sources that might provide money to support your research, and how to work with those sources once you find them. Now we can turn to the grant proposal document itself, but before examining the specific components and details of that submission (in Chapter 9), we want to emphasize how important it is to keep "the big picture" of the proposal in your mind throughout the writing process. Our suggestions for maintaining that broad perspective are grouped here in five clusters, each of which will be addressed within this section.

- First, it is important to understand how this type of proposal differs from those prepared in academic settings for theses and dissertations.
- Second, as you write the document it is useful to consider the standards that the evaluators (reviewers) are likely to have foremost in their minds as they read your proposal.
- Third, as you identify, shape, and describe your research question it is wise to remember that it represents the single component of the proposal that holds the greatest potential to elicit a positive or negative response.
- Fourth, the quality of your proposal document, both as rhetorical prose and as an example of your command over sound mechanics, will be noticed by reviewers and will become a part of the calculus through which they make a final determination about the wisdom of supporting your research.
- Fifth, and finally, you must understand that who you are and where you will perform your study can make a difference in how reviewers weigh your proposal.

The Application and the Proposal

Grant applications are formal appeals for the award of support, usually in the form of money, for undertakings by individuals or groups, in which the grantor has special interest. In the strict sense, a grant is an award of funds without a fully defined set of terms and conditions for use, whereas a contract is a work order from the grantor in which procedures, costs, and funding period have been established in explicit terms. Thus, the majority of what are commonly called grant proposals are applications for award of a contract—not a grant.

The grant request contains two components: (a) the application for funding and (b) the proposal that explains in detail what activities the funds—if granted—will support. Ordinarily, no distinction is made between the application and the proposal, the terms being used nearly interchangeably. Nevertheless, in any application for the support of inquiry, embedded amid budgets, vitae, descriptions of facilities, and plans for dissemination of results is a research proposal. To that extent, then, the preparation of a grant application requires most of the same skills as the preparation of any plan for research.

The differences between the research proposal embedded in the grant application and the dissertation or thesis proposals described earlier in this text rest primarily in the need to (a) conform to the grantor's format, which may be specified in considerable detail; (b) present the study in a way that will have maximum appeal to the granting agency; and (c) master skills required to complete other parts of the application that are not directly part of the research plan itself. In all other respects, a sound research proposal is a sound research proposal, whether in a grant application to the National

Institutes of Health, a philanthropic foundation, or in a dissertation proposal placed on your advisor's desk.

When your grant application is received by the funding agency, it will be read and evaluated—a process designated by the generic term *review*. In most cases, the application will be processed in company with other proposals that are competing for support during the same funding cycle (such cycles are often yearly but may involve either longer or shorter time spans). In reviewing the submission, the agency or foundation will employ a set of procedural steps to accomplish both the evaluation and, ultimately, the final determination of which submissions will be funded—and at what level of support.

Such review procedures vary enormously and range across a wide spectrum of complexity. At one end of the range are small organizations that may employ a single reader to evaluate and make recommendations. At the other extreme are the large federal agencies and substantial private foundations that utilize teams of both research peers and in-house specialists to complete a multistage review. If you are considering preparation of a grant proposal, it is imperative that you acquire a meticulous understanding of what kinds of review procedures you may encounter. The best source for that information, of course, will be the organization to which your application is directed.

Although review procedures vary in complexity and rigor, almost all are looking for the same things: probability of success and the quality of the proposal. These criteria are:

1. *Significance*: Is answering the research question important? Will it advance knowledge? Will it have an impact on the field of study in terms of scientific knowledge and/or practice?
2. *Approach*: Is the methodology sound and appropriate to answer the research question proposed?
3. *Innovation*: Does the proposal describe novel concepts, approaches, or method? Is the study original and creative?
4. *Investigator*: Is the researcher well trained and does he or she have the capability to conduct this study?
5. *Environment*: Is the laboratory, clinic, or field area appropriate to conduct the study? Does it have unique features that would enhance the probability of the success of the study? Does the institution or some other agency support this study?

Estimating the probability that you can conduct the research project to a successful conclusion and determining the value and integrity of your proposed study are the most important steps in the proposal evaluation process. It matters not how interesting the proposed research appears to be if the research is not successfully completed. The probability of your success is estimated by the reviewers by studying your training, experience with research, and the environment in which you work. The value of the research proposed is determined by the clarity and forcefulness of your arguments for the significance of your question, as well as by how well that inquiry is served by the proposed study design.

The Cornerstone: The Research Question

The foundation of a successful research proposal is the research question(s). If the answer to the question(s) that the investigator is proposing is important to the funding agency, fills in gaps in knowledge, will inform theory and potentially impact many people or institutions, then the reviewers will have the initial interest required to examine the proposal with thorough attention to detail. If, on the other hand, the first reaction to reading the research question is "so what?" then no matter how skillfully the proposal has been written, it will not receive a positive response.

Thus, considerable time and effort should be expended on shaping and developing your explanation of the research question. Once the general idea is in place, however, you can begin the process of determining who might be interested in the answer. The nature of your question will be the primary agency for directing you to appropriate sources from which to request funding. And once you have targeted one or more potential grantors, the next step will be to study the criteria they use to evaluate proposals.

Quality: The Proposed Research and the Research Proposal

Although the credentials of the investigator are extremely important, no amount of "star power" on the part of the investigator can compensate for an unimportant research question or a poorly designed study. First and foremost the research question should be important in terms of its comprehensiveness and generalizability. How many people in how many disciplines and professions will be affected by the findings? How generalizable will the results of this study be to research questions in other disciplines and professions? Will the potential findings have a significant impact on the practice of professions such as education, medicine, or rehabilitation practices? Second, is the proposed research innovative? All other things being equal, a proposed study that appears to have the possibility of breaking new ground, changing commonly used research techniques, or articulating new mechanisms to explain phenomena will rise to the top of the funding list.

A second critical quality rests in the proposed design and methodology. Reviewers will study those elements with particular intensity because it is in design and method that flaws can sabotage the results. For example, it is axiomatic that reviewers always give close attention to the descriptions of potential participants in the study to ensure that the subject characteristics, quantity, and selection procedures are appropriate. Also, you can assume in advance that readers will focus on any proposed tests or evaluation procedures to ensure that they are current and appropriate.

Finally, as you might anticipate, the statisticians among your audience will pore over the proposed analyses to be certain that they maximize the potential for providing a fair and rigorous test of your questions or hypotheses. If your study is qualitative in nature it is dangerous to assume that the same rules will not apply to your proposal. As an example, we have observed that reviewers have now come to scrutinize plans for data analysis with much the same intensity that was once reserved for operations with numbers and quantities!

This, then, brings us to the third salient characteristic of a proposal that will attract attention from reviewers. Because all three of us have served as reviewers, it is possible for us to say with absolute assurance that this aspect of proposal preparation is among the most underestimated and, therefore, potentially most lethal aspect of the whole research proposing process. To put it in the most direct terms possible, if the person asking for a grant does not take the time and trouble to produce a properly written document, then why would you trust them to be any more careful and prudent in producing research?

And to be equally frank, yes, by the words "properly written" we mean both the standard mechanics and the adequacy of expression. There is no excuse for either to be less than perfect. The world is not crowded with good editors for scientific text, but they are out there. Find one, use him or her with great care, and cherish his or her service with a respectful level of compensation!

Lest we be misunderstood, this is not an appeal for elegance of expression or flowery representations of ideas. In fact, our appeal is for exactly the opposite. Brevity is more than the soul of wit, it is the direct route to the reader's understanding—and a reviewer's affection! Be concise, even diagrammatic whenever possible, and (as they say) eschew all obfuscation in the form of popular jargon.

The Investigator and the Environment

Review committees want to feel comfortable in awarding a grant of financial support to an applicant. The committee members want to be assured not only that it is in their best interests to learn the answer to the research questions posed, but that the proposed study design is excellent, and that the researcher is competent to perform excellent work. The best evidence for competence is a series of publications on the topic to be researched. This has come to be known as a track record in the area of the proposed study. What record of accomplishment will you present to attract and reassure those who review your proposal?

Investigators are more likely to satisfy the review committee that their skills are adequate to perform the study if they include in their research applications citations to their previous work, documentation of pilot studies, and in some instances, if the application procedures allow, reprints of previous research reports that verify the reliability of the technique proposed or support the directional hypotheses to be employed.

In contrast, when the grant applicant is a complete neophyte, or has not completed a study in the specific area of the proposal, the committee will be forced to depend on less direct evidence of competence and, in the end, mostly on faith and intuition. It is understandable, however, that committees more often avoid such speculative investments and select instead proposals from individuals with a solid track record.

Because a strong track record is so important, successful young grant recipients usually are those who have worked through their doctorate in a very active research program and who also have had postdoctoral appointments. In both the doctoral and postdoctoral programs, these new researchers have had ample opportunities to practice proposal writing and to publish with seasoned mentors.

Factors other than track record are weighed by review committees (most notably the quality of the research question and proposal), and novice researchers or veteran investigators entering new areas do sometimes receive grant support. First-time grant proposals do not automatically encounter a hopeless situation that demands a display of previous research by a new investigator trying to obtain funding for the first research project. The fact remains, however, that previous performance weighs heavily among those factors that determine success.

The track record has been emphasized here not to discourage the beginner but to lend support to a suggestion that we have found particularly useful. A good way to build a track record is to begin with limited aspirations and seek limited funding from local sources, particularly from the institution in which the researcher works or from a professional association that encourages student research by providing small grants or awards.

The other important element of determining the likelihood of successfully conducting the research study is the environment in which the research is proposed. Reviewers are interested in the support that your college, university, or agency can provide you in the conduct of this study. Support includes space to conduct the study, equipment and computer resources, and availability of consultants such as statisticians, technicians, agency administrators, community leaders, and faculty in other departments who do research related to your proposed study.

Dealing With Rejection

Whether you submit a grant application to a governmental agency, a philanthropic foundation, or a business-affiliated organization, your proposal will be considered in competition with other proposals. Such grant competitions are mediated through review processes that, whatever unique form they may take, ultimately must involve human readers. This fact produces an implacable rule. *What is not noticed is not funded!*

Amid the sometimes formidable stack of proposals, the document that does not catch the eye—and thus the reward of closer attention and consideration—cannot compete on the formal criteria associated with quality of design and congruence with the grantor's priorities. For that reason, in Chapter 9 we will give close attention to the abstract and introduction that must present whatever attractive elements are contained in the research question and proposed methodology.

Beyond the problem of failing to attract attention, the reasons given for rejection may be phrased in diplomatic terms—but they boil down to a surprisingly small set of simple and all too familiar defects. Because in purely statistical terms the probability of rejection for any given proposal is high, beginners in the grant application business have good reason to note the common causes of "submitted, but not funded" with considerable care. As shown in Table 8.4, reasons for rejection fall into four general categories: mechanical, methodological, personnel, and cost-benefit. Having made the point that most rejections are the consequence of proposals that are unsound—often in simple and obvious ways—it is important to note that perfectly sound proposals sometimes are rejected. Research of considerable merit may remain unsupported for no reason that can be associated with the quality of the proposal. Because that unfortunate circumstance is a possibility, rejection must not be taken, by itself, as evidence of a fundamental defect.

Even if none of the reasons listed in Table 8.4 appear to be operative, a proposal simply may not appeal to a reviewer who must make a choice among equally strong contenders. Under such circumstances, subjective

Table 8.4 Common Reasons Grant Proposals Are Rejected

Mechanical Reasons

1. Deadline for submission was not met.
2. Guidelines for proposal content, format, and length were not followed exactly.
3. The proposal was not absolutely clear in describing one or several elements of the study.
4. The proposal was not absolutely complete in describing one or several elements of the study.
5. The author(s) took highly partisan positions on issues and thus became vulnerable to the prejudices of the reviewers.
6. The quality of writing was poor—for example, sweeping and grandiose claims, convoluted reasoning, excessive repetition, or unreasonable length.
7. The proposal document contained an unreasonable number of mechanical defects that reflected carelessness and the author's unwillingness to attend to detail. The risk that the same attitude might attend execution of the proposed study was not acceptable to the reviewers.

Methodological Reasons

8. The proposed question, design, and method were completely traditional, with nothing that could strike a reviewer as unusual, intriguing, or clever.
9. The proposed method of study was unsuited to the purpose of the research.

Personnel Reasons

10. As revealed in the review of literature, the author(s) simply did not know the territory.
11. The proposed study appeared to be beyond the capacity of the author(s) in terms of training, experience, and available resources.

Cost-Benefit Reasons

12. The proposed study was not an agency priority for *this* year.
13. The budget was unrealistic in terms of estimated requirements for equipment, supplies, and personnel.
14. The cost of the proposed project appeared to be greater than any possible benefit to be derived from its completion.

factors determine decisions, often leaving no wholly logical explanation for the action. The following section, dealing with resubmission, was prepared with the unfortunate authors of such proposals in mind.

A rejection, especially of the first grant proposal submitted for review, should not be unexpected. It is commonplace in academic life—especially for the novice. Just as it takes practice to learn how to perform any complex task truly well, it takes practice to write successful proposals. So long as the applicant has an opportunity to profit from evaluative feedback, each attempt should produce significant improvement.

Most review systems contain some provision for feedback as part of the rejection process. Often in the form of a summary review of the proposal, the feedback statement will list the main criticisms and may even suggest changes. The applicant should consider each point. If resubmission is contemplated, the new draft should address as many criticisms as possible through revisions, clarifications, and strengthened rationales. In addition, when the interval of time between the original submission and the notification of results has allowed collection of new pilot data, or acquisition of additional support from new publications, these should be added.

The easiest critique to respond to is one that raises questions about design or methodology. There the decision is simple: Either provide better substantiation for the choices made or change them. The review, however, may center not on what was proposed, but on the way the proposal was written. In some instances, the absence of any substantial comment about the research question or methodology provides the indirect suggestion of a failure to communicate clearly. In other cases, the reviewers may have understood the proposal perfectly but had reservations about a nontechnical issue such as the ratio of costs to potential benefits. In the former, rewriting for clarity is the needed response. In the latter, the only recourse is either to strengthen the rationale for the importance of the study or to find ways to cut the budget.

In terms of devising a response, the more difficult critiques often are those related to the proposed research questions themselves. In such cases, the author must read the reviewers' comments with great care so as to ascertain whether the questions raised are matters of logic and research design, or really are subtle expressions of doubt regarding the value of the research.

Sometimes reviewers signal their disinterest in the proposed study in implicit ways, because they don't want to make the blunt assessment: "I don't think this is worth doing!" In some (unfortunate) cases, this lack of candor leads to a furious hunt for every possible flaw in the proposal (no matter how minor) that can be used to rationalize rejection. That problem in the review process was one of several that led recently to a thorough reexamination of the NIH system. Although that evaluation did not lead to radical changes, it did produce a new question to which every reviewer must respond in his or her critique: How likely is it that the proposed project will make an original and important contribution?

Negative answers to that question offer strong evidence that the problem lies with your research question(s). The only useful responses are either to argue the matter by improving the rationale or to make significant modifications in the question(s) you want to ask. All of this suggests that in making use of reviewers' comments on your proposal, you must read both the lines and between the lines.

Once responses to criticisms have been advised, the rejected proposal may be resubmitted. An accompanying letter should note that the document is a resubmission and request return to the original reviewers. When this can be accomplished, it both makes use of familiarity and avoids a new set of criticisms based on a fresh set of perceptions (and biases). Proposals can be resubmitted several times, provided that feedback encourages revision and there is no suggestion from the grantor that the topic of study is inappropriate.

A telephone or e-mail conversation or meeting with the individual who coordinated the review process often can clear up ambiguities in the feedback provided with rejection. Such assistance is particularly useful when it is not clear whether the primary cause for rejection lay with the research question or with the substance of design and method. Further, a willingness to listen attentively, and politely, can serve to invite suggestions for improvement. In some instances, a staff member at the agency or foundation may even give the applicant an estimate of the probability for success in a resubmission. The applicant should remember, however, that staff members can only provide their personal judgment on such matters. In most cases, final decisions will be made by others.

After the first shock, a rejection should not inhibit the urge to find support for a worthy study. With critical feedback in hand, a stronger proposal can be drafted. The research that you do to substantiate revisions, the contemplation that you go through to clarify your plans, and the discussions that you have with colleagues about your proposed revisions all make you more knowledgeable about your topic and able to express your plans in an understandable and convincing manner. Most people who spend considerable amounts of time writing proposals, receiving grants, and reviewing proposals will tell you that more often than not, funding goes to the tenacious. It goes to the tenacious because in the trying and the editing, the researchers become much better at writing proposals, and the proposals become worth funding.

Planning and Preparation: Taking the Long View

As the competition for sizable grants has grown more rigorous over the years, the amount of detail involved, the advanced planning required, and the complexity of the proposal process itself have dramatically increased.

Shaking the money tree is not the straightforward process it was when funds were more plentiful and applicants fewer. Once, one or two weeks was the norm for the time invested in a proposal of modest scope. The same documents now may take several months of preparation. Nevertheless, if your goal is to make productive scholarship a part of your career, there is no alternative. Particularly if you want to prosper in a university setting, grant writing will be an essential skill for survival.

When the proper steps are taken in a logical sequence, and adequate time is allotted to complete each step, the worst that can be said about proposal writing is that it is exacting and sometimes tiring work. The best one can say is that it is challenging, and even stimulating. This is particularly likely to be the case when investigators in complementary areas of expertise collaborate in the process of developing the research plan. The interaction that occurs among colleagues during formulation of research questions and study design can be educational—and a powerful motivation for sustained effort. When a grant application has been completed, the proposal critiqued by expert reviewers, and the project approved for funding, there is more than just the satisfaction of success. There is the excitement of doing what the money can pay for—a good study.

Special Tips (and Encouragement) for Students

We would like to finish this chapter by providing some special tips and encouragement to students. It should not come as a great surprise, having read about all of the trials and tribulations of finding research support that have been discussed to this point, that few students apply for grants, awards, or fellowships with which to support their research—whether theses, dissertations, independent study projects, or pilot investigations. This pattern of lost opportunities arises from several sources.

First, overwhelmed by the difficulties of planning a study and writing the necessary proposal, students find it difficult to imagine they will surmount those hurdles, much less organize an application for funding. Second, faculty advisors often do not encourage students to apply for research grants. Indeed, there sometimes is an unspoken (and, perhaps, unintended) discouragement in professional attitude that holds student research to be something less than the "real thing" and, thus, not worthy of the effort involved in a search for funding. And third, given the temporal rhythms of graduate programs, the tasks of planning and writing a grant application many months in advance of a project or thesis often must compete with the more mundane but, nevertheless, essential work of attending classes, preparing for examinations, and meeting assistantship obligations. In that context, the speculative possibility of a grant to support next year's research is disadvantaged by the certainty of what is due on Monday morning.

In the larger universities, of course, where students are part of a research community and often involved directly in funded laboratories or long-term, grant-supported projects, they usually are socialized into a tradition that defines planning for funding as an essential (and routine) part of preparation for a study. Those fortunate interns and research assistants, however, represent a minority of the students proposing and executing research studies. Each year, literally thousands of students do research without funding of any kind, often without any programmatic expectation that they should learn how to search for support, and, even more unfortunately, often without strong faculty encouragement to believe they can succeed.

Why You Should Seek Funding

We believe none of the excuses for failing to seek financial support are valid. At the least, all of them are shortsighted and work to the disadvantage of students' education, as well as the quality of their research (and personal lives). We urge you to consider a very different and far more positive perspective on the possibilities for finding money to support your research.

Graduate departments in which faculty insist that their students not only begin early but also persist in grant seeking throughout their program invariably receive a disproportionately large number of student research grants—year after year. Equally important, such programs are providing training that is essential for success in any career that involves the conduct of research.

Beyond the personal and institutional discouragements noted above, there is the pervasive belief that money for student research is not available, or, at least, is in such scarce supply as not to justify the great investment of time required to make application. We hope that the special section labeled *Resources Earmarked for Student Research* in the early part of this chapter has convinced you otherwise. We can sum up the matter by revealing the three most important and pervasive prejudices that operate to deny students money for support of their research projects. Granting agencies give money only to those who have: (a) applied for it, (b) completed the application forms exactly as specified in the instructions, and (c) submitted the applications before the indicated deadline.

Those who ask, can get. Those who do not ask, do not get.

Think about it.

Admittedly, there are other forces at work. For example, it is true that available funds are unevenly distributed across disciplines. It also is true, however, that many awards are made on the basis of factors unrelated to the field of the proposed study (e.g., gender, race, nationality, institution, church affiliation, military status, vocation, organizational membership, geographic location, etc.). It is true that having received a previous grant often is a substantial advantage. It also is true, however, that "trading up" to a higher level of support can be facilitated by very small prior awards ($100–$200) such as those that often are readily available from sources within your own institution.

It is true that those who review applications are always more favorably disposed when something in the proposed study fits closely with the primary interest of the granting agency. It also is true, however, that when funding is considered *before or during* the preparation of the proposal (as opposed to being an afterthought upon completion), small adjustments in purpose, sample, site, method, or justification often can create just such an attractive match without significant sacrifice of the main research objective.

Begin at home. Virtually every campus has some procedure for making institutional funds available for research by graduate and undergraduate students as well. Locate that office and find out what is possible. The awards typically are small, but everything helps, and even a small success gives you a track record and, thereby, a significant advantage in subsequent applications, both inside and outside the institution.

Do it now. The famous three laws that govern the value of real estate ("location, location, and location") have their analogue in seeking research grants ("timing, timing, and timing"). The typical student novice completes his or her proposal, gains approval from advisors or the graduate school, and turns up at the local equivalent of the Office of Research Affairs with a 50-page document asking, "Where can I get money to do this?" In several respects, that forlorn inquiry already is too late. It is too late to influence the development of the proposal, and thus the best opportunity to make the study attractive to a potential source of funding. More to the immediate need, however, it probably is too late (or far too early) for the cycle of grant deadlines.

The average decision time for student grant applications lies in the range between three and nine months (a long wait often made longer by the time consumed in preparing the application), and 65% of application deadlines for student research funding fall between November and March (85% if you add October and April). Missing the deadline, for many agencies, means a full year delay before a new cycle of application consideration begins.

Network with students in other fields. A strong trend in research funding has developed to favor proposals that employ an interdisciplinary approach to inquiry. This is particularly true of federal agencies and applies in virtually all academic fields. Student projects that take advantage of that disposition not only will have improved chances for an award but also will offer preparation for participation in a collaborative research format that trainees are sure to encounter upon graduation. In many institutions, graduate school regulations even have been adjusted to accommodate theses and dissertations that involve interdisciplinary collaboration.

Whether through informal contacts or by more structured means, discussion with students in other disciplines often reveals that interesting theoretical and practical problems have parallel representation in several fields of inquiry. Some of those represent opportunities for the kind of research that reaches beyond the confines of a single perspective—and that can attract support for students who are ready to "think outside the boxes" of their disciplines.

Get ready to present yourself. In this book (Chapters 2 and 9), we have given much advice about academic résumés (vitae), and once again we urge you to maintain one (or several, to serve different purposes) in up-to-date condition. Keeping your records on a computer makes this a simple task. Almost all grant applications require submission of an academic résumé. In addition, many sources of support for student research ask for a "personal description" or "personal profile." This more closely resembles the essay that is part of application for admission to many colleges and universities.

Particularly in the case of pre-dissertation doctoral students, granting agencies know they cannot expect a track record of previous funding success or research publications in support of the candidate. Their emphasis, then, must be on selecting student applicants who know why they are in school, why they are engaged in research training, and what really interests them in the world of scholarship. With that in mind, and sufficient time to do multiple drafts and obtain editorial assistance, you can craft a profile that portrays you as a good bet for any agency's investment of grant money. Having those tasks of self-portrayal done in advance simply makes it easier to assemble a prompt and effective response when opportunity knocks.

Attention, undergraduate juniors and seniors! A surprising number of fellowships are designated for award to "entering students" in programs of graduate study (a specification that appears to be increasingly popular among foundations). In fact, some graduate fellowships specify that the applicant must not have accrued any graduate credit hours. Of course, not all these awards include the potential for funding student research, but of those that do so, virtually all will go each year to individuals who identified the opportunity and made application during their junior and senior years of undergraduate study.

Even tuition awards for graduate study may position you to receive priority consideration when later applying (to the same agency) for money to support a program-related investigation. If you think you may be destined for a graduate degree, exploring possibilities for financial support (including research funding) prior to receiving the baccalaureate is just sound career planning.

In some cases, part-time students can apply. Applications from part-time students and individuals without any current institutional affiliation are welcomed by some granting agencies. In such cases, it often is accomplishments rather than awarded degrees that must carry the day. Demonstrating the capacity to do the work of inquiry is always a central concern, however, and here, as elsewhere, strong and specific recommendations from scholars who themselves are active in the area of your proposed study may be a requirement.

No matter what the result—count your blessings. Searching for money with which to support your study, thinking about your proposal in terms of the perspective taken by funding agencies, and completing the tasks involved in making a sound application are all learning experiences that have lasting value. Among the senior scholars we know, there is not one with a strong record of obtaining research funding who has not also acquired some rejections along the way. It would be surprising if that were not true of student scholars as well. The verdicts "Approved, but not funded" and just plain "Not approved" are hard to digest. We would not try to persuade you otherwise. But going through the process can leave you smarter, more determined, and more likely to succeed on the next try. If rejected, apply, apply, again!

9

Preparation of the Grant Proposal

Once the funding agency is selected, the first step is to obtain its published guidelines for submitting proposals. Those guidelines must be followed meticulously. We will discuss here some of the considerations to be dealt with in writing typical grant proposals. For each specific proposal, however, the information in this chapter should be used in conjunction with the guidelines that are provided by the funding agency.

As noted in Chapter 8, your grant proposal will have two main parts, the application for funding and the description of the proposed research. The document will contain most of the same major divisions (see Table 9.1) that we have discussed throughout this book; however, many of the sections in a proposal for funding will have a slightly different focus, and it is those differences that will be discussed in this chapter. We will not repeat all of our earlier advice with regard to the preparation of generic research proposals. Where the primary emphasis or relative importance of a section is somehow particular to the process of seeking a grant, however, we will address that difference and suggest appropriate ways to respond. For that reason, you may find it helpful to begin by briefly revisiting Chapter 4.

Using Planning Models and Flowcharts

Preparation of a grant proposal, particularly the application aspect of the proposal, often involves dealing with a number of individuals in administrative

Table 9.1 Components of a Research Proposal Designed to Obtain Funding

Application for Research

- Biographical information about the investigators
- Facilities, equipment, and other special resources available
- Staffing and consultant needs; Support services
- Budget

Research Plan

- Abstract
- Purpose, Specific Aims, Hypotheses
- Background and Rationale
- Significance of the Study
- Preliminary Studies/Progress Report
- Research Design/Methods
- Data Analyses

Time Frame

Dissemination of Results

Appendices

- Pre-prints
- Reprints
- Photographs/Images
- Letters of Support

units external to the author's home base. Securing institutional review and approval, soliciting peer reviews, obtaining letters of support, accumulating budget information, gathering commitments from personnel, assembling vitae, and obtaining multiple signatures all serve to complicate the process. In addition, a set of sub-deadlines emerges that must be met in serial order. For all of those reasons, the demands of even a relatively straightforward grant proposal may quickly exceed the capacity of managerial habits that are entirely adequate for the limited scope of theses or dissertations.

In response to the complicated network of steps that is demanded in developing many grant proposals, there now are a number of commercially available management models, flowcharts, and step-by-step guidebooks intended to assist the proposal author. It is our judgment, however, that for most novices who are attempting to stay organized the best option may be to devise a do-it-yourself flowchart into which assigned responsibilities and deadlines can be inserted.

Once a potential study has been defined and an appropriate funding agency has been identified, the items listed in Table 9.2 can be supplemented as required by local circumstance and then be displayed with boxes and arrows to form a flowchart of sequenced events with appropriate time frame warnings. This will be, in effect, your own personal time frame, one that precedes the one you submit with the grant proposal.

Table 9.2 Tasks to Complete and Important Deadlines

First response to the request for proposals
Deadline for preproposal or letter of intent
Deadline for submission to Human Subject Review Committee
Deadlines for obtaining administrative signatures
- Chairperson
- Dean
- Office of Sponsored Research
- Director of Computer Services

Deadline for submission to a typist
Deadline for submission of proposal
Award announcement date

Major Development Steps

Obtain guidelines
Contact program officer at funding agency
Select primary authors of proposal
Prepare abstract or preproposal and submit letter of intent
Contact units for support of collaboration
Obtain initial administrative approval
Complete review of literature
Determine study design
Prepare first draft of proposal
Prepare abstract of proposal
Initiate human subjects review
Initiate internal peer review
Prepare budget
Revise proposal based on feedback
Collect vitae
Obtain letters of endorsement
Obtain written assurances from support sources
Complete institutional forms
Type final document
Obtain administrative signatures
Duplicate proposal
Submit proposal

If you are preparing your first grant application you will do well to remember the corollary of Murphy's Law that states, "Everything takes longer than you expect (*especially* in the summer)." Where one must depend on the voluntary cooperation of busy individuals in other parts of the university, several weeks of cushion are wise when establishing any set of target deadlines. This is particularly true of the typing and collation of the proposal. Even though the document is produced on a computer, the typing alone may take several weeks. Invariably, it will take longer than you anticipate. Because a proposal is a large project often involving 50 to 100 typed pages, all of which must have no errors at all, it will take the office staff away from other duties for unacceptable periods of time if typing is left to the last minute. Further, each draft must be checked and double-checked. The end result is a process that makes Murphy's corollary look positively optimistic.

The application portion of the proposal generally begins with several pages of standard forms that the funding agency uses for all requests for financial support. Those application pages are the "business and management" section of the proposal, so that is where we too shall begin.

The Application

The first forms in the document sequence are often referred to as the "cover pages." At minimum, they contain identification of personnel, biographical information, inventories of available resources, and checklists of funding agency requirements. Much of this is required for the management and record-keeping processes of the granting agency and, in some cases, the submitting institution as well. Collectively, the cover pages constitute what may be called the Application section of the proposal. Components of the Application section of the proposal are shown in Table 9.3.

The Budget

The budget, which is a direct reflection of the reason you are writing the proposal in the first place, will appear as part of the Application section. This is the place for you to request the money that you need to conduct the study. The first question that comes to mind is, "How much may I ask for?" First, calculate what it will take (a) to pay yourself, staff, and consultants, (b) to purchase and maintain needed equipment, and (c) to provide the materials necessary to conduct the study. Your ensuing request for financial support should be carefully estimated, supported by evidence, described in straightforward prose, and, of course, be absolutely honest in every regard.

Table 9.3 Application "Cover" Pages

Some or all of the following forms, in some combination and perhaps in different orders, are generally required by funding agencies.

- Brief description and identification number of the proposed study, and names of the principal investigators and the submitting institution
- The budget
- List of all personnel to be involved and time allocated to the research project (normally includes investigator(s) and both staff and student assistants)
- Biographical information for primary research personnel
- Sources of research support (current funding, collaborative facility arrangements, shared-equipment contracts, and other resources)
- Human subjects documents approved by the administration

Next, find out the size of the typical award from your target agency or foundation and, unless you have a compelling reason to do so, try not to allow your total request to exceed that figure. If your estimate of needed support is substantially larger than that of most funded proposals, you may have to consider scaling down your request. Funding agencies and foundations certainly are not populated by creatures of habit, but those who review proposals generally must respond to internal policy demands that represent more than the simple influence of tradition. You can be sure that any substantial departure from an agency's norm for funding will receive close scrutiny and require persuasive argument.

It is sometimes rumored among grant seekers that applicants should pad their budget with some excess equipment and unneeded staff so that the reviewers will have something to cut. This is a foolish notion, because it results in the applicant having items in the budget that are not well justified, which ultimately gives the impression of a casually constructed proposal. You do not want the award committee to begin haggling about whether your study is worth the money!

Although not a part of the plan for research, the budget is a central element in a grant application. It will be examined in detail by almost all members of a typical review committee, even those who may not be involved in evaluating the adequacy of design and methodology. Most application formats require that the budget be presented in brief form with no more than a page or two of appended explanation. Experience suggests that such explanation should be provided, even if it is not specifically required. Of course, if an agency specifically directs you not to provide supplemental pages of budgetary detail (as is the case with many of the NIH Proposal Guidelines), then you should comply with that instruction.

Staff and Consultant Needs

Personnel for whom funds are requested must be essential to the study. The most effective way to demonstrate such need is to specify responsibilities for each position. If project personnel must have special certification or advanced degrees, it is important to tell why a less qualified person could not be used. Enough information should be provided to support the level of commitment required in each position, whether for full-time or fractional employment.

The staff needed for the research project should be carefully planned and kept as small as possible while maintaining services essential to conducting the study. The staff budget generally is what drives up the cost of a research project more than any other expense. Along with staff salaries, fringe benefits (medical insurance, retirement) and raises for each year of the grant often must be included. The beginner at this game quickly finds that the costs of a project escalate rapidly as more personnel are involved. We suggest that in your first grant application, the proposal for staffing should be constructed with a frugality tempered only by feasibility.

The staff requirements should be explained in detail, and if a staff member's function dwindles in the second or third year of the grant, then that member should be phased out. Reviewers will note a conscientious effort on the part of the applicant to execute the project at a minimal cost. Positions to be funded for the full term of the study that are described with only global explanation of responsibilities and no indication of compelling necessity also will attract attention—all of it negative.

Consultants, too, should be requested sparingly and only when absolutely needed. When they are required, the need should be explicitly stated. If a particular consultant is to be employed, the applicant should explain exactly why, and a letter of agreement from that individual should be included in the appendix.

Supplies and Equipment

All purchases of expendables and equipment must be documented with care. Review committees will appreciate efforts to give exact and realistic estimates of supplies needed, rather than the use of speculative round numbers. For instance, if videotape is to be used, rather than simply requesting $1,125 for the purchase of tapes, the applicant should explain that 15 minutes will be required for 100 subjects on three occasions without possibility of reusing tapes. That will require one tape per four subjects, or 75 tapes at $15 each, and a final budget item of $1,125.

If funds for the purchase of large equipment are requested, the budget justification section might explain what attempts have been made to obtain the items by other means and why alternative plans such as rental or sharing will not suffice. Generally, funding agencies are not well disposed to requests for extensive equipment purchases, particularly large and expensive items. The position of many agencies is that the university or college should supply all the basic and less specialized equipment, as well as most large items that represent expensive and permanent investments in research capability. Young researchers may be dismayed to discover that their university takes the opposite position, holding that the investigator should obtain all expensive equipment through extramural funding, but that is the way things often work in complex institutions.

Funding agency officers understandably would rather award money to applicants who already have the necessary equipment. In that manner, the agency's funds go directly into research activities rather than into building the capacity to perform research. The only counter to this logic is to make the best possible case for why the applicant's facility is ideally, if not uniquely, positioned for the conduct of the particular study—given the addition of the needed equipment. Another helpful strategy here is to point out any substantial ways the applicant's institution is contributing to the proposed study in terms of equipment, facilities, or personnel assigned to the project. Agency and foundation personnel like to feel that research projects, especially large ones, are a joint university/agency enterprise, and they tend to resent any implication that they constitute a goose that lays golden eggs.

Budget lines for ancillary items such as analysis of data, preparation of manuscript, and communication with consultants may be permitted under the rules for application but should be explained with great care. Personnel assigned to the study must be justified with detail that goes well beyond the standard of boilerplate descriptions. In the golden days of research when grants were plentiful, consultants sometimes were included in proposals simply because they were professional colleagues who had an interest in the study. Today, there is a very different and much higher standard. Every person who is proposed for remuneration under the grant must be demonstrably necessary to completion of the study.

Even when the need clearly is imperative, caution is required in preparing your explanation for any budget line involving personnel. If, for example, a consultant is needed because the investigator has not mastered a technique that is central to execution of the study, a reviewer may wonder why the grant should be awarded to someone who is deficient in skills commonly used by workers in the area. There may be a perfectly reasonable explanation for such a circumstance, but that must be established by what appears in the proposal.

Travel Costs

Only travel that is necessary to complete the research and, perhaps, to disseminate the findings should be included in the budget. If travel to specific conferences for the purpose of presenting reports from the study is included, the nature of the audience should be indicated and the relevance of the conference to the study demonstrated. More important, the relevance of the conference attendees to the interests of the funding agency should be specified. Generally, funding agencies will specify in their guidelines the amount of funding they are willing to provide for travel.

Indirect Costs

Another major item in a proposed budget will be indirect costs (overhead) that the funding agency will pay directly to the institution. At this time, such costs vary widely and are negotiated by the agency and the university or college. Many large institutions have a standing policy on overhead for all grant contracts. Indirect charges presumably reimburse the university for the cost of building maintenance, processing of paperwork, utilities, and so on—all items that would have to be purchased by the grantor if the study were conducted at a facility owned by the agency. Other budget items dictated by the institution are insurance, retirement, and medical benefits for project personnel.

It is wise to take an early draft of the budget to the university officer in charge of grant negotiations and obtain assistance on all items not directly dictated by the nature of the research process itself. After several projected budgets have been written, the mysteries associated with indirect costs and employee benefits may become understandable to the researcher, but a substantial amount of grief can be avoided on the first effort if competent help is solicited and advice faithfully followed.

As a corollary to budget making, if a significant amount of support is to be provided by the institution in the form of new inputs of equipment, facilities, supplies, or personnel, the researcher must be sure to obtain formal approval from appropriate officers before proceeding too far into development of the proposal. In many institutions, agreements from the chairperson, the dean, and a vice president (at minimum) are required for such expenditures. It is best to have the needed commitments in hand well in advance of the point at which the proposal will be submitted for review.

Research Support

Any large project will require an extensive array of support services from the university and any other agency or unit of the community that is involved with the project. In the Application section, the *Research Support* pages are used to describe all of the resources that are available to support and even amplify the success of the research project. Whatever the source, the need for support services should be anticipated, both by providing for them in the budget and by ensuring that what must be purchased exists and can be delivered. If a computer graphics person is needed, one should be located to ensure that such a person is available. Specialized computer services may be required from the computer center, or unusual transportation capacity may be needed, such as the need to move video equipment into the community. These types of research support must be located and confirmed as part of preparation for the proposed study.

All space needs such as laboratories, offices, or work areas for technicians should be anticipated. If the research project is going to tax any unit of the university or agency, the people involved should be consulted prior to the final draft of the proposal. Examples include proposed projects in which a large number of telephone calls will be imposed on secretarial staff, frequent mass mailings will have to be prepared for postal pickup, or dozens of students must be asked to wait in hallways immediately outside classrooms. The applicant should try to anticipate any hardship that the project may create for anyone. Proactive planning of that kind can prevent or reduce logistic problems that generate ill will among support service staff—with inevitable consequences for execution of the study.

When support from other sources is critical to the research project, the researcher should obtain a written agreement from the appropriate officials specifying that the needed service will be available. This letter of agreement may be included in the appendix of the proposal. For example, if a specific school classroom is essential to the project, a letter reserving that facility for the specified period of time should be obtained. It would be an unpleasant surprise, for instance, to begin use of an environmentally controlled laboratory for a year-long study and then discover that it will not be available for the final two months of the project. Every precaution must be taken to prevent such disastrous accidents.

Other important resources include matching grants, facilities, shared equipment, subcontracts, and even unpaid consultants from within the submitting institution or from other collaborating entities. For example, some funding agencies are willing to provide matching or "challenge" grants that become active if the proposal is funded. These are highly attractive to other funding agencies, because it enables them to increase the funding impact of the contributions of all the agencies. Substantial "start-up equipment funds" provided by the institution to a new faculty member indicate a high regard for his or her research program and an anticipation that the investment will be repaid in the form of a successful project. The institution may agree to "match" the budget of the proposed research if it is funded, or provide a specified number of technicians or staff in the event of a successful grant application.

Such arrangements lead us to observe that there may be precursor actions for you to consider well in advance of writing a grant proposal—actions that might enhance your chances for funding. First, if your research agenda requires expensive equipment, very specialized testing and research spaces, or unusual time or travel requirements, negotiate those, or a large part of them, as a part of your recruitment package for a first appointment as a researcher/faculty member. That way you will not have to include these basic pieces of equipment or space in your request for funding.

If you are a graduate student with no more than doctoral and postdoctoral dreams, it may seem a long stretch to imagine yourself as a target for academic recruitment, but there is good reason to begin carefully thinking about how you will want to manage such eventualities. Academic dreams have a way of becoming sudden realities, and the advantage always goes to the prepared.

A second form of advanced planning is to scout around locally to identify potential sources of matching grants or offers of "in-kind" contributions that will substantially improve your total level of research support. In-kind contributions are provisions of personnel, data gathering, or data analysis that you would normally have to pay for with a grant. Examples are the offer of (a) a school system to allow a half-day staff member to go through student records and tabulate specific incidents that are pertinent to the proposed study, (b) a medical center to provide a secretary with sufficient time each week to call, make appointments, and organize patients for testing in your study, (c) a hospital to analyze blood samples from the participants in your study without cost, or (d) a city recreation department to provide reserved parking spaces, scheduled instruction, and free t-shirts for every participant in your study. All such in-kind contributions can be documented and entered into your proposal as research support. Budgetary supplements of that kind can exert a strong positive influence on how both you and your application are regarded by reviewers.

Biographical Sketch: Presenting Yourself

It would be easy to underestimate the importance of this section in a grant application. For many readers, vitae represent one more instance of the paperwork syndrome that infects higher education. Personal résumés are required, accumulated, and ignored in a host of bureaucratic functions. In consequence, unlike people in the business world, academics often pay relatively little attention to the construction and maintenance of attractive and utilitarian vitae. Too often they are allowed simply to accumulate the residues of scholarly and professional life without much thought to order, economy, or impact.

Two common misapprehensions prevail about the vitae section of a grant proposal: (a) that they are not read by reviewers and (b) that as a pro forma section, vitae have little influence on the success of a proposal. Both notions are untrue. Reviewers do read vitae. In fact, many reviewers, especially those who are not the primary reviewers, read vitae before they read the body of the proposal. Not only do vitae influence the judgments of reviewers, but they are the one part of the proposal in which the applicant can be aggressive by taking the argument for competence directly to the reviewer's attention.

The résumés of individuals to be associated with the proposed study are the primary vehicles for arguing that the interests, training, and experience of the investigator(s) make support of the study reasonable. The techniques for mounting this offensive are not difficult to imagine.

1. If a format is provided, follow it exactly by revising, reformatting, and reprinting every résumé. If a format is not provided, invent one that will best serve the strategies noted below.
2. Keep vitae reasonably short by being selective about content. Reviewers will be looking for the obvious items, and these should be given prominence:
   a. Publications that directly relate to and support the proposed study (as many as possible, while staying within page limitations)
   b. Publications in areas related to the study
   c. Receipt of previous grants in the area of the proposal
   d. Receipt of research grants in any area
   e. Involvement in a similar study whether funded or not
   f. Evidence of relevant training completed (e.g., postdoctoral study in the area)
   g. Unpublished papers or conference presentations in the area of the study
   h. Completed pilot studies in the area of the proposal

   Take care to give emphasis to current items. Any entry more than five years old should be deleted unless it continues to be cited in current work or helps to establish a track record that is directly related to the study. Don't burden the reviewer with irrelevant clutter.

3. Use a uniform format that is divided into subsections with prominent headings. The idea is that the format should make it impossible for a hurried reader to miss material that supports the competence of the investigator. Only research publications should be listed, but if there are few, cite relevant research presentations or abstracts to supplement the publications. Be certain, however, that the research publications are listed first, and that abstracts and presentations are categorized as such. It is not wise to list all of these in a mixture, leaving the reviewer to assume that you do not know the difference between an abstract and a publication, or that you are trying to give the impression of a great deal of scholarly productivity by intermingling these three types of activity.
4. Use short sections of text to provide detail about the exact nature of previous research experiences relevant to conduct of the proposed study. You are not limited just to the citation of dates, project titles, and publications. Tell the reviewers exactly what they need to know in a form that makes the task as easy as possible.

Where a team of investigators is to be involved, it will not always be easy to prepare an effective vitae section. Individualism in personal matters is a common trait among productive scholars, and the vita is correctly regarded as a personal matter. Nevertheless, this is a place where flexibility and some cooperative effort will improve the chances that all participants will enhance their vitae—by winning a grant to support their scholarship. Of course, if the funding agency requires all vitae to be reported in exactly the same format, by all means do so.

In summary, an ideal biographical sketch should provide a strong impression of:

1. thorough research training in the area of the proposed research,
2. consistent and cumulative productivity, and
3. recent publications that support the need for the proposed research.

Letters of Endorsement and Confirmation

It may be useful to demonstrate that respected scholars with expertise in the area of the proposed research regard the project as worthy and the investigator(s) as competent. For this purpose, letters containing such endorsements may be included in the proposal (usually in an appendix). In addition, letters that confirm the cooperation, support, or participation of organizations or individuals essential to the study, or testify to the availability of needed facilities or equipment are placed either at the end of the application portion of the proposal or in the Appendix, depending on the funding agency's guidelines. If, for example, schools and teachers or students are to be involved in any way, it is essential to have evidence that the appropriate authorities have reviewed the proposal and are prepared to participate if the grant is awarded. Similar brief notes may be required from the directors of service units such as the computer center, university health services, or laboratory facilities in other departments.

The documentation provided in such cases need not be elaborate or extend beyond what seems essential. The goal is simple. The applicant must provide some assurance that the staff, facilities, support service, and subject populations envisioned in the proposal will, in fact, be available.

The Research Plan

The overall organization of your research plan will tell a story about your organizational ability. Accordingly, you should give the organizational plan some serious consideration before you begin the writing process. Although we have mentioned this before, it bears repeating—try to think like a reviewer as he or she reads your proposal. What organization of the information will be clearest to someone who has not been immersed in this project as long as you have? If your project has three independent parts to it, then you may want to organize the research design, instrumentation, and procedures as three large sections. Conversely, if the instrumentation and procedures remain the same throughout two or three different experiments, then you may want to describe the instrumentation in the first study, and then describe only the different aims and procedures in following studies.

Because every research project is unique, only you will recognize the best organization for your study. Remember, though, that happy reviewers are positive reviewers. If they have to wade through paragraphs and paragraphs to find the details of interest, they will not be happy, and it is likely that they will quit trying.

Many investigators, upon receiving the evaluation of their proposal, exclaim, "The reviewers said that I did not provide an explanation of how participants would be assigned to groups, but it was right there on page 33. They just didn't read my proposal carefully enough!" That complaint, however, is almost never valid. The reviewers have read it, but simply didn't notice it, or, more likely, couldn't "get it" when they did read it!

In a well-organized and completely transparent procedures section, reviewers will not be able to miss an item of critical importance. That sort of writing and organization takes time to produce and an exceptional level of sustained attention. Excruciating? Sometimes it can be so. It remains, however, that effective explanations of research procedures also can make the difference between funding and no funding.

Abstracts

The word "abstract" normally designates a brief summary of a larger document. Most published research reports are accompanied by such an abstract. Written in the simple past tense, abstracts of that sort are histories of work already accomplished. In contrast, the abstract of a grant proposal is written in the future tense and summarizes work that will be done. When, as sometimes is the case, the abstract is prepared before the full proposal has been developed, it more properly should be called a *prospectus* or *preproposal,* but such niceties of language are seldom observed.

The abstract is prepared early in the proposal development process and serves several purposes. First, an abstract may focus the thinking of individuals developing the proposal by establishing a clear and explicit goal to which all subscribe. Second, a concise prospectus can be used for internal purposes to obtain preliminary administrative approval or to solicit support and cooperation from other units. Third, many agencies now require submission of a "letter of intent," that, in essence, is a one- or two-page abstract of the proposed study. These letters are screened by a panel of judges who rank them on pre-established criteria. Authors of the best abstracts are invited to submit full proposals for the second phase of consideration. Under such conditions, everything rests on the abstract, and no argument is required to convince the proposer that only the best effort will serve. Finally, whatever its function, an abstract prepared prior to development of the full proposal must be revised to maintain perfect congruence with the evolution of that document. Proposals that do not match the abstract may receive short shrift from reviewers.

Whether or not a letter of intent is required, the abstract that is submitted with the full proposal bears a disproportionate share of responsibility for success or failure. Often limited to a single page, these few paragraphs are what the reviewer will read first. When the bottom of that page has been reached, the reviewer must have a clear impression of the study's objective, method, and justification. If the proposal is to have a fair chance to succeed, however, another impression must have been communicated—that the study contains something of special interest, something that will sustain the reader's attention through the pages that follow. Thus, the abstract must accomplish the dual tasks of providing a concise picture of the study while also highlighting its unique characteristics.

Although different funding agencies require different lengths and types of abstract, in almost all cases there is no space for throwaways. Each word and sentence must convey a precise message to the reader. If a point is not essential to an understanding of the study, it is better left to the main body of the proposal. Because the abstract is a one-way, one-shot communication, absolute clarity is essential. No matter how well a point may be explained in the body of the proposal, if the reader is confused by the language of the abstract, the game may have been lost.

For both of the reasons noted above, economy and clarity, the watchword for writing the abstract is *plain language*. Avoid any special constructs that require definition. Don't coin new words in the abstract. Avoid slogans, clichés, and polemical style. Keep adjectives to a minimum and omit flowery descriptions. Finally, remember that the sure sign of an amateur is to apply banalities such as creative, bold, or innovative to one's own ideas. Reviewers, unlike graduate advisors, cannot ask what "that phrase" means at the next committee meeting, may not take the time needed to puzzle through a convoluted sentence, and certainly will not bother to figure out what probably is intended by an imprecise description.

Although one can assume that reviewers are both literate and familiar with the research process in broad terms, for the purpose of the abstract it is unwise to assume more. Writing beyond the technical competence of the reviewer can be fatal. The best rule is to imagine that you are explaining your study to an intelligent layperson. By eliminating specialized language and reducing esoteric constructs to their essential components, you can make the abstract intelligible to individuals with a wide range of scientific backgrounds—without appearing to write down to any reader.

The typical format required for a proposal abstract includes the following major elements:

- General Purpose
- Specific Goals/Aims
- Research Design
- Methods
- Significance (Contribution and Rationale)

Some agencies may also request that the applicant organization (institution), estimated cost (total funds requested), and beginning and ending dates be included in the abstract. It is important to read the instructions and provide only the information that is explicitly requested. Word count requirements almost always will limit communication of your own ideas, so space is at a premium.

The specific goal section should begin with a statement of what will be accomplished, presented in the format of a research question or testable hypothesis. For example: "The objective of this study is to determine whether the use of foam cushion inserts in athletic shoes will affect the incidence of heel injuries in runners. Frequency and type of injuries over a six-month period will be compared for two groups of 100 subjects of varied age, weight, ability, and sex, one group using shoes with cushion inserts, and the other using uncushioned shoes."

One of the hardest things to do is to reduce objectives and methods to just a few sentences. This difficulty probably contributes to several common errors found in the abstract section of the proposal.

Confusion of objectives with procedures: Taking a survey may be a procedure to be employed in the study, but it is not an objective of the study.

Confusion of objectives with the problem: The conditions that make the study important, in terms of either practical application or contribution to knowledge, are better discussed in the section on significance. In the occasional case in which some appreciation of significance is required before it is possible to understand the objective, the best course of action is to create a short subsection titled "The Problem" as a lead-in to the statement of the objective.

Attempting to specify more than one or two major objectives for the study: Save all sub-objectives for the body of the proposal. Do everything possible to help the reviewer focus on what is essential.

Failure to be explicit: The best insurance is to start with a conventional phrase that forces you to write about your intention. "The objective of this study is to. . . ."

The clarity and precision with which your objectives are presented may control how carefully the reviewer attends to the subsequent section on procedures, but it is axiomatic that *agencies fund procedures, not objectives.* How research is performed determines its quality, and it is here that the competence of the reviewer as a research specialist in the area of your proposal will be brought to bear. If the reviewer finishes the one or two paragraphs of this section with a clear, uncluttered idea of what will be done, a full appreciation of how those actions will accomplish the objectives, and a positive impression of what is intriguing or particularly powerful about your approach, your proposal will receive a full hearing.

The significance of the proposed study should be identified in modest but precise terms. Nothing can serve so quickly to make a reviewer suspicious of the merits of a study as to encounter some Chamber of Commerce enthusiasm in the abstract. In plain language, indicate how achieving your objective would be of value to someone, could improve some service, would fill a gap in an evolving body of knowledge, or would permit the correct formulation of a subsequent question. It also is appropriate, and often effective, to indicate a specific example of how a finding from the study might provide human benefit, even at some point subsequent in time or technical development. It is best, however, not to drown the drama of a simple example in needless embellishment. Finally, when deciding which potential consequence should receive emphasis in supporting the significance of the study, it is useful to examine the existing interests and commitments of the funding agency. When the proposed study takes some of its importance from a potential for contribution to ongoing projects of the grantor, there is a powerful argument for special attention to the application.

As the abstract is developed, revised, and given its final review prior to submission, it will be helpful to remember one of the somber facts that has emerged in this age of endemic grantsmanship. Some heavily burdened review committees are so large that it has become necessary to delegate to a subcommittee the task of giving full proposals a complete reading. The remaining reviewers thus may not see more of a given proposal than the cover sheets, the abstract, the author's vita, and, perhaps, the budget. That prospect should be sufficient to encourage attention to producing the best possible abstract.

Specific Aims and Hypotheses

We devoted a considerable amount of space in Chapter 1 to the development of purposes and hypotheses. At this point it might be helpful to revisit those pages. In that chapter we emphasized that the purpose and accompanying rationale were the most important aspects of any proposal. Even the research plan itself is no more than a careful explanation of how the purposes of the proposed study will be fulfilled.

In proposals for financial support, the purposes of the study are often characterized as "aims." Reviewers expect aims to be specified in terms that are clear, specific, and concise. The sequence of intended outcomes should be logical, and if hypotheses are appropriate for the study they should be nested under the appropriate aim. If reviewers reach the end of the Specific Aims and Hypotheses page and are unclear about what the research is about, they will likely quit trying to understand the rest of your proposal. Accordingly, this page should receive your closest attention, be revised as many times as required to achieve perfection, and be "tested" on willing colleagues who understand the topic of study. In fact, it sometimes is useful to inflict your account of aims on a friend who does not have expertise in the subject. If he or she cannot understand the general idea of the proposed research, then perhaps you need to revisit your effort one more time.

Background

To serve as background, reviewers need a very brief and highly focused description of what is known about the research topic. They also need an economical overview of any particular line of research that led to the proposed study.

In some cases, your description of the study's background should also include what is not known, and show precisely how the proposed study will respond to that gap in the body of relevant knowledge. Unlike the case for dissertation and thesis proposals, it is probable that only one or two pages will be allowed for the entire Background section. It is here that some prior practice at composing condensed versions of complex information will serve you well.

The Background section must also make the case that the research topic is continuing to attract the interest of scholars, that previous studies have produced gains in our understanding, that the aims of the proposed study constitute an obvious next step, and, ultimately, that the step is feasible. To write persuasively about such complex and closely interlocking matters will demand the best of your writing skills. As we recommended in Chapter 4, the place for most authors to begin will be with an outline of the important points that displays only a few descriptive phrases under each heading. If outlining simply is not in your repertoire, a diagram or flow chart may accomplish the same purpose but be better aligned with your way of conceptualizing complex relationships. For example, Specimen Proposal 3 in Part III of this book makes use of several diagrams to reduce lengthy text and compress complex relationships into a simple graphic representation.

Significance of the Proposed Research

Most funding agencies require a section in the proposal that has a title such as "Significance of the Proposed Research." Funding agencies are accountable either to the public or to their benefactors for the expenditure of their funds. Try to read your proposal from the *funding agency's perspective*. What will the agency's reviewers think is important? What aspect of your proposal will attract their attention and interest? Reviewers seek the applicant's best forecast as to the usefulness and importance of the results of the proposed project. That prediction will assist the review committee in determining the cost-benefit ratio of the proposed study.

The significance of the investigation might be explained in terms of synthesizing information from several research areas or by showing how the findings might be applied to human services. It could be impressive to point out that the findings might enable development of other types of research that previously had been impossible.

The significance section need not be long, but it should be carefully reasoned, emphatic, and should address concerns specific to the mission of the funding agency. Also, you do not need to use the word *important* anywhere in this section, because the reviewer already knows that you think the topic is important. The crucial objective for this section is for the *reviewer* to realize that the study has the potential to matter in ways that go beyond the pages of the final report.

Procedures and Methods

Research Design

A key component of the methods section in a grant proposal—a section that differs sharply from its analogue in a dissertation or thesis proposal—is a one-paragraph description of the research design. This paragraph orients the reader to the nature of the study and serves as an introduction to the methods section. The design section of the study is not the same as the analysis section. That is, the design is not necessarily described by the analyses to be employed, although terminology from those operations might be used.

The reviewer needs to know immediately whether the study is quantitative, qualitative, or mixed in format. Further, the particular type of design should be made explicit, as with studies that are descriptive, correlational, quasi-experimental, or experimental. In addition, it is essential to display the main parameters of the design. In the case of an experimental study, for instance, that would include the dependent and independent variables. Below is an example of such a paragraph.

> This study will use a quasi-experimental design, in which the dependent variables are simple and choice reaction time, and the independent variables are age, disease status, practice time, and test session. The design is a between-within fixed model design, where the between factors are age (young/old) and the disease status (control/stroke patients). The within factors are practice (2/4/6 weeks) and test session (pre-practice/post-practice). Simple and choice reaction time, the two dependent variables, will each be represented by their means and variances.

Participants

In studies of humans, the exact nature of the participants often is critical to interpretation of the findings. For example, if the research question of a study is whether age impacts a motor or cognitive performance, the reviewer will certainly want to know why a particular age or age range was proposed for selection of participants. As you describe the participants in your study, you should consider all the possible characteristics of humans that could possibly affect your results. Some of these are shown in Table 9.4.

Table 9.4 Characteristics of Participants That May Be Relevant to Study Outcomes[1]

Who?
- Gender
- Age
- Geographical location
- Type of institutional affiliation
- Professional/vocation affiliation
- Medical history
- Physical status/history
- Practice status
- Experience as a participant
- Social status
- Education
- Economic status

How Many?
- Total number
- Gender numbers
- Group numbers (Control/Experimental)
- Group assignments

How Selected?
- Availability/cooperativeness
- Convenience/random
- Agreements made
- Payments or other forms of reciprocity
- Procedures for assignment to groups
- Procedures for assignment to conditions/treatments

[1]Some or all of these characteristics may be relevant to the proposed research.

Tests and Measurements

This section is a major portion of the proposal for funding, and should describe the tests and measurements to be used precisely and in detail. In addition, explain why these particular tests and measurements were selected. For example, if you use a standardized telephone interview outline, what is its source, its validity, and why did you choose it over others? Reviewers assigned to your proposal will be experts on the topic you intend to study, and they will want to be assured that you are using state-of-the-art techniques and equipment, and that all of those are being utilized appropriately. Provide reliability and validity descriptions for all.

The format that you use to present this information will also convey an immediate impression about your organization skills, so close attention should be paid to the logic of sequences. Many times the nature of equipment and techniques is complex. Do not hesitate to use diagrams, pictures, or flow charts to explain equipment or procedures. Our own Figure 3.1 in this text and Figure 1 in Specimen Proposal 1 are examples of how such graphics may be used with complex processes. Such representations may save hundreds of words and provide the relief of immediate clarification for a beleaguered reader.

Procedures

After the abstract and the section containing objectives or aims, the section dealing with procedures is the next in importance for determining the ultimate destiny of your application. In this section you must concisely describe the *processes* by which you will conduct the study. The way tests are administered, observations are made, and data are recorded can make a great deal of difference in the outcome of the study. For example, testing participants on a cognitive task in six 10-trial blocks with 30-second rest intervals vs. 20-trial blocks with 10-second rest intervals may produce entirely different results from some participants.

Clearly, such procedural (design) elements are a crucial part of the research process. But, much less esoteric and seemingly routine details also can influence the quality of the data obtained. For example, how participants are greeted and oriented to their role in the study, how they are moved from test to test or from one location to another, who presents treatment protocols, and how data are recorded also can constitute important parts of the research process.

You can be assured that reviewers will examine all of the details related to data collection. If they do not complete that effort with a clear picture of your procedures, the consequence may be sadly predictable. They will not support your project for funding.

Analyses

This section includes both the details of data preparation and the procedures that you intend to follow to analyze the data. How will the data be reduced? For example, if you obtain three attitude scores, what will your dependent variable be—the highest score, the middle score, or the average score? Be precise. Will you use transformations of the data? If so, why, and which transformations will you use? If you plan to combine data from subscales, which subscales will you use, and why?

If the analysis of data is to be statistical, you may wish to review pages 82-87 in Chapter 4. Likewise, if the study involves analysis of qualitative data, you may wish to review our suggestions on pages 115–116 in Chapter 5. Organize the analysis section so that individual steps are described in a logical order. As we suggested in Chapter 4, a good way to lay out the analysis section is by specific aim and/or hypotheses, so that you can show how each analysis relates to each aim and hypothesis. Figure 4.2 is a good example of being explicit about the analyses. Table 9.5 is an example of how one analysis section might be organized. And, as mentioned earlier, if you are not a statistical expert, plan for a consultant to assist you with the development of your analysis section well before any data are collected. You also could imagine an analogous outline for a qualitative study focusing on ongoing analysis during data collection, analysis after fieldwork is completed, and procedures for trustworthiness and credibility.

The Time Frame

Many proposals submitted to government agencies are for funding of a three- to five-year project. The application and review process is so time-consuming that the applicant, the institution, and most agencies find the investment of such energy and time best returned if the award is for a substantial period. It is usual practice, therefore, to require development of a time frame that states explicitly when specific parts of the project will be completed. Such schedules serve a variety of purposes and may follow the Research Plan in the proposal, or be placed in the Appendices, depending upon the funding agency's guidelines.

The first and most obvious use of a projected time sequence is that it keeps the investigator and all personnel on schedule throughout a long period (see Figure 9.1 for an example). When all members of the research

Table 9.5 Generic Example of Analysis Section Outline[1]

1. Subject Characteristics
   a. Descriptive Statistics —all dependent variables
      i. Means, individual and group variability

   b. Distribution Statistics
      ii. Stem & leaf analysis, Box plots

   c. Missing data procedures

2. Specific Aim 1 (DVs A, B, C)
   a. Hypothesis 1: Group Comparisons
      i. MANOVA (2x4, IVs A and B)
         (1) assumptions, probability levels, power tests
         (2) Post-hoc tests – (probability levels)
      ii. Effect sizes

   b. Hypothesis 2: Within-group Comparisons
      i. Intra class correlations, variance
      ii. Significance testing
      iii. Effect sizes

3. Specific Aim 2
   a. Hypothesis 1: Stepwise regression analyses
      i. Criterion variables: DVs D and F; predictor variables: participant characteristics A and E, followed by other continuous variables
      ii. Assumptions, probability levels for entry, power tests

   b. Hypothesis 2: Discriminant function analysis
      i. Criterion variable: IV C; predictor variables—DVs C-F
      ii Assumptions, probability levels, power tests

[1]This example outline uses generic dependent variables (DVs) and independent variables (IVs). This outline can be adapted to any purpose, modifying the number of aims, hypotheses, analyses, DVs, and IVs to match your study.

team have a copy of the time frame, and when it is posted in offices, work rooms, and laboratories where every member of the team will have daily exposure to its terms, the effect is to encourage steady application of effort.

A second function of the time frame is to forecast for project personnel, reviewers, and the funding agency the probable contents of each progress report that must be filed (usually annually or biannually over the term of the funding period). Every reader of the time frame can clearly see what events will transpire prior to each progress report and, consequently, will expect a full and punctual accounting. This serves also to discourage investigators

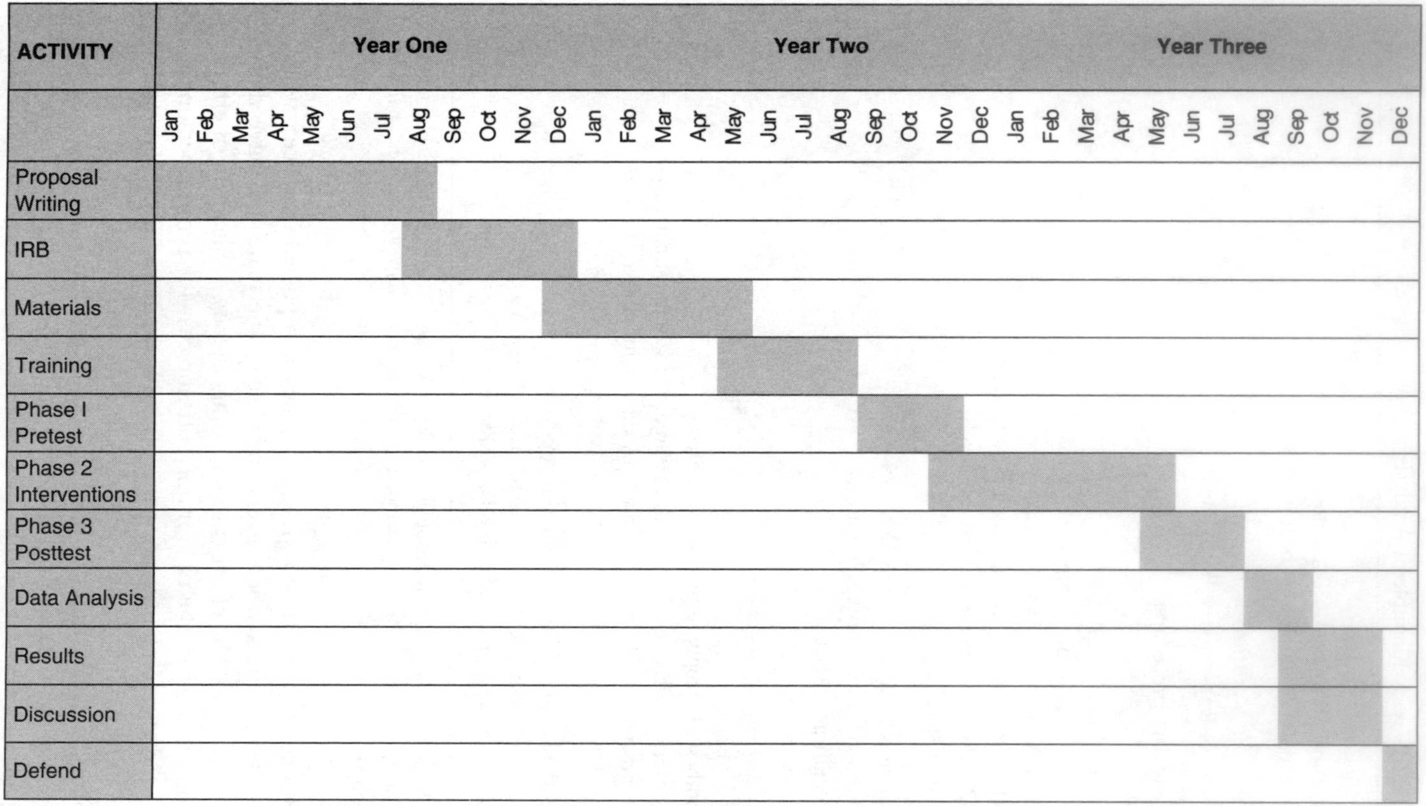

ACTIVITY	Year One												Year Two												Year Three											
	Jan	Feb	Mar	Apr	May	Jun	Jul	Aug	Sep	Oct	Nov	Dec	Jan	Feb	Mar	Apr	May	Jun	Jul	Aug	Sep	Oct	Nov	Dec	Jan	Feb	Mar	Apr	May	Jun	Jul	Aug	Sep	Oct	Nov	Dec
Proposal Writing																																				
IRB																																				
Materials																																				
Training																																				
Phase I Pretest																																				
Phase 2 Interventions																																				
Phase 3 Posttest																																				
Data Analysis																																				
Results																																				
Discussion																																				
Defend																																				

Figure 9.1 Example of a Time Frame

SOURCE: From the dissertation proposal "Effects of a home-based physical activity program implemented by a trained caregiver on the physical function of community-dwelling older adults" by Sandra Graham, The University of Texas at Austin, 2006. With permission of Sandra Graham.

from procrastinating and later finding themselves having to accomplish an enormous amount in the month or two that precede a progress report.

A third function that the time frame serves is to document the need for three or five years of funding for the project. If the reviewers can see in the time frame that every month is filled with work to be done, then it is clear that a project of the length proposed is necessary.

Fourth, if the time frame is well documented in the accompanying text it will enhance the reviewer's understanding of the entire project and will further emphasize the applicant's organizational skills. A well-conceived time frame, in which each part of the project is estimated in terms of its onset and duration, goes a long way toward convincing the reader that the applicant knows the area, the methodologies to be used, and all other aspects of the project. If it is well written, it also will buffer several possible criticisms—that the project cannot be completed in the amount of time proposed, that it does not require funding for the full period of time requested, or that the investigator does not understand the topic well enough to know how long certain procedures will take.

The time frame should include the schedule for hiring personnel, ordering equipment and supplies, putting equipment and facilities into operation, meeting with personnel from other institutions or agencies, acquiring subjects for the investigation, data accumulation, data analysis, and report write-up. In preparing the time frame, the applicant should try to think through and record every step that will be taken in the project. Then how long each step will take must be estimated. The more people are involved in a step, the longer it probably will take. The applicant may even find, after working out the time schedule, that the project will take longer than anticipated, and the funding period must be extended. If such extension makes the project too expensive, it may be necessary to consider a more economical design. For all these reasons, the time frame should be carefully worked out before the budget is made final.

Most grant proposal applicants end up developing two time frames: one is a personal schedule that includes the amount of time necessary to get the proposal through all levels of the approval process, while the other is an abbreviated time schedule for the actual research data collection, analysis, and write-up, to be included in the application portion of the proposal. The personal time frame, shown earlier in Table 9.2, although not submitted with the final proposal, should aid the writer in understanding the full scope of commitment that will be necessary. The research project time frame provided in Figure 9.1 simply equips the reviewer with an understanding of how long each process of the research project and write-up will take.

Positive responses to grant proposals are facilitated when the research plan is accompanied by a thorough and realistic time frame. Further, as we will assert at the close of this chapter, any study, whether funded under a grant or financed out of the investigator's own pockets, is likely to be significantly strengthened by laying out a flowchart that schedules the salient events—step by step.

Dissemination of Results

The final step that transforms the personal act of inquiry into research is dissemination of a report into the public domain. Peers must review procedures if results are to be treated as reliable knowledge, and if the knowledge is to be useful it must reach those who can make application. For those two reasons, many funding sources stipulate that plans for dissemination of results must be included as part of the proposal.

The rules here are reasonably simple.

1. Tell which results will be reported.
2. Indicate which audiences you intend to reach.
3. Specify how you plan to disseminate the final report.
4. Be specific in citing which particular conferences, development projects, local, state, or federal agencies, publications, or colleagues working on related research projects will be informed about the results from the proposed study.
5. Think carefully about how to reach appropriate audiences and go beyond the requisite printed report and ubiquitous paper at a national conference.

Appendices

The appendices generally contain documents that are required for approval, such as confirmation letters of support, but that are not explicitly necessary to understand the proposed plan of research. The process of selecting proposals to fund often is done by a number of reviewers, only two or three of whom (primary reviewers) may read your proposal in detail, including the appendices. In fact, in some agencies, the appendices are sent only to the primary reviewers. The remaining reviewers, who have other proposals that are their primary responsibility, may read only the abstract and specific aims, and then do no more than scan the methods section. They will not have a copy of the appendices unless they specifically request one from the selection process manager.

Examples of documents included in proposals for financial support that are not usually available in students' proposals are reprints of previous research reports that are relevant to the proposed research, manuscripts of *accepted* research reports, and color data photographs or images from pilot studies of the proposed research. Internet URLs are sometimes also included for reviewers who wish to view more and clearer images or video clips of various aspects of a proposed research project.

Summary

If you have read this far, we hope you are not immobilized by the thought of undertaking the tasks described. Difficult and demanding as proposal writing can be, the rewards are clear and attractive—a well-planned study and money to make it go. In addition, each time a proposal is written, the process becomes easier. Many of the proposal sections, when written on a computer, simply can be updated and reused in subsequent documents. For example, each time a proposal is submitted, the basic description of a university research facility simply is updated and reprinted. The same process can occur with a particular methodology used in a line of research. Once the description has been written, it need only be updated and then recycled. Grant gathering, like any other intellectual skill, improves with practice. You may never find yourself regarding the process as enjoyable, but the challenges are always fresh, and the sense of satisfaction in completing a good presentation can be a significant reward. Better still, of course, is obtaining the money.

PART III

Specimen Proposals

The specimen proposals presented in Part III of this text were selected with several intentions. First, we wished to display proposals that involved both different research paradigms and different designs for data collection and analysis. Second, we thought it desirable to present proposals that dealt with research topics drawn from a variety of disciplinary and professional areas. Third, we wanted to include examples that illustrate the difference between traditional proposals for academic theses and dissertations, and a proposal with which a graduate student might seek grant support for their research project. Fourth, and finally, for the purpose of helping you appreciate the developmental process through which most proposals must pass, we have included exemplars ranging from an early draft to a document in polished and final form. Ultimately, all four proposals presented in Part III did accomplish the purposes of their respective authors—the award of a research grant or approval by a graduate committee and subsequent successful completion of a thesis or dissertation.

We suggest that you examine more than one of these documents. Not only do the proposals involve different research designs, but they also illustrate alternative formats and writing styles. You can obtain valuable ideas about the preparation of your own proposal by observing how the demands for communicating a research plan were met in different contexts for inquiry.

For the first specimen proposal, we selected an experimental study in which the investigator lays out a plan to examine the impact of psychological factors on training programs for strength and flexibility in older women. Second in order is the proposal for a doctoral dissertation involving qualitative research.

In that document, the author proposes to examine high school students' perspectives concerning teachers who had a substantial impact on their school careers. For the third specimen proposal, we selected a study in which the investigator lays out a plan to utilize the Internet to conduct an online survey of the practices of front-line pediatric nutrition care providers for children with HIV infection. The final specimen in Part III is a grant proposal soliciting funding for support of a doctoral dissertation. The study plan described there involves the examination of how higher education in prison affects certain inmate characteristics.

Each of the proposals has been edited for the present purpose and is accompanied by our comments. The specimens used here are not presented as models that are perfect in every respect. Instead, we selected proposals that not only served the intentions noted above but also did so in a manner that reflected a consistently high level of general academic competence. As you will see, the proposals also involve both interesting research topics and sometimes challenging problems of study design.

Our accompanying comments are not intended to be critical of the authors. It is easy to play the role of "Monday morning quarterback" when you have ample time and none of the pressures associated with preparing a proposal. The purpose of our commentary is only to draw your attention to crucial elements within each research plan, as well as to underscore the strengths and limitations of presentation in each proposal document.

As you formulate your own critique of the proposals, it will be useful (and a point of fairness) to remember that each of the authors, having now gone through the process of presenting his or her plan for formal approval, and then having actually performed the proposed study, surely could make changes that would improve both the proposal document and the investigation it presents. Hindsight always is 20/20. As you read, give special attention to the points at which clarity or persuasiveness seems to weaken, and then think carefully about what changes in the text might better have served the author. Look also for proposal functions that are executed with precision, economy, and grace—and then go and do likewise!

PROPOSAL 1

Experimental Study

Strength Training in Older Women: Does Self-Efficacy Mediate Improvements in Physical Function?

Readers are reminded that, as indicated in the Introduction to Part III, this document has been edited for the purposes of both economy of presentation and consistency with the proposal functions presented in Chapter 1. Text that elaborated on particular concepts (once introduced) has been abbreviated, and several sections that were essential in the original proposal but are redundant for our present purpose have been omitted entirely. What remains here, however, is a faithful reproduction of the basic elements in the author's plan for the proposed study.

Introduction

The fastest growing segment of the population in North America consists of those individuals 65 years of age and older (U.S. Bureau of the Census, 1992).

AUTHORS' NOTE: This proposal, used with permission, was prepared in 1996 by Shannon Mihalko as part of the dissertation research requirement in the Department of Kinesiology at the University of Illinois at Urbana-Champaign under the direction of Edward McAuley. The dissertation was completed and accepted by the graduate school. Dr. Mihalko currently is an associate professor at Wake Forest University.

This rapid increase among the aging population, often coined "the greying of America," is forecast to continue over the next 60 years (Brown, 1992), resulting in an attendant concern for the health, functional ability, and quality of life in this population (Brown, 1992; Katz, 1983). Although older adults are living longer, whether they are living more functional, active, and independent lives is debatable. This ability to function independently has become an increasingly crucial public health concern (Katz, 1983; Phillips & Haskell, 1995).

> The author has begun by introducing the general social context of the problem to be addressed by the proposed study. The first paragraph was designed to pique readers' interest by discussing changes in our population—changes in which they are (or will soon become) participants. Because everyone expects (or, at least, hopes) to grow older, most readers will be interested in the topic and will be drawn along as the author now shifts from gentle introduction to definition of the concepts that form variables within the study. The discussion in the following paragraphs introduces two of those key concepts, age-related decline in physical function and the role of muscular strength and balance in physical performance by older individuals.

Attempts have been made to identify those factors that are associated with the marked declines in age-related physical function. Physical activity and exercise participation have been consistently identified as important components in improving the quality of life and level of functional independence among aging populations (McAuley & Rudolph, 1995). However, the percentage of aging individuals who participate in physical activity and exercise is extremely low, with epidemiological estimates indicating that less than 40% of those individuals over the age of 65 participate in any leisure time physical activity, whereas over 40% of these adults maintain a sedentary lifestyle (Stephens, Jacobs, & White, 1985).

In addition to declines in physical activity, aging has also been associated with decrements in muscular strength, balance and movement impairments, increased incidence of falls, and reductions in performance of activities of daily living (ADLs) such as eating and drinking, dressing, washing, using the toilet, getting in and out of bed, rising from a chair, and moving about a home (Fiatarone et al., 1990; Goldberg & Hagberg, 1990; Katz, 1983; Tinetti & Speechley, 1989). It has been argued, however, that those physiological deficits often directly associated with aging are not solely due to the process of aging, but are largely a consequence of decreased participation in physical activity (Goldberg & Hagberg, 1990; Mihalko & McAuley, 1996). In fact, substantial physiological gains have been demonstrated in those aging individuals who engage in regular physical activity and exercise (Blair et al., 1980; Fiatarone &

Evans, 1993; Goldberg & Hagberg, 1990; McAuley, Courneya, & Lettunich, 1991; McAuley & Rudolph, 1995).

Research examining the influence of regular physical activity and exercise on physiological parameters has demonstrated a wide range of effects on various physical conditions, including the reduction of coronary artery disease and all-cause mortality (Blumenthal et al., 1991), increased levels and performance of ADLs (Gill, Williams, & Tinetti, 1995; Katz, 1983; Phillips & Haskell, 1995), increased muscular strength and endurance (Fiatarone & Evans, 1993; Mihalko & McAuley, 1996), improved balance and movement parameters (Tinetti & Speechley, 1989; MacRae, Feltner, & Reinsch, 1994), and more recently, a decreased risk for falling (Province et al., 1995).

In the next paragraphs, the construct of "falling" is introduced as a vital concern in the health of elderly persons. Notice how the author first connects the construct to elements in the previous paragraphs, and then extends it by explaining how physical abilities may affect activities of daily living.

Falls and fall-related injuries warrant particularly thorough investigation as they are major contributors to mortality, morbidity, and institutionalization in older adults and therefore pose a significant threat to the aging population (Dunn, Rudberg, Furner, & Cassel, 1992; Rubenstein & Josephson, 1992; Sattin, 1992). A landmark study by Province and colleagues (1995) examined the relationship between exercise levels and the relative risk of falling and fall-related injuries in adults 60 years of age and older. Their meta-analysis of epidemiological evidence was generated using the database from Frailty and Injuries: Cooperative Studies of Intervention Techniques (Ory et al., 1993). Specifically, the risk factors for falling and fall-related injuries identified as primary outcomes in this meta-analysis included functional status, ADLs, muscular strength, balance, and previous falls. Province et al. (1995) report that assignment to interventions that had exercise as a component was associated with a decrease in the incidence of falls among the elderly.

In previous studies, researchers have compared fallers and nonfallers in an attempt to identify the major risk factors for falling. Throughout that literature, the most consistently identified risk factors that are implicated in falling are lower-extremity strength, levels of balance, and those variables associated with abnormal gait patterns (MacRae et al., 1994; Tinetti, Speechley, & Ginter, 1988). In fact, impairment of gait and balance have been associated with a significant three-fold increased risk of falling, whereas decreases in lower-body strength appear to increase the odds of falling five fold (MacRae et al., 1994).

In aging populations, clear differences have been shown between fallers and nonfallers on parameters that are potential risk factors for falling, as well as on future behavior and lifestyle changes and restrictions. In fact, 26% of those aging individuals who have fallen in the past year report avoiding common daily physical activities (Tinetti et al., 1988). Another study found that fallers report a 41% decrease in physical activity over a one year period as compared to 23% in nonfallers (Vellas, Cayla, Bocquet, dePemille, & Albarede, 1987). Other risk factors for falling consistently identified throughout the literature include: use of an assistive device, such as a walker or cane; impaired vision and a decrease in nocturnal vision; arthritis; self-reported limitations in mobility; and limited levels and/or dependence in ADLs (Langlois et al., 1995).

Activities of daily living are the most common referents of physical disability and functional dependence among the elderly (Guralnik & Simonsick, 1993; Phillips & Haskell, 1995) and are often characterized by primary self-maintaining functions such as bathing, dressing, transferring from bed to chair, climbing stairs, and walking a short distance (Guralnik & Simonsick, 1993; Katz, 1983). This decrease in physical functioning accompanying aging leads to an inability to live independently, often referred to as frailty, and is typically characterized by the need for assistance in carrying out one or more ADLs (Guralnik & Simonsick, 1993). Approximately one-fifth of the U.S. population who are 65 years of age or older require assistance in at least one ADL and may therefore be classified into this category (Guralnik & Simonsick, 1993). Furthermore, 23% of older adults living at home have difficulties with ADLs, and about 10% require help to complete these activities.

The inadequate assessment of functional ability among aging populations suggests the need to make wider use of physical performance tests that are truly relevant to the sample under consideration. In addition, it is necessary to incorporate the assessment of specific tasks and movements that are required to perform these common daily activities (Gill et al., 1995). Typical ADLs, such as climbing stairs, reaching up, and lifting objects, require the involvement of the three components of physical fitness often referred to as "muscular fitness" (American College of Sports Medicine, 1993, p. 48), which includes muscular strength, muscular endurance, and flexibility (Czaja, Weber, & Nair, 1993; Phillips & Haskell, 1995). That is, those aging adults who have undergone declines in muscular fitness are likely to report more difficulty in performing many daily activities, which often require lifting, carrying, pushing, and/or pulling capabilities (Czaja et al., 1993).

The first sentence of the next paragraph provides a brief summary of the previous discussion (signaled by the opening phrase, "The recent literature suggests . . ."). The closing sentence, however, serves a very different function. It draws the reader's attention to a fact that will become the basis for the proposed study—we do not know whether increased strength reduces falls directly, or through other intermediate agencies (which nicely explains the rhetorical question posed in the title). Using that question as a means of transition, the author promptly turns to a discussion of the relationships involved. Note that the author is in the enviable position of being able to use some of her own preliminary work as part of the supporting literature.

The recent literature suggests that participation in physical activity, specifically resistive strength training programs, is associated with increases in the three components of muscular fitness (Fiatarone et al., 1990, 1994; Frontera et al., 1988). In turn, an increase in muscular strength has been associated with improvements in the levels and performance of ADLs (Fiatarone et al., 1990, 1994; Mihalko & McAuley, 1996) with a potential reduction in the risk for falls and fall-related injuries among aging individuals (Binder et al., 1994; Lipsitz et al., 1994). Whether the association between increased strength and reduced risk for falls and enhanced ADLs is direct or whether it can be explained by other underlying mechanisms is not, however, known.

One logical explanation for the association between greater levels of muscular strength and improvements in ADLs and reductions in the risk for falls and fall-related injuries may simply lie in the resultant changes reported in muscular strength levels from participation in strength training programs. That is, improvements in muscular strength among aging adults via participation in resistive strength programs may directly lead to an increase in ADL levels and performance, as well as to a decrease in falls and relative risk for falls. In fact, preliminary findings (Mihalko & McAuley, 1996) offer limited support for this hypothesis. Here, Mihalko and McAuley (1996) report that strength improvements accounted for 10.3% of the variance in ADL performance following an 8-week high intensity strength training program. Future research to corroborate these findings is necessary, however, since the assessment of ADLs was based on proxy reports rather than on actual performance levels.

In addition to the resultant muscular strength gains consistently reported as a function of participation in resistive strength training programs, it is possible that ADLs and the relative risk for falling may also be influenced by enhanced balance. For example, significant increases in balance parameters were demonstrated by frail older adults participating in an 8-week low to moderate intensity group exercise program (Binder et al., 1994). Moreover, this program led to significant increases in lower extremity strength, gait speed, and self-reported physical function in this cohort of older individuals at risk for recurrent falls. Therefore, increases in balance parameters may indeed augment the influence of physical activity on independent functioning and falls prevalence.

In conjunction with the findings on the effects of physical activity on physical function and the performance levels of ADLs, it has been suggested that a decrease in ADLs is often reported as a result or consequence of falling. More recently, however, reductions in ADLs and decreased performance levels in a variety of ADLs have been investigated and identified as potential risk factors for falling among the elderly (Langlois et al., 1995). Additionally, several investigators (Tinetti & Powell, 1993; Tinetti et al., 1988) have argued that fear of falling may be a significant predictor of falling among aging adults and perhaps, a more considerable problem than falling itself. Fear of falling is typically defined as an enduring concern about falling that results in restriction of daily activities and functioning (Tinetti & Powell, 1993). Specifically, fear of falling, after adjustment for age and gender, has been significantly associated with increased frailty and depressed mood, restriction of mobility and social activity, and decreased satisfaction with life (Arfken, Lach, Birge, & Miller, 1994). The prevalence of falling and the impact of fear of falling on physical activity levels and ADLs among the elderly is overwhelming. In fact, approximately 50% of those older individuals who report fear of falling avoid performing common ADLs and voluntarily reduce their levels of physical activity (Tinetti et al., 1988).

In the paragraph below, the author arrives at the primary destination for all that has preceded—the key concept of self-efficacy and the suggestion that it may be one of the intermediary mechanisms for functional change. Retrospectively, the step-by-step logic of the presentation now becomes transparent, and the author's thoroughness in that process is impressive. Our only concern is that subsequent to mention in the title, the key concept (self-efficacy) is named for the first time on page 10 (of the original proposal document) and has been preceded by a great deal of dense and heavily documented prose. A reader might be excused if he or she began to wonder "What is all of this leading to?" a number of pages before arriving at the answer.

There are several strategies for circumventing that problem. One would have been to introduce a sketch (perhaps a simple graphic model) of the primary components, in their ordered relationships, at a much earlier point in the introduction, saying, "This is the tour we are going to take, here are the primary stops en route, and there is the destination." With that guide in mind, this lengthy "introduction" might not seem at all arduous (or mysterious).

Another strategy simply would have been to reduce the density of the presentation. Although, for example, the many multiple citations probably were familiar and reassuring to advisors who were specialists in the field of gerontology, limiting the documentation to just one or two items for each major point might have served the purpose of the proposal just as well—and allowed other readers to move more easily through the less encumbered text.

Although fear of falling may decrease ADLs and limit physical functioning in older adults, little is known about the determinants of fear of falling or the potential mediating constructs between fear of falling and physical function. As actual behavior is often hard to predict based upon self-report measures of such constructs as fear and fear of falling (Tinetti, Mendes de Leon, Doucette, & Baker, 1994), this fear of a lack of control over one's ability to perform ADLs without falling is often conceptualized and assessed by perceptions of self-efficacy (McAuley, Mihalko, & Rosengren, in press; Tinetti & Powell, 1993, Tinetti et al., 1994). As proposed by Bandura (1977, 1986), self-efficacy is a situation-specific sense of control, concerned with the beliefs or convictions that one has in one's capabilities to successfully engage in a course of action in order to satisfy situational demands. It has been argued (McAuley et al., in press; Tinetti & Powell, 1993) that Bandura's (1986) social cognitive theory offers a potentially useful framework for explaining and understanding the relationships between physical activity, fear of falling, prevalence of falls, and ADL performance levels. Specifically, a social cognitive perspective posits that self-efficacy perceptions may mediate the underlying changes reported in ADLs and falling among active older adults as compared to their less active counterparts.

Purpose

In the statement of the purpose below, after identifying the broad conceptual framework of the study, the author lays out two objectives for the investigation. The first involves comparison of two treatments in terms of consequent changes in the parameters of strength and balance. The second objective, however, is to examine the relationships of those *physical* changes to *behavioral* (performance) changes measured by tests of ADLs and falls, as they may be mediated by *psychological* changes (in perception) measured by tests of self-efficacy and fear of falling. As complicated as that may sound, in operational terms it still boils down to a basic two-treatment group experiment with pre-post measures of six primary variables.

It might also have been helpful at this point if the author had indicated the order in which the variables are presumed to act. For example, does increased strength lead to improved ADL and thus higher perceived self-efficacy? Or might it be that increased strength itself leads to a greater sense of self-efficacy, and thereby to improved ADL performance?

The present study incorporates a social cognitive framework to examine the relationships between the changes in muscular strength and balance resulting from participation in a high-intensity strength training program, and the subsequent changes in efficacy perceptions, ADL performance, fear of falling, and prevalence of falls among an aging cohort. In an attempt to examine the mediational role played by self-efficacy in changes in ADL performance and falling, this study had two primary objectives: (a) to compare strength and balance changes in aging individuals who were assigned to either a high intensity resistive training program or a stretching and flexibility control group and (b) to determine whether unique variation in changes in ADL performance and falls could be attributable to changes in perceptions of control or self-efficacy.

Rationale

> Although reasons for conducting the study have been woven throughout the introduction, here the author directly addresses the question "Why do this study?" Notice that for the purpose of clarity, the author neatly separates practical and theoretical reasons for conducting the study. The ensuing discussion leads through a series of cumulative steps to the final assertion that although the evidence for each link is as yet inconclusive or incomplete, there is a sufficient chain of evidence to support testable postulates (hypotheses) about the complex relationships involved.

There are both practical and theoretical reasons for conducting this study. From a practical perspective, knowing the benefits of the two exercise programs and how they differ in improving the physical capabilities of the participants will provide information that is valuable to those working with the elderly. If one exercise program is substantially better, then physicians and exercise specialists will have a basis from which to make informed decisions.

On a theoretical level, this study will integrate a number of the percepts of efficacy relevant to falling, ADLs, muscular strength capabilities, balance parameters, and physical function and provide information on the factors that decrease fall prevalence and reduce ADL limitations. Moreover, we will better understand whether increases in physical capabilities, such as muscular strength and balance, lead to enhanced perceptions of personal control relative to successful performance of common ADLs without falling. This in turn may provide information on whether self-efficacy influences balance and improves subsequent function, and reduces the fear of falling. Although there is little empirical evidence to date that links physical activity and fear of falling (McAuley et al., in press), improvements in balance (Binder et al., 1994), increases in muscular strength (Fiatarone & Evans, 1993), and a reduction in falls risk (Province et al., 1995) have been consistently related to physical activity and exercise among aging individuals.

Considerable evidence exists to support the hypothesis that successful participation in exercise and physical activity programs can serve to enhance perceptions of mastery or personal control (Ewart, Stewart, Gillilan, & Kelemen, 1986; McAuley et al., 1991; McAuley, Lox, & Duncan, 1993; Rodin, 1986). Specifically, McAuley and colleagues (McAuley et al., 1991; McAuley et al., 1993; McAuley, Courneya, Rudolph, & Lox, 1994) have demonstrated that participation in exercise programs not only increased levels of self-efficacy, but these perceptions of personal capabilities also led to increased adherence to exercise regimens and subsequent physiological responses to exercise. Thus, a social cognitive perspective (Bandura, 1986) would postulate that fear of falling, falling, and ADL performance are multiply determined by physiological, cognitive, and behavioral influences. It is possible, therefore, that perceptions of self-efficacy may mediate the relationships between physical activity participation and subsequent physiological improvements with changes in ADL performance, fear of falling, and prevalence of falls among aging individuals. The results of this study will substantially contribute to our understanding of these complex relationships.

Hypotheses

The complex nature of this study, including both the presence of six major variables (some of which were divided into subvariables) and the need to examine several different types of relationships through complex statistical analyses, required the author to establish a large number of hypotheses. For clarity of presentation and ease of use, these were grouped into six logically derived clusters. For economy, we have included here only representative examples from each cluster, indicating in each case the number of omitted hypotheses.

Notice that each hypothesis is specific to a particular relationship. For example, in Hypothesis 1a the relationship between type of training and pectoral muscle strength will be subject to a statistical test of significance (some form of analysis of variance is implied). When the analysis is complete, the investigator will be able to accept or reject the hypothesis. By returning to our discussion of hypotheses in Chapter 1, you can examine the adequacy of each hypothesis. For example, is the hypothesis unambiguous? Does it express a relationship? Is an appropriate statistical test implied?

You will also notice that some of these hypotheses are directional—positing whether the relationship might reasonably be expected to be positive or negative. The presence of previous research suggesting such directionality encouraged the hypothesis and, in turn, allowed the author (in at least some of the instances) to choose a more powerful statistical test (one allowing greater confidence in its prediction) than the test used with hypotheses that lack specified direction (null hypotheses).

Specifically, it is hypothesized that:

1) Participation in a resistive strength training program will result in greater increases in muscle strength than participation in a stretching and flexibility program, for the:
   a. pectoral muscles
   b. latissimus dorsi muscle
   c. deltoid muscles

> We have omitted six additional hypotheses. For hypotheses such as these, it is acceptable to write a separate statement for each measurement to be taken, or to nest a series of such items within a single stem statement. For the sake of economy and simplicity, we prefer the latter (as above).

2a) Participation in a resistive strength training program will result in greater decreases in fear of falling than participation in a stretching and flexibility program.

2b) Participation in a resistive strength training program will result in greater increases in Falls Efficacy Scale scores than participation in a stretching and flexibility program.

> We have omitted two additional hypotheses.

3a) Participation in a resistive strength training program will result in greater improvements in self-reported ADLs than participation in a stretching and flexibility program.

4a) Change in muscular strength is negatively related to changes in fear of falling.

4b) Change in balance is negatively related to change in fear of falling.

> We have omitted four hypotheses here.

5c) Change in muscular strength is positively related to change in ADL performance.

5d) Change in balance is positively related to change in ADL performance.

6a) Change in muscular strength will predict a unique percentage of the variance in residualized change in fear of falling scores, in addition to that predicted by change in balance and self-efficacy.

We have omitted eight additional hypotheses.

In many proposals, sections would be included at this point to deal with a variety of special needs within the proposal. Among the most common of those are definitions of technical terms that are particular to the investigation or to the author's supporting arguments, and specification (for the reader) of the limitations and delimitations that must be remembered when considering the proposed study. An alternative to segregation of such material into separate sections, and the option used in this proposal, is to weave the information into the text as it becomes relevant. The first time a technical term such as "self-efficacy" was used, for example, it was followed by a brief definition, adjusted to the nature of the immediate context.

The choice between segregating or distributing definitions is partly a matter of personal style and partly a matter of logistics (length of the manuscript, number of unfamiliar terms, complexity of definitions, etc.). If a great number of definitions are required, for example, it is easier to revisit one that appears in a single listing than to locate it embedded somewhere in many pages of text. On the other hand, establishing subsections of the proposal (such as "Limitations" or "Definitions") just for the effect of having them serves no useful purpose. The primary consideration should be to make it as easy as possible for your readers to follow the presentation. In what follows, you can observe exactly how the author has elected to handle these matters.

In the original proposal, the following review of the literature required 42 pages of text. Following the brief introduction, we have selected the first and last subsections of that review for use here. A lengthy middle subsection has been omitted. Because it employed exactly the same review format, for our present purpose the reader will not be deprived by the deletion. A brief summary does provide an overview of the entire review at the close of the section.

Literature Review

This chapter begins with an epidemiological overview of aging in the United States, in order to highlight the importance of examining the aging process from a public health perspective. This section will also discuss the relationship between aging and physical function from an epidemiological perspective, with an emphasis on declines in physical activity and common activities of daily living (ADLs), as well as on the increase in falls risk and prevalence of falls associated with age. This section will be followed by a review of the effects of physical activity and exercise on function in the elderly, highlighting the role played by muscular strength in falls risk reduction and ADL enhancement. Finally, the last section of the chapter will be devoted to reviewing the potential underlying mechanisms proposed to mediate strength effects on these parameters of physical function, with particular emphasis on the role played by self-efficacy perceptions.

> At the close of the first paragraph, the reader knows three things about the review of literature in this proposal. First, although citations to prior studies have appeared throughout the sections serving as introduction and definition of primary variables, the concepts involved are sufficiently complex (and numerous) to require further explanation. Much of that discussion will involve the use of research literature to support the design of hypotheses and selection of methods of measurement. Thus, for this proposal it became reasonable to have a separate section designated as "Literature Review." Second, an overview of the chapter indicates that the review will be divided into three subsections—an important step in helping readers to organize their understanding of the author's discussion. Third, and finally, the author has specified the relationships to be treated within each subsection: (a) aging, health, and physical function; (b) physical activity, strength, and physical function; and (c) the mediating role played by self-efficacy between strength and physical function.
>
> This might be an appropriate point for you to revisit our discussion of preparing a literature review (Chapter 4). In particular, we draw your attention to the way the author uses the literature to support an ongoing discussion that is the central element in this section. This contrasts sharply with the ineffective strategy of writing text that consists of long (and boring) sequences of studies only occasionally punctuated by reference to their implications. Note also that extended detail is provided only when appreciating a particular point truly is essential to understanding the proposed study. As we argued in Chapter 4, if the discussion is developed with care, much of it can consist of general statements, with the reader always having the option of recourse to relevant detail in the cited references. As you read, also note how the use of topical subcategories and the insertion of headings and subheads eases the task of following the review. The addition of summaries for each subsection (as well as to the chapter as a whole) makes it difficult to lose track of information despite the complex set of variables involved.

Epidemiological Perspectives

Aging in the United States

According to the United States (U.S.) census (U.S. Bureau of the Census, 1992), individuals over 65 years of age make up the fastest growing portion of the population in North America and have often been referred to as the "fastest growing minority" (Mackintosh, 1978). In fact, the proportion of Americans who are 65 years of age and older is estimated to double from 12%

in 1986 to 24% in the year 2020, reaching an approximate number of 52 million (Tinetti et al., 1988; U.S. Bureau of the Census, 1992). Specifically, population estimates have indicated that this segment of the U.S. population had grown to 31.2 million in 1990, as compared to 26 million in 1983. This "greying of America" is projected to continue over the next several decades (Brown, 1992), raising several concerns relative to public health issues and health care economics. The overall impact of this rapid increase in the aged will influence such factors as reduction in independent functioning, the need for long-term care and institutionalization, and ultimately, the prevalence of chronic disease and all-cause mortality and morbidity (Schneider & Guralnik, 1990).

Approximately 85% of those individuals 65 years of age and older have one or more chronic health conditions, with an additional 42% demonstrating limited functional capabilities (Katz, 1983). At present, the proportion of elderly individuals who are functionally impaired and/or dependent is alarmingly high and estimated to grow exponentially in the future (Tinetti et al., 1988). For example, one-quarter million hip fractures that are attributable to osteoporosis and indirectly associated with falls and fall-related injuries are reported each year in the U.S. among those individuals 65 years of age and older (Smith & Tommerup, 1995). As the population ages and becomes more functionally dependent, the impact on society, health care, and the public health system could be overwhelming (Bokovoy & Blair, 1994).

Aging and Physical Activity

As the population of older adults continues to increase, the question posed by many is not how long can we prolong life, but how can we extend the quality of that life? Indeed, although individuals may live longer, they may not necessarily lead more active, happy, and independent lives. In fact, several recent conferences and workshops on health, physical activity, and exercise have emphasized the importance of physical activity and an active lifestyle, recommending the evaluation and promotion of physical activity and exercise among elderly populations (Bouchard, Shephard, & Stephens, 1993; Haskell et al., 1992; King et al., 1992). Additionally, the U.S. Department of Health and Human Services (1990) has identified physical activity and fitness as one of the major objectives for the health of the nation, as well as an integral component of overall health promotion. Specifically, the U.S. Public Health Service (1990) has incorporated the importance of prolonging independent living among this elderly population in their national health objectives for the year 2000, with an emphasis on increasing the rate of participation in regular, appropriate physical activity to 40% in individuals 65 years of age and older.

Although considerable evidence exists to suggest that physical activity and exercise are essential ingredients in successful and healthy aging, the participation rates of North Americans in physical activity and exercise programs are disappointing. Analysis of the 1990 Established Populations for Epidemiologic Studies of the Elderly data base indicates that 26% of the study population were incapable of walking 1/2 mile, 42% were unable to complete heavy household chores, and 19% were not able to climb stairs (National Institute on Aging, 1990). Furthermore, the Healthy People 2000 report indicates that more than two out of every five older adults over 65 years of age lead a basically sedentary lifestyle (Public Health Service, 1990). In fact, disability resulting from this sedentary lifestyle is twice as common in those individuals 75 years of age and older, as compared to those aged 45 to 64 years (U.S. Department of Health and Human Services, 1990).

Unfortunately, many older individuals who attempt to participate in an exercise or physical activity program withdraw prior to the onset of apparent health benefits (Dishman, 1990). This finding is not surprising in light of the well-documented attrition rate in exercise programs, with approximately 50% of those adopting a new exercise regimen dropping out within the first 6 months (Dishman, 1990). With the apparent increase of older populations in which inactivity and decreases in independent functioning are prevalent, the importance of determining the impact of these declines in physical activity on functional status, morbidity, and mortality is abundantly clear.

Aging and Physical Function

Activities of daily living. In addition to the established age-related declines in physical activity, aging has been associated with decreases in various components of functional independence (Gill et al., 1995; Katz, 1983; Simonsick et al., 1993; Tinetti & Speechley, 1989), with a considerable segment of the community-dwelling population over 65 years of age reporting limitations in some aspect of physical functioning. For many older adults, a reduction in levels and performance of ADLs has a greater impact on the preservation of independence in day to day functioning than the chronic diseases from which these reductions had originated (Ensrud et al., 1994). For example, an estimated 10% of nondisabled adults aged 75 and older who are still living in the community become dependent in ADLs each year (Gill et al., 1995) and approximately one-quarter of the aging population has difficulty performing activities that are deemed necessary for self-sufficiency (Dawson, Hendershot, & Fulton, 1987; Young, Masaki, & Curb, 1995).

The ability to live independently in the community is extremely difficult for those older adults who need assistance with or simply cannot perform common ADLs (Czaja et al., 1993). Specifically, 23% of adults aged 65 years and older have difficulty with activities such as bathing and dressing, and about 10% require assistance with these ADLs (Dawson et al., 1987). In addition, 27% need help with more complex activities, such as meal preparation and housework (Guralnik & Simonsick, 1993). Furthermore, there is a high rate of home accidents and falls among this aging population, possibly resulting from the inability to perform these common self-maintaining functions (Czaja et al., 1993; Province et al., 1995; Sattin, 1992). With the present and anticipated aging of the population, the increasing proportion of older adults who are functionally limited or dependent is not expected to diminish in the near future (Manton, Corder, & Stallard, 1993).

Falls Risk and Prevalence. One of the most serious and prevalent sources of medical problems and functional decline among aging individuals in the U.S. is represented by falls and fall-related injuries (Tinetti et al., 1988). Approximately 30% of those individuals 65 years of age and older suffer a fall each year, with about 50% of these elderly persons sustaining more than one fall in this time period (Tinetti et al., 1988; Tinetti & Speechley, 1989). Consequently, serious injuries result from approximately 10% to 15% of these falls (Tinetti et al., 1988). Specifically, 0.2% to 1% of these falls result in hip fractures, 5% in fractures at other sites, 5% in soft-tissue injuries and dislocations, as well as long periods of immobility due to the inability to regain normal gait patterns (Kiel, O'Sullivan, Teno, & Mor, 1991; Sattin, 1992; Tinetti et al., 1988).

Falls and fall-related injuries pose a major threat to aging individuals and are a significant contributor to morbidity, mortality, institutionalization, and immobility (Dunn et al., 1992; Sattin, 1992; Tinetti et al., 1988). Unintentional injury is the 6th leading cause of death among those adults 65 years of age and older. These unintentional deaths may be largely attributable to falling, especially for those over the age of 85 years (Province et al., 1995; Sattin, 1992). Additionally, these falls have a considerable social, medical, and economic impact.

Several paragraphs have been omitted here that offer details concerning the threats to individuals, as well as the social and public health implications, presented by injurious falls in the elderly.

Summary

This section reviewed the epidemiological literature on aging. People are living longer but that does not necessarily mean they have an optimal quality of life. Many age-related declines occur with aging. Over one-quarter of elderly adults have problems doing simple daily tasks, and accidental death, particularly from falling, is a problem in the aged. Participation in physical activity programs has been recommended for the elderly (U.S. Public Health Service, 1990) and may help reduce the accidental deaths and improve the quality of life.

The middle subsection of the literature review, "Exercise Effects on Function in the Elderly," has been omitted here.

Self-Efficacy Theory

Overview of Self-Efficacy Theory

Self-efficacy theory is a social cognitive approach to behavior that views behavior, physiology, cognitions, and environmental influences as interacting determinants of one another (Bandura, 1977, 1986). Self-efficacy is defined by Bandura (1977) as a belief in one's capabilities to overcome successfully the demands of a situation, in order to achieve a desired outcome. Self-efficacy theory offers a common mechanism by which individuals posit control over their own behaviors and act as self-referent mediators, rather than simply passive acceptors of environmental stimuli and external forces (Bandura, 1977, 1986). Efficacy cognitions are specifically related to the activity or behavior in question, and subsequently, efficacy cognitions are apt to change in response to changing environmental stimuli.

Increases in efficacy perceptions are likely to occur as a function of positive mastery experiences, whereas decreases in efficacy perceptions are likely to result from negative experience and failure. By definition, self-efficacy cognitions are concerned with the convictions or beliefs that an individual possesses about his or her ability to meet the situational demands of a particular course of action successfully (Bandura, 1977, 1986). It is important to emphasize that it is not the skills an individual actually possesses that are important, but rather what judgments the individual forms relative to what can be accomplished with these skills.

Those individuals who have high levels of self-efficacy tend to exert more effort, approach more challenging tasks, and persist longer in the face of obstacles and barriers. For example, an elderly individual who has high levels of self-efficacy for performing ADLs without falling is more likely to engage in ADLs within a variety of environmental situations. On the other hand, when faced with aversive stimuli, those individuals who have low self-efficacy perceptions tend to attribute negative experiences internally, drop out of the activity altogether, and report higher levels of depression and anxiety (Bandura, 1982). In the context of falling, aging individuals who have low levels of efficacy for maintaining balance and believe they are at risk for falling are more likely to avoid these situations altogether.

Omitted here are several paragraphs that describe self-efficacy theory, its derivation within the social sciences, appropriate forms of measurement, and domains within which it successfully has been applied.

Although self-efficacy theory has been applied to numerous domains, perhaps the domain of most promise is that of health behavior and specifically, that of exercise and physical activity (McAuley, 1992a). Research in the exercise domain has demonstrated that not only is self-efficacy enhanced by the exercise experience, it is also a determinant of physical activity in both diseased (Ewart et al., 1986) and asymptomatic populations (Dzewaltowski, 1989). This has been documented in large-scale community studies (Sallis et al., 1986), and in training studies (McAuley, 1992a; McAuley et al., 1991).

This last section of the review of literature describes the aspects of the research literature that are most relevant to the present study: the relationships among the dependent and independent variables.

Self-Efficacy Theory, Physical Activity, and Exercise

Exercise behavior has been recognized as a complex, dynamic, and multiply determined phenomenon (Dishman, 1990) with a host of determinants having been identified as having a potential direct or indirect effect on exercise participation and adherence (McAuley, 1993). Self-efficacy theory (Bandura, 1997, 1986) offers a promising and sound theoretical framework from which to examine the interplay between biological, psychological, social, and environmental processes in the specific domains of health, exercise, and physical activity (Dishman, 1990; McAuley, 1992b). A majority of the researchers examining self-efficacy theory within the domains of exercise and physical activity have focused on efficacy as a determinant of exercise behavior (McAuley, 1992b, 1993).

In a large-scale community study, findings indicated that self-efficacy perceptions had influenced the adoption and maintenance of exercise behavior in a general population (Sallis et al., 1986). Further support for this relationship was found in a study of older individuals recovering from acute myocardial infarction (Ewart et al., 1986), who participated in a 10-week exercise rehabilitation program involving either jogging and circuit weight training or jogging and volleyball exercise regimens. The number of minutes above or below the prescribed intensity level that subjects participated in the jogging activities, as well as the strength gains accrued in the weight training program, were predicted by levels of self-efficacy. That is, subjects who had higher levels of lifting-efficacy had greater strength gains at program end. Those subjects with greater post-treadmill efficacy also had an increase in their perceived ability to carry out tasks involving arm or leg strength, such as home activities and stair climbing (Ewart, Taylor, Reese, & DeBusk, 1983).

Ewart and colleagues (Ewart et al., 1983, 1986) have demonstrated that those individuals with increased self-efficacy are more likely to participate in a variety of life behaviors than their less efficacious counterparts (McAuley, 1993). With respect to exercise and physical activity, aging individuals who have a greater sense of efficacy in their physical capabilities will adopt a more active lifestyle and perhaps attempt more ADLs and therefore decrease their levels of functional dependence. In turn, an increase in physical activity and exercise participation may serve as a further means to enhance efficacy perceptions. Indeed, exercise and physical activity have been investigated as a mechanism by which efficacy cognitions concerning one's physical capabilities may be enhanced. In this vein, it appears that efficacy cognitions about one's physical capabilities may be altered and modified by chronic exercise and physical activity (Ewart et al., 1986; Holloway, Beuter, & Duda, 1988).

Most of the research in this area has examined the effects of physical activity on self-efficacy cognitions among middle-aged and older males and females, with the majority of studies involving clinical populations (McAuley, 1993). For example, Kaplan, Atkins, and Reinsch (1984) examined the effect of a walking program on efficacy perceptions for compliance to future exercise programs in subjects with chronic obstructive pulmonary disease (COPD). Not only was efficacy significantly correlated to program compliance among this clinical sample, but efficacy was also found to generalize to other related ADLs. In addition, exercise efficacy was found to be a predictor of five-year survival in another sample of patients with COPD (Kaplan, Ries, Prewitt, & Eakin, 1994). Thus, participation in strength training programs may not only lead to enhanced efficacy for strength training, but also to compliance with future strength training endeavors and increased efficacy for strength-related ADLs.

Here we have deleted several pages presenting additional examples of the pervasive and interactive relationships among exercise, affective responses, and efficacy expectations.

Across these studies, it has been shown that both chronic and acute bouts of physical activity and exercise consistently influence efficacy cognitions in a positive manner (McAuley, 1993, 1994). Through many subsequent studies, evidence has accumulated to support these two related aspects of the exercise/self-efficacy relationship. That is, self-efficacy has consistently been both a reliable predictor of exercise behavior, as well as a positive outcome of the exercise experience. In addition, several studies exist that examine the mediational role played by self-efficacy in affective responses (Bozoian, Rejeski, & McAuley, 1994; McAuley & Courneya, 1992). From this standpoint, physical activity is theorized to influence affective responses indirectly through its positive influence on efficacy perceptions.

Summary

Self-efficacy theory posits that behavior, physiology, cognitions, and environmental influences interact with each other. Self-efficacy theory may help explain the relationships between strength and falling variables in adults. An extrapolation of the findings suggests that self-efficacy plays a mediational role in the relationship between physical activity, falling, and physical function since self-efficacy perceptions have proven to be reliable predictors, outcomes, and mediators of various behaviors, and hold particular promise in the domains of physical activity and exercise. Thus, muscular strength changes that result from participation in a strength training program may both directly and indirectly affect ADLs, falling, and fear of falling, via increased perceptions of efficacy for strength-related activities.

The transitional section with the heading "Summary of the Proposed Study" serves as a bridge between the preceding review and the following explication of method, measures, procedures, and analyses. Notice how this overview begins with a backward reference to the main points established in the review, then moves through a clear statement of the rationale for the study, and then closes with a paragraph that directs attention forward to what will be accomplished by the study (now to be proposed in detail).

Summary of the Proposed Study

Participation in a high intensity strength training program leads to increases in muscular strength and fitness, as well as improved balance parameters (Binder et al., 1994; Ewart et al., 1986; Phillips & Haskell, 1995). Resistive strength training, once thought to be potentially hazardous for an elderly population, has recently been acknowledged for its role in not only enhancing physiological outcomes, but also in improving ADL performance, reducing the risk for falls and fall-related injuries, and enhancing the quality of life among elderly persons (Brown et al., 1990; Czaja et al., 1993; Ewart et al., 1986; Fiatarone & Evans, 1993: Goldberg & Hagberg, 1990; Phillips & Haskell, 1995). Whether the role played by increased muscular strength in the reduction of falls risk and ADL enhancement is direct or whether it can be explained by other underlying mechanisms has yet to be determined.

Bandura's (1986) social cognitive theory offers a theoretical framework for examining the underlying mechanisms that may explain the effects of resistive strength training on falls prevalence, fear of falling, and ADL performance in aging individuals. Specifically, a social cognitive perspective (Bandura, 1986) would suggest that fear of falling, falls prevalence, and ADL performance are multiply determined by physiological, cognitive, and behavioral influences. Furthermore, it has been argued that self-efficacy perceptions may mediate the relationships between physical activity participation and subsequent physiological improvements with changes in falls prevalence, fear of falling, and performance of ADLs among elderly persons. From a public health standpoint, the integration of the percepts of efficacy relevant to muscular strength, balance, ADLs, falling, and overall physical function may serve not only to decrease falls prevalence and reduce ADL limitations, but may ultimately lead to enhanced physical, economic, and social independence in our aging society.

The present study employs a social cognitive framework to determine the direct and indirect effects of changes in muscular strength and balance resulting from participation in either a high intensity strength training program or a stretching and flexibility program on ADL performance and falling among a sample of older adults, via self-efficacy. As a result, the mediational role played by self-efficacy will be more thoroughly examined and the direct and indirect effects of strength training on ADL performance and falls prevalence will be determined.

Method

> For most proposals, the brief introductory paragraph is a perfect model of economy and clarity. It provides a simple overview of major steps and then directs the reader to a figure that meshes the individual parts into a diagram of actions in sequential order. In what follows, you will see three other proposal strategies that are worthy of emulation: (a) previous studies by the author, her advisor, and other members of her research group

are used as pilot work to support particular decisions; (b) references are cited that provide reliability and validity information for each measure; and (c) complex details, test forms and protocols, and other supplementary materials are placed in appendices.

This study addresses the effect of strength and stretching/flexibility exercise programs on a variety of variables. In addition, the underlying mechanisms that influence increases in ADLs will be investigated. Participants will be recruited and then oriented to the study. Pretests will be conducted on behavioral, physical, and psychological measures. Participants will then be randomly assigned to one of the two exercise programs. At the conclusion of a 12-week exercise program posttests will be administered to all participants. A diagrammatic overview of the design is presented in Figure 1.

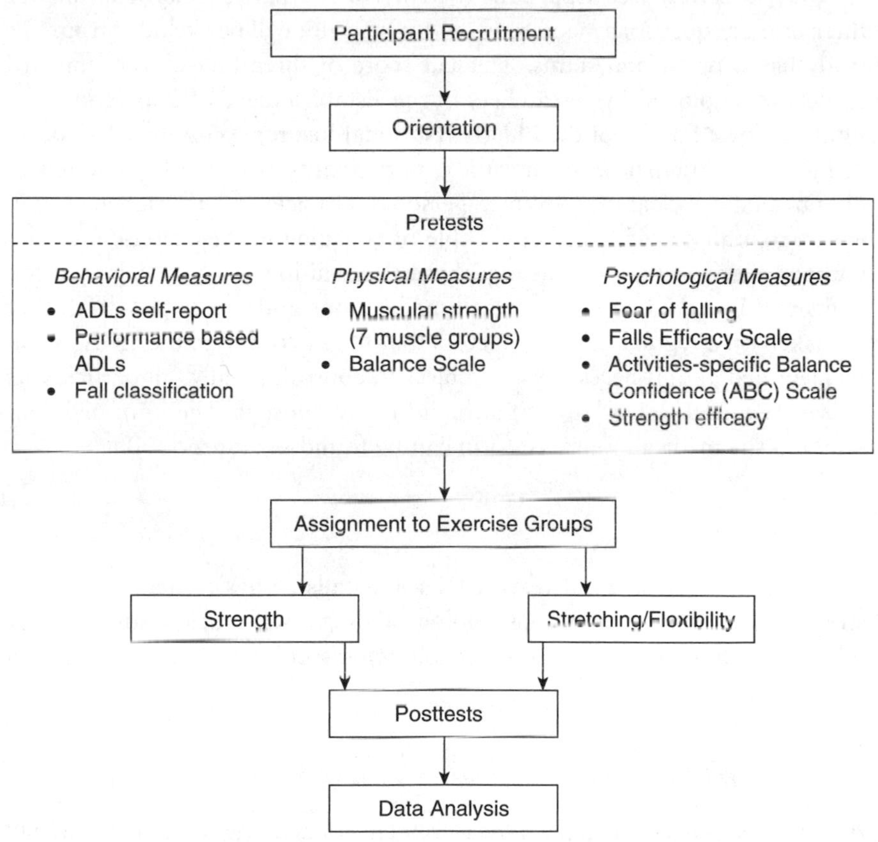

Figure 1 Overview of the Research Design

NOTE: ADL = Activities of Daily Living.

Participants

Participants for this study will be recruited via local newspaper and radio advertising. In addition, the addresses of retired faculty and staff from a midwest university will be obtained from an established mailing list and these individuals will be sent an invitation to participate in this study. Potential participants will be asked to respond to further announcements which will explain the time commitment and outline the general aspects of the exercise programs. All participants will be required to satisfy the following inclusion criteria: (a) females 60 years of age and above, (b) no history of regular physical activity over the past six months, (c) free of any medical condition that might be further exacerbated by regular physical activity, and (d) an adequate level of functional mental status. In order to screen for declines in the latter criterion, participants will be asked to complete the Pfeiffer Mental Status Questionnaire (Pfeiffer, 1975), which is a short measure and will be administered by interview (see Appendix A). Any of the participants who answer three or more questions incorrectly on this measure will be excluded from this study based on mental status. This cut score of three follows the standard guidelines employed by researchers in the Establishment of Populations for Epidemiologic Studies of the Elderly (National Institute on Aging, 1990).

Prior to participation in the study, participants also will be required to obtain medical clearance from a personal physician. This clearance will ensure that all participants are capable of engaging in the exercise program without aggravating any existing physical condition. Additionally, participants will be asked to document all medications and prescribed drugs that are taken on a regular basis, so factors that might confound the analysis can be identified and, if necessary, be subject to control. Finally, all participants will be required to read and sign the informed consent. The informed consent and the medical clearance form can be found in Appendix B.

Measures

Measures of behavioral (e.g., ADLs and falls), physical (e.g., muscular strength and balance), and psychological (e.g., self-efficacy and fear of falling) function are described in the following sections and can be found in the appendices.

Behavioral Measures

Activities of Daily Living. Two measurements of activities of daily living will be administered in order to investigate both subjective and objective ADLs.

With the current emphasis on performance-based testing (Elam et al., 1991), rather than on proxy assessments (Mihalko & McAuley, 1996), the use of both measures will allow for a comparison between the assessment of ADLs based on performance and self-report (See Appendix C).

> The comparison indicated above certainly would be possible, and it might be of considerable practical interest. As a general rule, however, it is unwise to introduce analytic operations that are not truly part of the study as proposed (which is the case here). If nothing else, such side excursions may distract readers from the main exposition and add little more than a further complication. There is, of course, the alternative of making it a genuinely functional part of the proposal.

The development of the self-report measure of ADLs (Mihalko & McAuley, 1996) was based upon ratings by nursing home administrators on the importance of various ADLs in the daily functioning of their senior residents. This adaptation of Lawton and Brody's (1969) Instrumental Activities of Daily Living Scale was constructed in an attempt to target ADLs relevant to functional ability among sedentary older persons. This measure includes the assessment of daily activities such as getting up and down from a chair or bed, getting in and out of a car, carrying and picking up objects, and reaching up overhead. Participants will be asked to rate 20 ADLs separately on a 5-point Likert-like scale, ranging from 0 to 4, (i.e., 0 = cannot do, 1 = can do with assistance, 2 = can partially do without assistance, 3 = can do independently with some difficulty, and 4 = can do easily). Therefore, scores for this measure of subjective ADLs can range from 0 to 80.

A performance-based assessment of ADLs also will be completed by each subject. Here, all participants will be asked to attempt four ADLs, including reaching up overhead, getting up and down from a chair, stair climbing ability, and walking speed. The first three ADLs will be scored on a 5-point Likert-like scale by objective raters, based upon strict scoring criteria which will mirror the scale of the subjective ADL measure. For example, when attempting to stand and sit from a chair, participants will be given a score of 0 if they need maximal assistance to stand or sit, a score of 1 if they can stand and sit with minimal assistance, a score of 2 if they can stand or sit after several tries, a score of 3 if they can rise to a complete standing position and then sit with the use of the arms of the chair, and a score of 4 if they can stand and sit easily. The last performance-based ADL will assess walking speed by asking participants to walk an 8-foot course with no obstructions (Guralnik et al., 1994). Participants will be instructed to walk at their usual speed with the aid of an assistive device if needed. Two raters will record times for two walking trials and an average of the four times will be used for analysis. For the walking measure, categories will be created in order to allow for a collective performance-based measure of the four ADLs (i.e., 5-point Likert-like scale). That is, participants who can complete the task will be assigned scores of 0 to 4, dependent upon the quintiles of time needed to complete the walking task, with the fastest times being represented by a score of 4 (Guralnik et al., 1994). The four categorical measures will be combined for a total ADL score ranging from 0 to 16. The analysis of walking speed as a variable will utilize average walking times (i.e., actual seconds).

Fall Classification. To address the limitations of past fall assessments which have typically employed one-item dichotomous, retrospective measures, falls will be assessed in this study by using computerized dynamic posturography, commonly referred to as the Equitest. The Equitest perturbs the support surface and visual surround in order to examine the contribution of visual, vestibular, and somatosensory components involved in balance control. Any time the subject steps forward or begins to fall, the computer stops the particular task and records the occurrence as a fall. Therefore, falls will be calculated by summing the number of times that the subject "falls" over the course of the test. Participants will be supported by a loose harness, thereby preventing the occurrence of an actual fall.

> When participants are to be exposed to a complex situation or activity such as that presented by use of the apparatus described above, it usually is wise to plan some form of acclimatization procedure. This would be the appropriate point to briefly sketch how that would be accomplished.

Physical Activity History. A short inventory assessing participants' exercise history will be employed to determine overall physical activity background (see Appendix D). Specifically, participants will indicate: (a) the number of days per week they typically engage in physical activity; (b) the number of minutes per session of physical activity; (c) the intensity level of each session, ranging from 1 "easy" to 4 "hard"; and (d) the mode of activity in which they typically engage.

Physical Measures

Muscular Strength. Levels of muscular strength will be assessed for the following muscle groups: pectorals, latissimus dorsi, deltoids, biceps, triceps, hamstrings, and quadriceps. The strength level of each muscle group will be determined by the individual's one repetition maximum (1 RM) or the greatest amount of resistance that the individual can move one time throughout a full range of motion about the working joint. For example, participants will attempt an exercise for each muscle group at a relatively light resistance, contracting with enough force to overcome the resistance, thereby causing the resistance to move throughout the full range of motion. The resistance will be progressively increased for each subsequent attempt until the subject is no longer capable of completing the exercise. The resistance level for the last completed attempt will be recorded as the subject's maximal strength score for that exercise and muscle group. In addition, the five muscle groups of the upper body, as well as the two for the lower body separately will be combined in order to have a composite measure of both upper and lower body muscular strength. Finally, the upper and lower body strength scores will be combined in order to calculate a composite measure of total muscle power or strength.

Balance. Level of balance ability has been shown to predict future falling risk and to be associated with enhanced ADL performance (Binder et al., 1994). Furthermore, exercise programs have been related to increased levels of balance among the elderly (MacRae et al., 1994; Nichols et al., 1995). Therefore, the Balance Scale (Berg, Wood-Dauphinee, Williams, & Maki, 1992) will be employed to assess participants' balance to determine whether the exercise program had a positive influence on balance levels and whether changes in balance are associated with improved ADL performance and reduced falls. The Balance Scale is a 14-item scale and will be scored by trained objective raters on a 0–4 metric with scores ranging from 0 to 56 (see Appendix E). In contrast to the limited one-item assessment of balance typically employed, this measure asks participants to execute 14 movements that are essential to daily living in the elderly, including turning, standing, moving and reaching, transferring, and standing on one foot. The Balance Scale possesses adequate psychometric properties (Berg et al., 1992) and McAuley et al. (in press) have reported an acceptable level of internal consistency ($\alpha = .90$).

Psychological Measures

Fear of Falling. Although fear of falling has been suggested as one potential cognitive mediator of the relationship between physical activity, ADLs, and falling among the elderly (Arfken et al., 1994; Tinetti et al., 1988), little is known about the determinants of fear of falling or the mediating constructs between fear of falling and physical function. In order to more clearly understand the construct of fear of falling, a single item 1–5 Likert-like scale will be employed and participants will be asked to indicate the degree to which they are afraid of falling (McAuley et al., in press). Using this format provides data regarding the fear of falling in a continuous form, as opposed to the more typical dichotomous assessment of the construct (see Appendix F).

Self-Efficacy. Research examining physical activity, ADLs, and falling requires a multi-dimensional approach. Therefore, a number of measures will be employed to assess self-efficacy cognitions relative to falling, balance, and strength capabilities in order to tap different aspects of physical function (see Appendix G). For all self-efficacy assessments, participants will indicate their degree of confidence on a 100-point percentage scale comprised of 10-point increments, with 100% indicating complete confidence and 0% indicating no confidence at all. Total strength of self-efficacy for each measure will be calculated by summing the confidence ratings and dividing by the total number of items in each scale, resulting in a maximum total efficacy score of 100. The assessment of self-efficacy in this manner is in accord with the recommendations proposed by Bandura (1986).

In order to assess one's perceived self-confidence for avoiding falls while performing common ADLs, the Falls Efficacy Scale (FES; Tinetti et al., 1990) will be employed. The FES is a 10-item scale which assesses participants' beliefs in their capabilities to carry out basic ADLs without falling. Tinetti and her colleagues (1990, 1994) have shown falls efficacy to be an important correlate of ADLs and physical function in a community sample of older persons and the FES to be a reliable and valid measure across a variety of samples. Adequate internal consistency ($\alpha > .90$) for the FES has been consistently reported in the literature (McAuley et al., in press; Tinetti et al., 1990, 1994).

The second measure of self-efficacy will be the Activities-specific Balance Confidence (ABC) Scale (Powell & Myers, 1995). In the rationale for the development and utilization of the ABC scale, Powell and Myers (1995) have argued that the FES consists of fairly global activities and may not be sensitive enough to assess balance confidence among highly functioning older adults. By employing both the FES and the ABC scale in the present study, a comparison can be made to determine which is the most sensitive measure for aging populations. The newly developed 16-item ABC scale assesses one's confidence in performing various ADLs without compromising one's balance and has initially promising psychometric properties, with adequate internal, convergent, and construct validity (Powell & Myers,

1995). In addition, the ABC scale taps a broader continuum of activity difficulty than the FES, including such items as picking up an object from the floor, standing on a chair to reach, and walking on icy sidewalks.

Again, we must caution readers that the introduction of "potentially interesting" comparisons that are not, in a direct sense, a part of the primary design of the study puts both economy and clarity at risk. Given the presentation here, it seems reasonable to believe that it might be fruitful to compare the sensitivity of the FES and the ABC Scale when used with the participants in this study—but that matter certainly can be pursued without including it in the proposal.

The last measure of efficacy was developed for the study and was geared toward the assessment of efficacy beliefs for successfully completing strength-related activities. This measure of efficacy will be employed to tap efficacy cognitions which are specifically related to the activity in question, i.e., strength training. The strength efficacy measure is comprised of 10 items, which assesses one's confidence in lifting objects which are increased in weight by small increments. For example, the first item will ask participants how confident they are in their ability to lift an object weighing 3 pounds for 10 times without stopping. Adequate internal consistency ($\alpha > .91$) was found for the strength efficacy measure in a similar sample.

Procedures

Orientation

An initial meeting will be held to inform potential participants about the randomly assigned exercise program, detailing time commitment, possible risks, and providing a description of the exercise protocol. Interested participants will then attend an orientation session conducted by the primary investigator one month prior to the exercise intervention. At this time, all interested participants will complete the informed consent, medical and physical histories, and the Pfeiffer Mental Status Questionnaire by interview. Immediately prior to the beginning of the intervention and before the balance and Equitest assessments are completed, participants will attend a second meeting at which the medical clearance will be collected and the remaining measures and physiological tests will be completed. Participants then will embark on a 12-week strength training or stretching and flexibility exercise program.

> Group administration of paper and pencil tests (and, particularly, psychometric instruments) always holds the potential for introducing unintended data artifacts. Both the context and the process must be planned with great care and executed with meticulous observation of protocol. Who will be present? How will instructions be delivered? How will responses be recorded? Will participants be allowed to ask questions? In what order will the measures be administered? Detail about all of this might be relegated to an appendix, but we would prefer to be reassured that the author has thought it through.

High Intensity Strength Training Program

The protocol for the strength training program will follow the guidelines developed by Fiatarone and her colleagues (1990, 1993) which entails working with 80% of the one repetition maximum (1 RM). Participants will begin the program at 60% of their 1 RM, which will be assessed at the second orientation session immediately prior to the exercise intervention. The resistance progressively will be increased to 80% of the 1 RM, as participants become accustomed to the training regimen.

A trained exercise specialist will demonstrate proper form and technique for each exercise as follows: incline bench press, front or side-lateral raises, one-arm bent-over row, biceps curl, triceps kickback, leg extension, and leg press. Specifically, each subject will be instructed to complete one set of up to 12 slowly conducted repetitions for each exercise. Once a subject can complete 12 repetitions at their pre-determined resistance level, the weight will be increased by one increment for the next exercise session. For example, if a subject can complete 12 repetitions at 5 lbs. for the side-lateral raises with proper form, the weight will be increased to 8 lbs for that exercise at the next exercise session. Exercise sessions will be held three times per week on alternate days, lasting approximately 30 minutes, and progressively increased from 60% to 80% of the 1 RM. In order to ensure that participants are indeed working at 80% of their current 1 RM, an assessment of maximal strength will be taken during the 6th week of training. Finally, exercise frequency for all participants will be recorded with a log of completed exercises for each session.

Stretching and Flexibility Control Program

The stretching and flexibility exercise group will serve as an attentional control group against which the effects of the strength training program can be compared. This group will meet on the same basis as the strength training group, be supervised by an exercise specialist, and therefore receive the same amount of attention and feedback. The focus of the stretching and flexibility program will be on general stretching and flexibility exercises for the whole body. For example, participants will be led by the exercise specialist in progressive exercises that involve constant, controlled, and steady stretches for all large muscle groups. Each stretch will be held for approximately 20–30 seconds and repeated 5–10 times. Furthermore, each session will start and finish with 5 minutes of warm-up and cool-down exercises. Finally, exercise frequency will be recorded by the exercise leader in an attendance log for all participants.

Post-Testing

Immediately following the 12-week exercise interventions, participants will complete all psychological, behavioral, and physiological testing. In addition, participants will be given guidelines for continuing their exercise programs upon cessation of the organized training intervention. For example, the use of resistive equipment and objects that can be found in the home for strength training exercises will be discussed.

Data Analysis

Repeated Measures Multivariate Analyses of Variance (MANOVA) will be employed to assess between-group and within-subject changes over time in the variables of interest as a function of the exercise interventions. Univariate and simple effects analyses will be used to follow-up any significant multivariate effects indicating between group differences. If any pre-treatment differences in the variables of interest exist between groups, the pre-treatment value will be used as a covariate in subsequent analyses. Multiple regression and correlational analyses will be employed to investigate the hypothesized relationships among the dependent variables of interest. Specifically, the analyses that will be employed to test each hypothesis are detailed below.

> The reader is reminded that only selected examples within the six clusters of hypotheses and subhypotheses have been presented in this abbreviated version of the original proposal. Notice how the three forms of analysis (variance, correlation, multiple regression) are organized around the hypotheses to which they are applied. This simple arrangement allows for both economy and clarity.

Hypothesis 1, 2, 3: A series of mixed model repeated measures MANOVAs with treatment groups as the between-subjects factor and time as the within-subjects repeated measures factor will be conducted to examine changes in (a) muscular strength and balance, (b) self-efficacy and fear of falling, and (c) ADL performance and falls prevalence, as a function of the exercise interventions.

Hypothesis 4, 5: Correlational analyses will be conducted to investigate the relationships between changes in muscular strength and balance with changes in self-efficacy and fear of falling (hypotheses 4a–d), and ADLs and falling (hypotheses 5a–d).

Hypothesis 6: Hierarchical multiple regression analyses will be employed to determine whether self-efficacy mediates the relationship between muscular strength and balance changes and changes in ADL performance, fear of falling, and reported falls. First, residualized change scores for the psychological variables will be calculated by regressing the post-program scores onto the pre-program scores. Although simple pre-post change scores are acceptable for those physiological variables that possess high reliability, such strategies are less advisable for assessing change over time in psychological variables (Cohen & Cohen, 1985).

The original proposal included a list of references as well as the appendices noted below.

Appendix A. Pfeiffer Mental Status Questionnaire

Appendix B. Informed Consent and Medical Clearance Forms

Appendix C. Self-report Forms for Activities of Daily Living and the Record Form for Performance-based Test of Activities of Daily Living

Appendix D. Physical Activity History Form

Appendix E. Balance Scale

Appendix F. Fear of Falling Scale

Appendix G. Falls Efficacy Scale, Activities-specific Balance Confidence (ABC) Scale, and Strength Efficacy Scale

PROPOSAL 2

Qualitative Study

Teachers Who Make a Difference: Voices of Mexican American Students

Note to the Reader

The proposal that follows involves use of the qualitative paradigm for research. If you are unfamiliar with this kind of inquiry, a prior reading of Chapter 5 will help you to understand the particular problems that were confronted by the author, both in planning the study and in preparing the proposal document. Also, if you are familiar only with experimental and quasi-experimental designs, Chapter 5 will explain what may seem to be unorthodox or, at least, unexpected ways of handling some elements in the proposal.

As you read, it will become increasingly clear that the author's assumptions about the nature of such matters as reliability, objectivity, validity, replicability, and generalizability differ from those made by investigators using traditional models of quantitative

---

AUTHORS' NOTE: The original of this proposal, used with permission, was prepared by Belinda J. Minor under the direction of Professor Linda C. Wing, in partial fulfillment of the requirements for the Ed.D. in the Graduate School of Education at Harvard University. The ensuing doctoral dissertation was completed and accepted by the Graduate School in 1997. Dr. Belinda J. Minor, the author, now lives in California where she is a school administrator and part-time lecturer in administration and teacher development.

science. Although those differences sometimes are more apparent than real, they do exist, and they are vital to an understanding of qualitative research.

The special and uniquely valuable powers of the qualitative paradigm reside precisely in how it is different from other forms of inquiry. Changing the starting assumptions allows research questions of a very different order. In the following proposal, for example, the author lays out a plan for asking a group of senior high school students what they think (and feel) about the role played in their school careers and lives by particular teachers. The research questions served by that process demand much more than just a reconstruction of events, or the simple collection and tabulation of opinions. The author is proposing to ask the participants to reflect upon and discuss the perceived influence of teachers on their motivation, academic achievement, persistence in school, and aspirations. Questions of that kind, and the data collected to serve them, are quite unlike what one expects to encounter in a quantitative study.

To the extent, then, that qualitative researchers start with different assumptions about some aspects of inquiry, their proposals likewise will display some distinctively different characteristics. At bottom, however, there must be rules of thought and procedure that ensure that qualitative designs represent research that is systematic, transparent, rigorous, and faithful to the demands of its own paradigmatic assumptions. Proposals for qualitative research must allow a clear distinction between what is science, on the one hand, and what is no more than careful reportage, thoughtful observation, or connoisseurship, on the other. To have different rules for inquiry is not the same as having no rules for inquiry. For that reason, some investment in prior study will make your reading of the following proposal both much easier and far more profitable.

Finally, we want you to take particular note of the fact that the following document was selected for use here because it represents a relatively early stage in proposal development. Authored by a well-prepared and highly motivated doctoral candidate, and guided by the efforts of a competent and helpful graduate advisor (and thesis committee), the draft seen here was subsequently revised into a final form that received prompt approval. For our readers, however, the particular advantage in this document lies in the opportunity to look over the shoulders of the author and her advisors during the process of tweaking the proposed study toward a finished level of polish. As they did, you can work at the task of considering additions and alterations that would further strengthen this strong initial effort.

Teachers Who Make a Difference: Voices of Mexican American Students

Title for Retrieval. Granted, this is no more than the tentative title used for the proposal, but this is the place to start crafting one that will best serve its primary function—retrieval of the study by potential readers. The present title clearly identifies the two groups that will be of central interest in the study: (a) Mexican American students and (b) teachers. Furthermore, although less explicit, the wording would lead most readers to assume that the study will involve quoted material of some kind from students, as well as an effort to describe the influence of their teachers. Brief, nonpedantic, and nicely phrased to seem intriguing, the title may well serve to attract interest—but can do so only for those who read it!

It is important to remember that in most instances titles are read only when they are retrieved from some type of index, whether from a data-based computer retrieval system or the pages of a hard-copy volume. When indexing is based on keywords assigned by the author (or a review specialist), an artistic title such as the one used here can serve perfectly well. When, however, the assignment of the study to index categories is done on the basis of the words actually appearing in the title, what was aesthetically attractive may become functionally inadequate.

In an index based on title words alone, persons searching for qualitative studies of student life in secondary schools, investigations of student achievement in high schools, research employing phenomenological interviewing with adolescents, or even studies that identify variables related to teachers' influence on students will be unlikely to retrieve this study. Research method, social context, school level, and primary variables may be inferred by the present title, but they are not actually specified.

By all means, write graceful titles. Be sure, however, to use the key words that might be important in placing your study into index categories where it can be noticed and retrieved by people who will find it valuable. Our suggestions in Chapter 6 concerning the design of titles should help you to achieve that goal.

A standard title page was omitted here.

Abstract

This qualitative study will describe and analyze the perspectives of Mexican American high school seniors regarding the influence of their teachers on their learning, school careers, and lives. The theoretical framework undergirding this study is "effective teacher" research. The research design will involve "in-depth phenomenological interviews" (Seidman, 1991), focus groups, and observations involving a primary sample composed of twelve Mexican American seniors. In addition, a second sample of six will be interviewed in a separate focus group as a "member check" (Guba, 1981). The research will have implications for educators who want to make a positive difference in Latino students' learning, achievement, and persistence to high school graduation.

Theoretical Framework. In terms of the definition we have used throughout this book, a body of research is not, in itself, a "theoretical framework." The author apparently is using the term in its more general sense to indicate the source of theoretical constructs and empirically based assertions that formed the background for designing the study and developing the proposal. It is our strong preference, however, to reserve use of the term *theoretical framework* to designate a particular (and formal) theory, or network of related theories, used as the singular vantage point for defining a research problem and interpreting the results of an ensuing study. In that sense, then, a theoretical framework is a coherent set of postulates about the nature and functioning of some aspect of the world.

In light of demographic changes in the United States—distinguished by a large, young, and quickly growing Latino[1] population—coupled with a sizable and historically persistent achievement gap between Latino and Anglo students, the education of Latino students is a critical challenge facing our nation. The life challenges confronting most high school dropouts, including higher unemployment and lower wages, as well as the costs to society are well-documented (Catterall, 1985; Fine, 1991; U.S. Department of Education, 1993).

Since the publication of *A Nation at Risk* by the National Commission on Excellence in Education in 1983, national leaders have exhorted schools to eliminate the achievement gap and to educate **ALL** children to high standards. However, many teachers today question their ability to meet this goal because they locate the source of students' academic and motivational problems in their families, peers, SES, culture, or personality and believe that, in relation to those external factors, their influence is of little consequence. (Clark, 1990; Grant & Secada, 1990; Harris, 1991; Hidalgo, 1991; Ladson-Billings, 1994; Paine, 1989). Tragically, teachers' self-perceived lack of "efficacy"[2] may actually contribute to the low achievement of students (Ross, 1945).

Research is needed to help educators appreciate and understand their influence on Latino students' achievement. Despite the numerous factors that are, or seem to be, beyond their control, teachers must understand how they can use that influence in ways that make a positive difference in their students' lives.

This study will contribute to meeting this need by responding to the following research questions: What do Mexican American[3] high school seniors report as the salient qualities and behaviors of teachers who have made a difference in their lives? How do the students characterize the influence of these teachers on their motivation, learning, academic achievement, persistence in schools, and aspirations?[4] In what other ways do pivotal teachers influence students' lives?

Abstract. Not all graduate schools require an abstract in a prescribed format for dissertation proposals (abstracts invariably are required in grant proposals). Because preparing an abstract for the proposal provides practice on a vital skill (see our discussion of abstracts in Chapter 9), we are much in favor of the requirement and suggest that you develop one whether or not it must be included in the formal document.

It would be inappropriate to attempt a critique of the proposed study as presented in the abstract. Technical details can be judged only in the context of their explanation. It is possible, however, to consider the adequacy of the abstract as a mechanism for accomplishing its primary purposes: (a) concise communication of the study's purpose, method, and justification; and (b) suggestion that the findings will contain something of special interest.

In this abstract, the writing is crisp, graceful, and a model of efficiency. The first paragraph touches on purpose, method, and justification without use of jargon or technical terms likely to be unfamiliar to reviewers. Use of footnotes and citations in an abstract would not be our preference, but they do serve a clear purpose in this instance.

Three of the five paragraphs are employed to develop further the argument for the importance of the study (and they do so very effectively). That, however, leaves only the opening and closing paragraphs to deal with a formal statement of the research questions—and methodology. Necessarily, a great deal must be left out. Two of the missing bits of information are among those that most reviewers will be looking for: (a) method for selection of participants and (b) means of data analysis. Although we can anticipate that those will be specified in detail in the subsequent body of the proposal, most readers will expect some brief indication in the abstract. Even if there is a word or page limit on the length of the abstract, that small expansion should not be difficult to achieve. The opening sentences of the third paragraph, for example, might be sacrificed without serious loss as a way of providing space to identify the type of participant selection and data analysis being proposed.

Headings. Note that the use of an underscored, centered heading is not allowed in either APA or Chicago style. Unless otherwise specified by the graduate school (as may have been the case here) or funding agency, it is best to employ one of the heading styles commonly used in the discipline or professional field of the study. The correct number and type of headings in the APA format, for example, can easily be established by consulting the "Level of Headings" section of the *Publication Manual of the American Psychological Association* (2001). Headings for subsequent sections of this proposal have been revised to conform to APA guidelines. Page numbers have been omitted in the table of contents.

Table of Contents

Table of Contents. A glance through the table of contents will remind you that when they are carefully done, as is the case here, tables of contents serve not only as a way of locating particular topics within the proposal document but also as an excellent outline of how the author has developed the presentation. The 14 appendices indicate that the author has tried to keep the main text as compact and tightly focused as possible, relegating supportive materials, complex technical definitions, forms, and protocol specifications to attachments that can be consulted as needed. Because virtually all readers will want to read the material on sample selection, however, we think it inappropriate to force them to flip back and forth between appendices (F and G) and the main text. Appendices are for non-essentials only.

Introduction

Background

This qualitative study will describe and analyze the perspectives of Mexican American high school seniors regarding the influence of their public school teachers on their learning, school careers, and lives. The theoretical framework undergirding this study is "effective teacher" research. The research design will involve "in-depth phenomenological interviews" (Seidman, 1991), focus groups, and observations involving a primary sample composed of twelve Mexican American seniors. In addition, a second sample of six will be interviewed in a separate focus group as a "member check" (Guba, 1981). The research will have implications for educators who want to make a positive difference in Latino students' learning, achievement, and persistence to high school graduation.

The Problem

In light of demographic changes in the United States—distinguished by a large, young, and quickly growing Latino population—coupled with a sizable and historically persistent achievement gap between Latino and Anglo students, the education of Latino students is a critical challenge facing our nation. Latino school enrollment has doubled over the past twenty years. Historically, Latino students have attained lower standardized test scores[5] and have suffered lower high school completion rates and substantially higher dropout rates than Anglo and African Americans (supporting data are presented in Appendix A). The life challenges confronting most high school dropouts, including higher unemployment and lower wages, as well as the costs to society are well documented (Catterall, 1985; Fine, 1991; U.S. Department of Education, 1993).

Since the publication of *A Nation at Risk* by the National Commission on Excellence in Education in 1983, national leaders have exhorted schools to eliminate the achievement gap, and to educate ALL children to high standards. However, many teachers today question their ability to meet this goal because they locate the source of students' academic and motivational problems in their families, peers, SES, culture, or personality and believe that, in relation to those external factors, their influence is of little consequence (Clark, 1990; Grant & Secada, 1990; Harris, 1991; Hidalgo, 1991; Ladson-Billings, 1994; Paine, 1989). Tragically, teachers' lack of self-perceived "efficacy" may actually contribute to the low achievement of students (Ross, 1994).

Research is needed to help educators appreciate and understand their influence on Latino students' achievement and to understand how they can use that influence in ways to make a positive difference in students' lives, despite the numerous factors that are, or seem to be, beyond their control (see Appendix B).

This study will contribute to meeting this need by responding to the following research questions: What do Mexican American high school seniors report as the salient qualities and behaviors of teachers who have made a difference in their lives? How do the students characterize the influence of these teachers on their motivation, learning, academic achievement, persistence in school, and aspirations? In what other ways do pivotal teachers influence students' lives?

Introduction. At first glance, this introductory section appears to be a repeat of the abstract, a tactic that, in itself, is perfectly acceptable so long as both texts perform the functions needed at each location. Closer reading, however, will reveal that there is additional detail here, along with changes in wording and references that lead the reader to supplementary material in the appendices—all signals that we now are in the main body of the proposal.

Research Questions. Because research questions are such a central element in a proposal, we would prefer that they always be listed and numbered in serial order, rather than be embedded in the text of a paragraph. That helps the reader treat them as separate entities, each with its own demands on methodology.

Sample. Although the main exposition is yet to come, this introduction will cause most readers to post caution flags at three points: (a) How will the sample groups be selected? (b) Will 12 participants be sufficient in number to serve the relatively ambitious goals presumed by the research questions? and (c) How can six *different* participants be used as a "member check" when use of that term generally is taken to mean a confirmation of data by the original sources?

Assumptions. Finally, there is an opportunity at this point to test your own ability to think through a problem in study design. In formulating the research questions above, the author has made a very important (though tacit) assumption about her participants and about their experience of schooling in particular. We do not refer here to the usual assumptions about method (participants will volunteer, will not deliberately mislead the interviewer, etc.), but to how the students understand what happened to them in school. As with Poe's famous *Purloined Letter,* what the author has assumed is so obvious (and taken for granted) that it easily becomes invisible. Examine the research questions carefully and watch for further clues as you read—and leave our "Postscript" following the proposal until you have finished a complete reading of the document.

Literature Review

Since the turn of the century, educational researchers have conducted many investigations in an attempt to find ways to define and measure attributes of "effective" education. Nevertheless, no consensus has emerged concerning either criteria for or quantification of that elusive construct.

Three lines of inquiry have dominated such studies: "effective schools" and "effective educational practices" research (Appendix C), and investigations of "effective teachers" (Appendix D). As explained below, the study proposed here will contribute to the latter.

Cruickshank (1990), in his comprehensive review of the effective teacher research from the 1890s through the 1980s, notes two historically distinct periods of research that differ theoretically and methodologically. Research on "effective teachers" conducted prior to 1960 used subjectively derived items and vague terms created by *administrators* to describe teacher attributes and behaviors that they *presumed* to be effective. That line of study proved to be of little value.

Subsequent to the publication of the so-called *Coleman Report*[6] researchers turned their attention to identifying specific teacher behaviors present or operative when pupils were succeeding (Cruickshank, 1990). Conducted in the 1970s and 1980s, these quantitative studies used a host of new instruments to record classroom behaviors, which were then analyzed systematically. The goal of this methodology was "to determine *how and to what extent teachers perform a group of precise actions and the extent to which performing these actions is related to other desirable attendant classroom events and/or pupil learning*" (Cruickshank, 1990: 68, emphasis in original).

Cruickshank's table of 85 "promising teacher effectiveness variables" (Appendix D), culled from his review of ten other reviews of roughly 2350 studies, shows few areas of agreement about which variables are most closely associated with effective teaching. Only eight of the 85 variables associated with effective teachers appear in at least four of the reviews. These are: clarity, attending/monitoring behavior, use of more pupil participation, equitableness of pupil participation, structure, time-on-task persistence and efficiency, use of feedback, and criticism (the latter is negatively related to effectiveness).

One of the main limitations of the quantitative methods used in the above studies is that, because they are based primarily on *observations,* they cannot explain why certain teacher qualities or behaviors appear to be associated with student outcomes. *The connections* between what students perceive teachers to be doing and the students' motivation and achievement remain invisible to the observer and must be speculated by him or her.

Researchers who study "effective teachers" do not disaggregate data to examine what kind of teachers and teaching are effective for students from different ethnic or cultural backgrounds. Instead, when they differentiate among student groups, they use terms such as "low SES," "disadvantaged," or "at-risk," to refer to poor students of all ethnic backgrounds. The extent to which these or other variables are salient for effective teachers of Latino students remains unclear. The proposed study will focus on "effective teachers" for one somewhat heterogeneous (Matute-Bianchi, 1991; Suarez-Orozco & Suarez-Orozco, 1995) cultural group: Mexican Americans.

One large, quantitative study not mentioned in Cruickshank's review (possibly because it involved teacher *ineffectiveness*) was conducted by the U.S. Commission on Civil Rights (1973), and focused specifically on the behaviors of teachers with Mexican American students. The investigators used the Flanders Interaction Analysis (Amidon & Flanders, 1963) to code observations of 494 classrooms in 430 southwestern schools. They found that teachers were "failing to involve Mexican American children as active participants in the classroom to the same extent as Anglo children." The study reported:

> Combining all types of approving or accepting teacher behavior, the teachers responded positively to Anglos about 40 percent more than they did to Chicano students. Teachers also directed questions to Anglo students 21 percent more often than they directed them to Mexicans (U.S. Commission on Civil Rights, 1973: 43).

Clearly, several of the behaviors of "effective teachers" were withheld from Mexican American students. Low Mexican American achievement was associated with teachers' inequitable treatment of the students, but the researchers inferred the causal connection between teacher behavior and student outcomes. In the proposed study, I will seek to explore the substantive connections between teachers' pedagogy (broadly defined to include all teacher behaviors including curricular choices) and students' motivation, learning, achievement, and aspirations by soliciting the perspectives of the students themselves.[7]

My study will join others that focus on "effective teachers" of students from culturally diverse backgrounds. These contain two new types of research: 1) *cross-cultural quantitative studies* of the numerous factors (including teachers) associated with student motivation and achievement (e.g. Gibson & Ogbu, 1991; Stevenson & Stigler, 1992; Suarez-Orozco & Suarez-Orozco, 1995)[8], and 2) *qualitative studies* in which methods from anthropology, sociology, and psychology are used to investigate the relationship between teacher attributes and culturally diverse students' achievement.

Literature. In this section, the author has elected to shift from the brief introductory sketch of the study directly to a discussion of related literature. Some readers might find that transition somewhat abrupt. At the least, they might reasonably feel that a more complete overview of the study would be helpful in understanding how prior research clarifies and supports formulation of the questions and design of the research methodology. We think that most readers, however, will be reassured when it becomes clear that the primary concern here is with explicating the research questions and the choice of a design from within the qualitative paradigm.

The author's primary purpose in these pages is to explain the nature and Importance of the research questions. That is, as was true in much of the abstract, the points developed in this section deal with describing the educational problems faced by Mexican American students, testifying to the importance of dealing effectively with those problems, explaining what research can and cannot tell us about that process, and asserting how the proposed study will contribute.

Organizing for Explicitness. We can suggest only that appreciating how the research literature supports these arguments might be made easier if the steps of logic were made more explicit. A simple strategy commonly employed in research writing would be to organize this same content around each research question in sequence. If preferred, common-language points of logical development (like those laid out in the previous paragraph) could be used to achieve the same end. Either strategy would serve to focus the reader's attention on one point at a time, yielding both greater clarity and a more powerful cumulative effect.

Our preference for such explicit organization notwithstanding, given the generally high level of exposition here we doubt that any reader will miss the primary points being made. The author is using a body of educational research to explain both a set of social problems and how results from the proposed study might be employed in resolving those problems. In what follows, she continues the argument that her investigation might do so in a unique and potentially powerful manner.

This body of qualitative research, while still in its infancy, describes some effective teacher behaviors—including listening to students, caring, and showing respect for students' cultures—that were not noted in prior research. Indeed, they are entirely absent from the list of 85 "teacher effectiveness variables" enumerated by Cruickshank. The four main foci of the qualitative research included:

Lists of citations have been omitted in the four categories below.

1) **Studies of dropouts.** These studies reveal, primarily through interviews with dropouts, how dropouts attribute much of their loss of motivation and lack of achievement to ineffective teaching practices. The students commonly identify these practices as: lack of caring and prejudice against students of color.

2) **Studies of minority teachers' memories of influential teachers.** These studies find qualities such as: caring that students learn, connection of material to "real life," not moving on in the curriculum until the students have mastered a lesson, pushing students to think, and getting to know students and their families, to be attributes of effective teachers for students of color.

3) **Studies involving interviews and direct observation of teachers whom parents, administrators, or colleagues deem "effective."** These studies find that, among other attributes, effective teachers of students of color use "culturally relevant" pedagogy.[9]

4) **Studies involving ethnographies and/or interviews with students, individually or in focus groups.** These studies, like the studies of dropouts, point to the importance of teacher "caring" and "respect" demonstrated through instructional practices that elicit and value all students' learning.

My study will fit into this fourth category, as it attempts to understand Mexican American students' experiences with and perceptions of the influence of teachers on their motivation, achievement, persistence in school, and personal lives.

Human Attribution. We draw your attention to the phrasing in the sentence immediately above. A literal interpretation would lead the reader to believe that the study, an inanimate entity, is going to do something that only an animate entity (a human, in this case) can achieve—that is, to understand something. We confess that this has become a commonplace of construction in writing about research, and it is likely to remain so. Nevertheless, we prefer to write as though it were the investigator, or the reader of the research report, who is described as performing human functions—understanding, exploring, theorizing, and the like. Thus, we would much prefer: "My study will fit into this fourth category, and through it I hope to better understand Mexican American students' experiences. . . ." As you can see, this is not a matter of correct or incorrect grammar (which might be resolved by application of the appropriate rule), but rather a problem in the logic of word usage. We think you will serve your readers better by attributing things attempted to the researcher and not the research.

The only study that has been directed exclusively at Latino students' perspectives on the influence of teachers on their learning was conducted by Gladys Cappella Noya for her Harvard Graduate School of Education doctoral thesis, completed in 1995. Noya's sample includes eleven adolescents, Puerto Ricans and Dominicans, whom she engaged in "collective discussions, writing, and individual conversations." She found the following themes to be salient:

> . . . participants' perceptions that teachers play a determinant role in the demarcation of the depth to which students are willing to bring themselves into the classroom; students' needs to express their doubts, questions and disagreements; and young people's appreciation for relationships among students and teachers which acknowledge human dimensions beyond their respective teacher/student status, and which transcend hierarchical boundaries (vii).

Noya found the students in her sample described effective teachers as teachers who demonstrate caring and respect for students. Participants indicated that teachers demonstrate respect by listening to student comments in class without cutting them off or humiliating them with criticism. Students particularly value teachers who can create a class climate in which everyone feels free to ask questions and think aloud. These teachers convey a sense that the students are intelligent people, capable of doing challenging work and contributing to the construction of knowledge in the classroom. They enable students to bring their whole selves into classrooms and reciprocate respect.

Further review of Noya's findings has been omitted here.

Although my research questions are similar to Noya's, and there will be some similarities in our methodology, the proposed study will differ both methodologically and theoretically from Noya's in some important ways. First, Noya's methodology employs what could be considered "extreme" sampling (Patton, 1990: 169). Her sample includes only students who have been rejected from traditional American public schools and students who attend a relatively affluent school on the island of Puerto Rico. My methodology will use "intensity sampling." "Using the logic of intensity sampling, one seeks excellent or rich examples of the phenomenon of interest, but not unusual cases" (Patton, 1990: 171). My sample will be drawn from students who attend a typical, urban public high school.

Second, Noya's analysis does not consider how students perceive teachers' responses to their ethnicity or social class. I intend to look for the relevance of these qualities for Mexican American students.

Third, and finally, since Latinos are not a unitary population, and Mexican Americans may have different experiences and perspectives from Puerto Ricans, due to differences in region, history, and culture (Suarez-Orozco, 1995), when contrasted with Noya's findings it is likely that there will be differences in my participants' perceptions of teachers.

Prior Research. Here the author has introduced a key reference—one with which her advisors are likely to be familiar. Not only does she make skillful use of the results to support her proposal, but she also carefully shows how her own work will make a distinctive contribution that moves beyond the prior investigation.

Method

The purpose of this research is to describe and analyze the perspectives of Mexican American adolescents regarding the influence of their teachers. The central questions are: What do Mexican American high school seniors report as the salient qualities and behaviors of teachers who have made a difference in their lives? How do the students characterize the influence of these teachers on their motivation, learning, academic achievement, persistence in school, and aspirations? In what other ways do pivotal teachers influence students' lives?

These phenomenological[10] questions will be investigated through a qualitative research design that will consist initially of interviews, focus groups, and observations of eighteen students. Lincoln and Guba (1985: 225) caution that "the design of a naturalistic study . . . *cannot* be given in advance;

it must emerge, develop, unfold . . ." (emphasis in original). Thus, where it seems reasonable to modify the design, I will do so and will "report fully on what was done, why it was done, and what the implications are for the findings" (Patton, 1990: 62).

A qualitative approach seems to be the most reasonable approach for answering these research questions, since qualitative research is especially appropriate for studies where little empirical research exists (Patton, 1990).

Site

The site for this study will be a typical urban California public high school with a predominantly Latino student body. I have chosen a predominantly Latino school because more than two thirds of the Latino students in the United States are educated in racially isolated schools (Appendix A-1). The site is in California because California has the largest Mexican American school population. I have selected a typical, rather than an "exemplary" school because I want to learn how individual teachers can make a difference to Latino students in average schools, under typical circumstances. I chose a high school as the site for the research because I am interested in interviewing seniors who are on the brink of "making it" through the system. Their insights concerning the influence of teachers on their motivation, learning, and achievement may have implications for supporting other students in ways that will lead to the same end result.

One site is sufficient for this study because a number of schools "feed" into the high school and the district has a voluntary school choice program. Thus, I can select students who previously have been exposed to a wide range of teachers and pedagogy while conducting my study at this site.

Repetition. Is it really necessary to repeat the research questions here? Probably not, but we think it a good idea, if only because some reviewers turn directly to the methods section.

Flexibility. Is the second paragraph note about necessary flexibility appropriate? Indeed it is in a qualitative study, though with a novice researcher the license to make such adjustments should be used with great caution. The apt quotation from Patton notwithstanding, in most dissertation proposals it is customary to make it explicit that changes will be made with the advice and consent of the primary advisor, or, for major changes, all members of the thesis committee.

Why Qualitative? The claim for appropriateness of the qualitative paradigm in the third paragraph might better be made by noting that only a qualitative study would be

(Continued)

(Continued)

appropriate for answering the questions posed. The argument that qualitative studies are best where little is known can too easily lead to the assumption that they are valuable only for "exploratory research." That view weakens the position of qualitative designs as viable alternatives for any question that they fit—irrespective of the status of existing knowledge.

Defining Terms. Based on our experience with many thesis committees, we can predict that if a distinction is made such as "typical" rather than "exemplary," someone will ask for a definition of what the author considers to be "typical." It might be better to deal with that here, rather than in an extemporaneous exchange with advisors.

Variability. Finally, we think that the author intends to assert that one school is enough because the substantial variety of teaching present in feeder schools will produce a participant sample within which a wide range of educational experiences are represented (that is an important condition for the study). Although the argument seems plausible enough, we would prefer that it be made even more explicit. It is important to avoid any implication that diversity of students somehow makes this "typical" school site more representative of urban high schools (a caution the author does make in a subsequent section of the proposal). Qualitative studies of this type do not have strong claims to external validity (transferability of results) as a consequence of sampling (although they may address the matter of generalizability through other means).

Appendix. As illustration for our earlier criticism of placing the explanation of sample selection in the appendices, you must now turn periodically to F and G while reading the following (and do so if you are to understand our comments on that section).

Sample

The primary sample will be twelve Mexican American high school seniors—the intensive focus of the study; a second sample will be six additional students who will participate in a focus group serving as a "member check" (Guba, 1981) at the end of the study. I will purposefully select participants from a pool of volunteers in order to include a range of diverse characteristics including: males and females, students from various academic programs, students of varying grade point average, and first and second generation Mexican Americans (see Appendix F). The primary sample of twelve students will each be interviewed three times. From these twelve, two will be selected for observation and six for participation in ongoing focus group interviews (see Appendix G).

I will interview twelve students in order to select a cross-section of students who have been exposed to diverse school and life experiences. While this is not a study specifically about gender, track, grades, or generation, prior research indicates that students' experiences and perceptions vary systematically along those dimensions (Gilligan, 1982; Oakes, 1995; Suarez-Orozco, 1995). Thus, those factors will serve as "sensitizing concepts" (Patton, 1991: 391) in my data analysis. By including a cross-section of students along these dimensions, I hope to discover perceptions of effective teaching that are either widely shared or that differ categorically (Patton, 1991: 172). By selecting six other students for focus groups, using these same criteria, I can test my findings from the interviews and seek further information.

Since Fine (1991) suggests that many students of color "make it" despite rather than because of teachers and the system, this sample of "achievers" will be asked to locate the influence of their teachers in particular. Acknowledging that other people and institutions impact their lives, I will be cautious and not presuppose a causal relationship between teacher influence and student success in school.

How Many? Unlike the situation in many quantitative studies, there is no reliable mechanism for definitively answering the question "How many participants are needed?" In the end, establishing that number will be a matter of judgment. There are, nevertheless, some simple rules of thumb (discussed below) that apply to conditions like those in the proposed study.

Complicating Conditions. As the author skillfully argues, one of the desired conditions is representation of particular characteristics in the sample group. Her citation of prior research to sustain that point is exactly correct. From a reading of Appendix F, however, we also know that maintaining a degree of balance among those characteristics is an additional requisite (for example, a final sample with only one out of twelve student participants to represent the "second generation" condition clearly would be unacceptable). This is further complicated by the fact that the author obviously wants to maintain an even gender balance (or close to it) as a priority consideration.

A Rule of Thumb. The applicable generalization here is that as the number of desired characteristics increases, and as you add provisions about maintaining a minimum balance of characteristics in the total sample, the number of participants also must be increased. Finding participants with exactly the right characteristics simply is too difficult; with a small, all-volunteer initial pool to draw from, it may be impossible. Given the specifications as proposed, we think it is highly unlikely that the desired sample can be obtained. Either the total number of participants must be increased (at least twofold would be our rough guess), or the number of characteristics decreased (or, ideally, both might be adjusted).

Building the Sample. Although it preempts the later discussion of selection procedures, it may be helpful here to insert our advice about how the desired sample (size and nature) might be obtained. We would urge the author to proceed with a pilot

(Continued)

(Continued)

recruitment at the study site. Using only the three original criteria for preliminary selection (senior class status, Mexican American, volunteer), the initial group of recruits thus generated can then be inspected for representation and balance of other characteristics. Given a pool of sufficient size, it may be possible to make necessary adjustments by purposeful selection of candidates for the final (now much larger) participant sample. If that does not produce a sample with the desired characteristics, the option of a further "targeted" recruitment effort may be available. Beyond those strategies lie the less desirable alternatives of simply working with whatever the call for volunteers has provided or of moving on to a different site for another try.

Access and Sample Selection

I have negotiated access to the district and school where I plan to conduct the study between January and March, 1996. In January, I will solicit participants by posting announcements and by staffing a sign-up table where students can meet me, ask questions, and fill out questionnaires. I will perform the contact work myself because, "building the interviewing relationship begins the moment the potential participant hears of the study" (Seidman, 1991: 37). I will select the participants from the pool of students who complete the demographic information questionnaire (copy in Appendix H) and whose parents co-sign the "informed consent" agreement (Kimmel, 1988: 67–76).

Because Latino students are not dispersed equally among the different academic tracks, and because the sample size is so small, the sample will not be "representative" of the school population, nor is it intended to be (Patton, 1990: 185). Using the logic of qualitative inquiry, I am more interested in understanding, in depth, the experiences and perspectives of diverse individuals than being able to generalize those experiences to a larger population. To be considered for participation, students must be seniors of Mexican origin who have attended schools in the U.S. for at least five years (so that they have experience with teachers in at least two U.S. schools).

Groundwork. It is clear that the author has done her homework, completing preliminary negotiations for access to the site and drafting informed consent forms and other study documents. All of that is powerfully reassuring to advisors. Again, readers are cautioned against any assumption that the sample is intended to represent the population at the school site or the student populations in urban schools more broadly. Planning to handle the recruitment tasks herself (on site) is particularly important, for reasons that she makes clear.

Appendix I. It is essential that Appendix I, "Interview Guide," be read in conjunction with the following. Again, the decision to relegate an item to the appendix has worked to the disadvantage of the reader.

Interviews

I will use in-depth, phenomenological interviewing with twelve Mexican American seniors as my primary method of data collection. Consistent with Seidman's (1991) model, each participant will be interviewed three times. Interviewing is important because, as Seidman (1991: 4) asserts, "If the searcher's goal . . . is to understand the meaning people involved in education make of their experience, then interviewing people provides a necessary, if not always completely sufficient, avenue of inquiry."

In order to provide context for understanding the participant's perspective, the first interview in Seidman's model (1991: 11) focuses on the person's life history. Using an "interview guide approach" (Patton, 1991: 288) consisting of open-ended questions (see Appendix I), I will focus on the participant's school life history, particularly on experiences with and perceptions of past teachers. The second interview focuses on the "concrete details of the participant's present experience in the topic area of the study . . . upon which their opinions may be built" (Seidman, 1991: 12). In this interview, I will ask students to describe their lives and the impact of current teachers. "In the third interview, participants are asked to reflect on the meaning of their experience" (Seidman, 1991: 12). In this interview, I will ask students to make connections between pivotal K-12 teachers and the students' motivation, achievement, persistence in school, and aspirations. I will also ask students to reflect on how teachers can support students in ways that will help them achieve at high levels in school and beyond.

Whenever possible, I will conduct interviews at least one day, but not more than one week, apart with each participant, in order to give them a chance to reflect on their experience and to build rapport over time. I will tape record each interview. A bilingual transcriber will assist me in transcribing half of the interviews. I will transcribe the other half and will listen to all the tapes as I proofread and annotate the transcripts. Immediately following each interview, I will record field notes (Patton, 1991: 239).

Seidman's Protocol. Although phenomenological interviewing in the format provided by Seidman is a complex process that requires considerable interpersonal skill, careful attention to detail, and a thorough understanding of purpose (as well as a period of guided practice), the brief description given here is entirely adequate. The Seidman book provides a definitive source, and most advisors are likely to be familiar with the methodology involved. Where that is not the case, prior conversations and provision of reference materials can serve as preparation.

Transcribing. Even though the author is bilingual (a point made clear in a later section of the proposal, along with her intention to give each participant the choice of language for their interviews), the use of a bilingual transcriber for a portion of that difficult task is a significant addition. Providing for an external check on her own technical skill reflects the author's concern for maintaining a high standard of accuracy in the data—again a signal her advisors will interpret as indicative of sound preparation and readiness for the long task ahead.

Cuing Participants. As some of you may already have concluded from a reading of Appendix I, however, there may be some reason for concern about the author's plans for collecting data from interviews. An interviewer's choice of words can signal the interviewee as to what sort of responses are anticipated. Given the probability that in the context of a friendly, nonthreatening setting, most participants already will have some inclination to say things that please the investigator, such signals can serve as a powerful source of data contamination. That being the case, advisors will be looking for clear assurance that the author intends (initially, at least) to avoid giving her participants any indication that teachers should be identified as influential figures in their lives.

Whether teachers are called "memorable," "influential," "pivotal," "effective," or simply "good," those descriptors constitute a clear signal about what is expected when students talk about teachers. Even a seemingly neutral word like "memorable" can give more direction than intended. Stimulating participants to reflect upon and talk about the behaviors of teachers whom they perceive as having fallen into any of those positive categories will be an appropriate (even necessary) part of the proposed study. *It will be so, however, only after students have, of their own unprompted volition, indicated that teachers were among the people "who made a difference" for them.* In other words, it is essential to first establish the "whether" of things, before moving on to the "how"!

It is easy to understand how the author, committed to the vital significance of the research questions, steeped in the supportive literature, and eager to move ahead to conducting the study, might have overlooked the possibility of cuing her participants. Happily, the correction is easily made—and there is clear evidence in the proposal that she knows precisely how to do it. To paraphrase points already made in her presentation: Don't rush, lead with comfortable, nonprescriptive questions that invite students to share their stories, listen intently and show interest in the participants' perceptions, be prepared to use follow-up probe questions when needed to clarify or extend a response, and move the conversation only gradually (perhaps in the later interviews) toward the explicit questions that address such matters as teacher effectiveness. The reader should note that a modest pilot trial of the interview guide might have sensitized the author to this matter before presentation of this preliminary draft for review.

Observations

I plan to observe at least two of the participants from when they rise in the morning until they prepare to retire in the evening. By spending a day with the students I expect to get a sense of how the students negotiate their home, school, and work cultures, which may prove helpful in contextualizing the information and ideas that they share about the relative influence of teachers in their lives.[11]

During the observations, I will take field notes (see Appendix J)[12] and will type them soon after, adding reflections. I will use the data from the observations to inform my follow-up questions in the interviews and to generate tentative hypotheses.[13]

> *Rationale.* We are not convinced that the proposed observations will contribute significantly to the data collected from other sources. Furthermore, given what is provided here (and elsewhere in the proposal) concerning observations, we are not sure the author herself has fully come to grips with the how and why of this part of the research plan. With an already demanding schedule of data collection, and the arduous tasks of transcription and analysis beyond that, unless more persuasive arguments can be advanced, we are inclined to advise dropping this component of the proposed study.

Focus Groups

"Groups are not just a convenient way to accumulate the individual knowledge of their members. They give rise synergistically to insights and solutions that would not come about without them" (Brown et al., 1989:40). Both Hidalgo (1991) and Noya (1995) conducted focus groups with Latino adolescents and found the sessions fruitful. In this study, I will facilitate two groups. One will consist of six of the twelve primary interviewees. We will meet as a group three times for fifty minutes to discuss the attributes of influential teachers. The six participants in the member check group will meet once for ninety minutes, near the end of the study, to critique the developing analysis.

> *Citing Sources.* The rationale for obtaining focus group data makes good sense in the context of this study (although for some readers a much more thorough definition of the term might have been required to make that apparent). That conclusion is furthered by documentation of the fact that the technique has previously proved to be a helpful supplement to interviews. Because the formation and facilitation of focus groups requires considerable knowledge, skill, and planning, we would normally expect to see both citations to standard works on the methodology (see Chapter 5 for our suggestions in that regard) and indication of successful pilot work by the author.

Trustworthiness

I have made several choices to increase "trustworthiness" (Lincoln & Guba, 1985) and to minimize common threats to validity. First, I have selected a school where neither students nor teachers know me to minimize threats to validity arising from personal bias towards participants or individuals named by them. Second, because I am bilingual, I will interview students in the language they prefer. The bilingual transcriber will assist me in checking the accuracy of both direct quotes and translations. Third, the final focus group will provide a member check on the analysis. Fourth, I will discuss the possible influence that my own life experiences—for example, as a former teacher and administrator who worked for ten years in southern California with many Mexican American students—may have on my perceptions and thinking. Finally, by conducting multiple interviews over time, as well as by using multiple data collection strategies, I will minimize the chances that the findings will be based on idiosyncratic data.

Getting It Right. It is impossible to define here the complex of criteria and procedures that qualitative researchers assign to the word "trustworthiness." Advisors will know what is intended, however, and a full discussion is not required in the proposal. At bottom, trustworthiness has to do with the quality (goodness) of qualitative research. Thus, in practical terms, it has to do with establishing and meeting criteria that will lead readers of the study report to conclude that the investigator "got it right"! Put another way, the term denotes a revision of traditional criteria such as reliability, internal and external validity, and objectivity, to make them more appropriate to the assumptions made in the qualitative paradigm.

What the author presents here are procedures that will serve the ends of trustworthiness, though we would have preferred that the several components of that construct be made explicit (the usual list of criteria includes credibility, transferability, dependability, and confirmability), and that proposed strategies for meeting each then be discussed in turn.

Coming Clean. We were particularly pleased to note that the author plans to write a thorough explication of the nature and potential influence of her own life experiences as they might be expected to shape the research process. Of course, that discussion will be necessary for an informed appreciation of the study's conclusions.

Confirming Data. We continue to be concerned about the proposed use of a secondary sample of students, who were not interviewed, for the purpose of conducting a "member check." The word "member" here ordinarily designates a full participant who has been selected to read through transcripts to ensure accuracy of what has been recorded, or to review products of data analysis such as thematic statements or individual profiles. The purpose is to detect any substantial discrepancies between the perceptions of the participants (whether as individuals or as members of a group) and the investigator's reconstructions and representations. If that is the intention, then it is the participants in the primary sample who will have to do the checking.

Ethical Concerns

Due to the nature of the research questions, the confidentiality provided through carefully maintained participant anonymity, and my status as a graduate student with no connections to the teachers or school administration, there is minimal threat to the well-being of the students in the study as a result of their participation.

My stance of "empathetic neutrality"[14] should minimize the threat of "interviewing as exploitation"—a process that turns others into subjects so that their words can be appropriated for the benefit of the researcher" (Seidman 1991:7) that presents ethical problems in some studies.

Protecting Participants. The author's careful reassurances notwithstanding, hard experiences have taught us that there *always* is the lurking possibility of harm to participants in a naturalistic field study. Anonymity never can be guaranteed, if only because it is not entirely under the researcher's control. In a typical school, everyone will know who the participants were and exactly what was asked in the interviews and discussed in the focus groups. Furthermore, if member checking is done, whatever is checked has the potential to become public and, correctly or incorrectly, will be assigned to every participant.

We think the author's arguments about meeting her ethical responsibilities are reasonable. Certainly, she has given us good reason to believe that she has given serious consideration to the problem and will protect her student collaborators in every possible way. We want to warn our readers, however, that ethical concerns are never simple. Careful forethought in planning and watchful diligence in execution are responsibilities that come with the decision to do qualitative research.

Limitations

As with all exploratory research, the findings of this study will be tentative. It is important to note: 1) the sample size and procedures for participant selection, while appropriate for a qualitative study, will not support generalization to a larger population of Mexican Americans, and 2) the relationship between students' perceptions of effective teachers and empirical measures of "teacher effectiveness" has yet to be defined. The exploration of that relationship is beyond the scope of this study.

Data Analysis

Because this is an exploratory study, I will begin coding data from interview transcripts, field notes, and focus group transcripts using coding categories (Patton, 1991:402) suggested by the literature (Appendix K), but I will actively seek patterns of data suggesting new categories as well. Using data organized and displayed in that fashion, I will return first to an analysis of interviews from individual participants, as separate cases, and only then consider the wider matter of cross-case analysis. Such a sequence of analytic steps would conform broadly to the suggestions of Glaser and Strauss (1967), who have argued that an understanding of individual cases (before they are aggregated in any way) is the best guarantor for theoretical assertions that are grounded in specific contexts and real-world patterns.

Next, because "a phenomenologist assumes a commonality in those human experiences and must use rigorously the method of bracketing to search for those commonalities" (Eichelberger, 1989: 6), I will read across interviews, noting similarities and differences. I will then use pattern coding (Miles & Huberman, 1994) to identify common themes. While I will examine the entire sequence of responses about teachers by each participant, I will also distinguish between those categories generated by the students themselves and those discussed in response to probes. For example, if the students mention nothing about the relevance of the instructional materials to their lives, I might inquire about the connection, but will indicate that I needed to probe to elicit that information.

Finally, I will prepare the data analysis by briefly describing each of the twelve participants, and, by using quotes from their interviews, will illustrate common themes as well as atypical responses. Transcripts from the focus groups will be utilized primarily as a check for confirmation of the salience of themes across the sample, although I will include in my thesis excerpts of dialogue that illustrate important points of consensus or disagreement. The observations will be used both to describe two of the participants more fully and to fill in descriptions of the school.

Keeping Honest Books. Again, the procedures indicated here (coding, category development and sorting of data, identification of patterns, construction of thematic statements, and writing grounded descriptions of participants) are all appropriate to the research task. We can add several suggestions that might strengthen the plan for analysis.

First, it would be wise to consider use of one of the software programs for computer-based analysis. This study will generate such an enormous volume of data that hard-copy manipulation will be difficult. Second, it will be important to maintain a record of the exact source for each category. Not only are advisors often curious about such origins, but the final report will require writing about category development long after immediate memory has dimmed. Third, if at all possible, do some pilot trials of the proposed analysis using interview data obtained from a source outside the study site (an appendix showing that this has been accomplished successfully always is reassuring to advisors). Finally, make it an absolute and inviolate rule not to ignore or discard data that run contrary to initial expectations, or that appear not to fit comfortably into emerging categories for analysis.

Conclusion

This country cannot afford, ethically or economically, to fail to educate another generation of Mexican American students. While many factors outside the control of teachers may contribute to underachievement of Latinos in the United States, evidence abounds to confirm that teachers can make a difference, although many teachers question their ability to do so. In the proposed study I will explore Mexican American high school seniors' perspectives on what constitutes effective teaching and their perspectives on the influence that teachers have on their learning, school careers, and lives. It is my hope that teachers who are actively seeking ways to support the achievement of their Mexican American students will find, in reading this thesis, useful ideas that will support them in their important work. It is equally my hope that teachers who have doubted their ability to make a difference in their Mexican American students' achievement may reconsider their thinking and pedagogy upon contemplating this research.

The Bad News and the Good News. We were delighted to find the author still enthusiastic and convinced of the potential utility in her proposed study. Sad to report, hardly anyone reads theses and dissertations—except other graduate students! The encouraging flip side of that rule, however, is that publications based on theses and dissertations can be aimed at any audience the researcher desires to reach—other scholars, professors, policy makers, and practitioners at any level of education—including, of course, the teachers specified above. All that is required is preparation of a sound manuscript based on the dissertation (or some part thereof) and selection of an outlet that reaches the intended readers. Good proposals lead to good studies, and when appropriately translated and disseminated, the findings from good studies can make a positive difference in any professional field.

A Postscript to this Proposal

The Rest of the Story. This proposal, already well along in its development, was reviewed by a committee of thesis advisors, the author was given feedback and suggestions for improvements, the revisions were executed, and with the full support of the advisory committee the study was approved by the Graduate School. In that process, some, though not all, of the points in our critique above found their way into the final proposal. Notably, the observation component suggested in the first draft was dropped (as was the "member check" procedure), the sample size was expanded (to 34 participants), the specifications for representation and balance of characteristics were maintained in somewhat less demanding form, and discussion of teacher influence was placed at the end of the interview sequence.

As for our earlier note about the author's assumptions concerning the role played by teachers in the cosmology of influences that shape student school experience, by this point most readers will have detected that her unspoken but pervasive expectation was that some (if not most) of her participants would nominate (and describe) particular teachers as pivotal individuals in their school careers. From the outset, we found ourselves asking, "What if none (or very few) of her participants nominate teachers as influential in making it through to graduation?" At the least, that outcome would limit a search for the attributes of effective teachers—although it might well open the door to other important insights into students' perceptions of what and who were helpful in their school careers.

Given her own life experiences, the personal values that are evident in the proposal, and the nature of her continuing professional development, the expectation that teachers might be salient was perfectly reasonable. Indeed, in the case of several of her participants, the assumption was verified by direct and persuasive testimony. Virtually every one of the study participants indicated that they had encountered one or more teachers whom they considered to have been excellent. Furthermore, they unanimously affirmed the proposition that the quality of teaching makes an important difference in school.

Nevertheless, the main direction of findings ran contrary to some of the assumptions apparent in the first draft of the author's proposal. *The majority of students perceived their education and lives to have been shaped more profoundly by their interactions with parents, friends, and peers than with teachers.* That her study was sensitive to just such a possibility, and that she was ready to make that unexpected and (thereby) particularly valuable finding the center of an articulate and powerful report, *were the consequence of thorough preparation, strong advisement, and development of a sound research plan through successive revisions of her proposal.* The lesson to be taken here is that effort expended on construction of a carefully considered proposal for qualitative research may well be as important as conduct of the study itself.

References

The reference list has been omitted here.

Appendices

Except F, G, and I below, all appendices have been omitted here.

Appendix F: Sample Selection

I intend to select a cross-section of students for participation in both samples, the primary sample of 12 and the "member check" group. Ideally, the sample will assume a form approximately like this:

Composition of Primary Sample

All must be seniors of Mexican decent who have attended US schools for at least five years.

1 male in honors classes

1 female in honors classes

1 male in "regular" classes who is getting good grades (A's & B's)

1 male in "regular" classes who is getting average grades (C's & F's)

1 female in "regular" classes who is getting good grades (A's & B's)

1 female in "regular" classes who is getting average grades (C's & F's)

1 male in "basic" or special education classes

1 female in "basic" or special education classes

1 male first-generation Mexican American

1 female first-generation Mexican American

1 male second-generation Mexican American

1 female second-generation Mexican American

Total = 12

Composition of Secondary Sample

All must be seniors of Mexican decent who have attended US schools for at least five years.

1 male first-generation Mexican American

1 female first-generation Mexican American

1 male second-generation Mexican American

1 female second-generation Mexican American

1 male or female honors student

1 male or female student in "basic" or special education classes

Total = 6

Appendix G: Sample and Methodology

Table 1

Participant Number	*Activity*
1	3 in-depth, individual interviews + one day observation
2	3 in-depth, individual interviews + one day observation
3	3 in-depth, individual interviews + 3 focus group interviews
4	3 in-depth, individual interviews + 3 focus group interviews
5	3 in-depth, individual interviews + 3 focus group interviews
6	3 in-depth, individual interviews + 3 focus group interviews
7	3 in-depth, individual interviews + 3 focus group interviews
8	3 in-depth, individual interviews + 3 focus group interviews
	Note: participants 3–8 will all be in the same focus group that will meet three times.
9	3 in-depth, individual interviews
10	3 in-depth, individual interviews
11	3 in-depth, individual interviews
12	3 in-depth, individual interviews
13	"member check" focus group interview
14	"member check" focus group interview
15	"member check" focus group interview
16	"member check" focus group interview
17	"member check" focus group interview
18	"member check" focus group interview
	Note: participants 13–18 will all be in the same focus group that will meet once for ninety minutes.

Appendix I: Interview Guide

Interview #1: Personal Educational History

I'm interested in hearing about your experiences in school, particularly with memorable teachers, starting back as far as you can remember.

(I will listen for and possibly probe: when the participants began school, where they attended, what level groups they were assigned to, memorable experiences, the kinds of grades they got, their attendance patterns, any changes in their experiences or feelings towards school or teachers over time, and descriptions of memorable teachers—especially good ones—and their influence; and anything else that the students share about their school history.)

Interview #2: Concrete Details of Present Experience

A) **Last time, you told me about your past experiences in school, especially with teachers, and I have found them very powerful. Today, I'm wondering if you could tell me about your life now—both inside and outside school.**

B) **What do you see as the influence of the teachers you now have on your motivation, learning, achievement, and plans for the future?**

(In addition to listening to the students' responses to these questions, I may ask them to describe the best teacher whom they have at present.)

Interview #3: Reflections on the Meaning of Experience

A) **Now that you've told me a lot about your school experiences, I'm wondering if you could help me see some of the long-term and short-term influences that you think specific teachers have had on your life.**

(Here I may follow up on stories from the previous interviews)

B) **What advice would you give to a teacher who really wants to make sure that all of the students in his/her class are both motivated to learn and able to succeed in school?**

(Here I will listen for students' connections between "effective" and "ineffective" teachers and their own motivation, achievement, persistence in school, and aspirations. If they have not already done so, I will ask them to describe to me the qualities and attributes of an effective teacher, either based on someone they know, or based on how they imagine one could be.)

Notes

1. In this proposal, the terms "Latino" and "Hispanic" refer interchangeably to people, or their descendants, who hail from diverse Spanish-speaking regions such as Mexico, Cuba, Puerto Rico, the Dominican Republic, and parts of Central and South America (Suarez-Orozco & Suarez-Orozco, 1995).

2. Teacher efficacy measures the extent to which teachers believe their efforts will have a positive effect on student achievement.

3. I will focus the study on Mexican Americans, rather than on all Latinos, as Mexican Americans constitute more than half of the Latino population. I have chosen to use the term Mexican American, rather than Chicano, Mexicano, or any other terms because many people prefer it (Nieto, 1992).

4. These terms ("motivation," etc.) will be defined by the students.

5. The validity of these tests for Latino students is questionable (Valencia, 1991).

6. U.S. Department of Health, Education, and Welfare (1966).

7. The relationship between students' perceptions of effective teachers and empirical measures of teacher effectiveness is unknown and is beyond the scope of this study.

8. These investigations focus primarily on "macrosystem factor," (LeCompte & Dworkin, 1991: 57) such as family variables and school structure (see Appendix B), rather than teacher effectiveness *per se,* thus they will not be discussed further here.

9. Definitions of "culturally relevant pedagogy" differ, but often include involving students' families and connecting subject matter with students' lives and experience in meaningful ways (see Appendix E-1).

10. Phenomenological inquiry focuses on the question: "What is the structure and essence of experience of this phenomenon for these people?" (Patton, 1990: 69).

11. My experience visiting homes of Mexican Americans while working for Migrant Education leads me to believe I will not have difficulty finding participants who are willing to bring me into their homes.

12. I will assign pseudonyms to the school and to teachers mentioned by participants.

13. The focus of this study is on *students' perceptions;* therefore, I will not observe teachers.

14. Empathetic neutrality: "Empathy . . . is a [caring] stance towards the people one encounters, while neutrality is a stance toward the findings" (Patton, 1990: 58).

PROPOSAL 3

Online Electronic Survey Study

The Nutrition Care Process of Front-Line Pediatric HIV Providers

The proposal is divided into three sections: *(a) Introduction, (b) Related Literature,* and *(c) Methods.* We have removed the author's designation of these divisions as chapters. Where page counts are given here, reference is to pagination in the original full manuscript copy of the proposal.

Introduction

That there may be an association between the adequacy of human nutrition and the effectiveness of immune functions has long been suspected. For example, centuries of empirical observations made during periods of famine have noted the reduction in individual resistance to epidemic diseases. More specific elucidation of the nature of that relationship, however, came only in the 20th Century with the development of nutritional immunology as an area of disciplinary inquiry.

The foundation for dramatic advances in our understanding of the role of nutrition in resistance to disease was established in 1968. In that year,

AUTHORS' NOTE: This proposal, used with permission, was prepared in 2003 by Pamela Rothpletz-Puglia as part of the dissertation research requirement in the Nutrition Education program at Teachers College, Columbia University under the direction of Isobel Contento. The dissertation was completed and accepted by the university. Dr. Rothpletz-Puglia currently is Director of Nutrition and Wellness at the Francois-Xavier Bagnoud Center of the University of Medicine and Dentistry of New Jersey.

members of an expert committee of the World Health Organization (WHO) produced a monograph entitled *Interactions of Nutrition and Infection.* In that document the authors asserted a definitive and straight-forward conclusion. Review of the evidence had indicated, unambiguously, that malnutrition and infection can be mutually aggravating and produce more serious consequences for the patient than either condition acting independently (Scrimshaw, Taylor, & Gordon, 1968).

In short, not only is there evidence of a clear relationship between nutrition and immune function, but the paths of that influence may run in either or both directions. As the following section will illustrate, subsequent research was to demonstrate the prescience of that earlier assertion by the WHO committee. In a wide range of studies, nutritional deficiencies and pathogenesis were found to be locked into deadly cycles of mutual influence.

Here, 6 pages of research review and the accompanying 45 citations have been deleted from the specimen proposal. In those pages the author lays out the preliminary basis for the question that will be at the center of her study. She begins with research that establishes the primary landmarks in nutritional immunology by explaining: (a) why the functions of immunity depend on nutrition, and (b) which nutrients have been identified as critical in that association. Then she shifts the focus to a particular instance of the nutrition/immunity interaction—the disease created by infection with human immunodeficiency virus (HIV). She also cites research that has tracked the nature of the cyclic interaction between disease and deficiency. Her introduction of the illustrative figure below makes this relationship perfectly transparent.

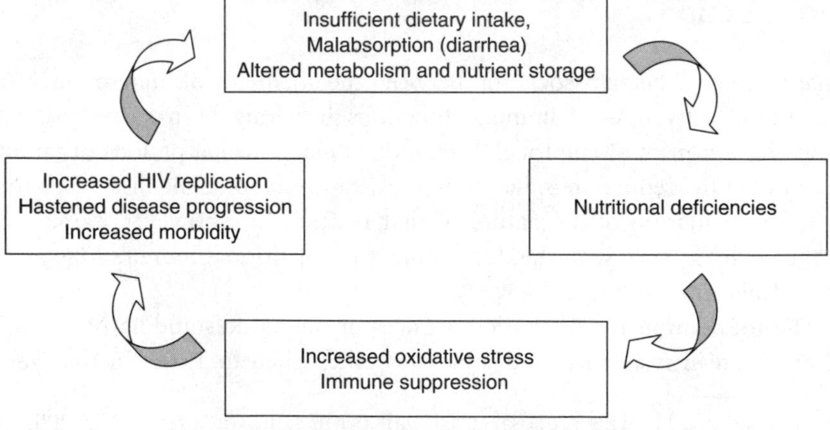

Figure 1 Cycle of Micronutrient Deficiency and HIV Pathogenesis

SOURCE: From Semba, R., & Tang, A., Micronutrient and pathogenesis of human immunodeficiency virus infection, in *British Journal of Nutrition, 81*, 181–189. Copyright © 1999, Cambridge University Press. Reprinted with permission.

Finally, the introduction moves on to explain the use of aggressive antiviral therapies for HIV infection—and the aversive nutritional consequences that they bring for the patient. The focus again shifts here to give central attention to medical management of therapeutic regimens in cases of pediatric HIV, and the growing need for specific guidelines covering the nutrition care process for children. Having established the trail of research evidence to that point, the author now turns to the general purpose and significance of the proposed study. As you read the next paragraph, note that the author (Rothpletz-Puglia) uses her own previously published research to help make a case for the study.

The emergence of new and increasingly complex problems in the long-term management of HIV therapy, and particularly the increased incidence of overweight, require an ongoing evaluation of the current pediatric HIV nutrition care process. Moreover, strategies for handling these emerging nutritional problems need to be incorporated into practice guidelines. The American Dietetic Association (ADA) is the professional group that develops the most widely used dietary guidelines or medical nutrition therapy (MNT) protocol for pediatric HIV/AIDS (Rothpletz-Puglia, Heller, & Morris, 2000). The ADA is in the process of updating the MNT guidelines with an emphasis on evidence-based recommendations. However, the scientific evidence may be years away or may never be pediatric-specific.

The need for nutrition management strategies has now been asserted and the concurrent ADA effort to establish guidelines has been introduced to the reader. With that groundwork in place, the author is ready to foreshadow the purpose of the study. Notice in the second paragraph how she neatly includes identification of an appropriate procedure for collecting data.

Meanwhile, a perspective that is conspicuously missing in the literature is that of front-line nutrition providers themselves. For example, what are nutrition providers doing in practice, despite not having evidence-based guidelines? The collective knowledge of front-line providers can provide information for understanding the process and characteristics of pediatric HIV nutrition practice in the era of highly active antiretroviral therapy (HAART).

In this study the analysis of pediatric HIV providers' reported nutrition care strategies will be reviewed and analyzed using ADA's Nutrition Care

Process Model as a framework (Lacey & Pritchett, 2003). In addition, the critical incident technique, an inductive job analysis strategy, will be used to elicit patient-provider characteristics.

> Even though not yet stated in formal terms, we now know the primary purpose of the study (to review and analyze provider's strategies), in what form the data will be collected (written reports from providers), from whom data will be collected (Pediatric HIV providers), the theoretical framework for the analysis (the ADA Nutrition Care Process Model), and the methodologies proposed for gathering data about patient-provider characteristics). That is a great deal of information to pack into two sentences. Such laudable economy could be furthered by introducing acronyms for items that will be repeated frequently in the following text, as, for example, by using NCPM for the ADA model, and CIT for the critical incident technique.

This Nutrition Care Process Model was developed to provide a framework for describing the characteristics of nutritional care and the series of actions nutritionists take to achieve a particular outcome. The Nutrition Care Process Model (see Figure 2) is a visual representation that reflects key concepts of each step of the nutrition care process The model is intended to depict the relationship among all the components as they overlap, interact, and move in a dynamic manner to provide the best quality nutrition care possible. As you can see, the four steps in the nutrition care process are nutrition assessment, nutrition diagnosis, nutrition intervention, and nutrition monitoring and evaluation. The Nutrition Care Process Model also identifies other factors that influence the quality of nutrition care, including the relationship between the client and the nutrition provider, the strengths that the nutrition professional brings to the process, and the environmental factors that influence the process.

The purpose of the Nutrition Care Process Model is to describe the spectrum of nutrition care practices that can be provided by nutrition professionals (Lacey & Pritchett, 2003). The model does have the patient-provider relationship depicted as the central core of this process, and provider characteristics such as knowledge, skills, and critical thinking are shown in an outer ring of this model, but these aspects of the model are not characterized further than having been labeled. This researcher had an interest in understanding provider characteristics in relation to quality nutrition care, and the critical incident technique was identified as a strategy for obtaining data that would make that possible.

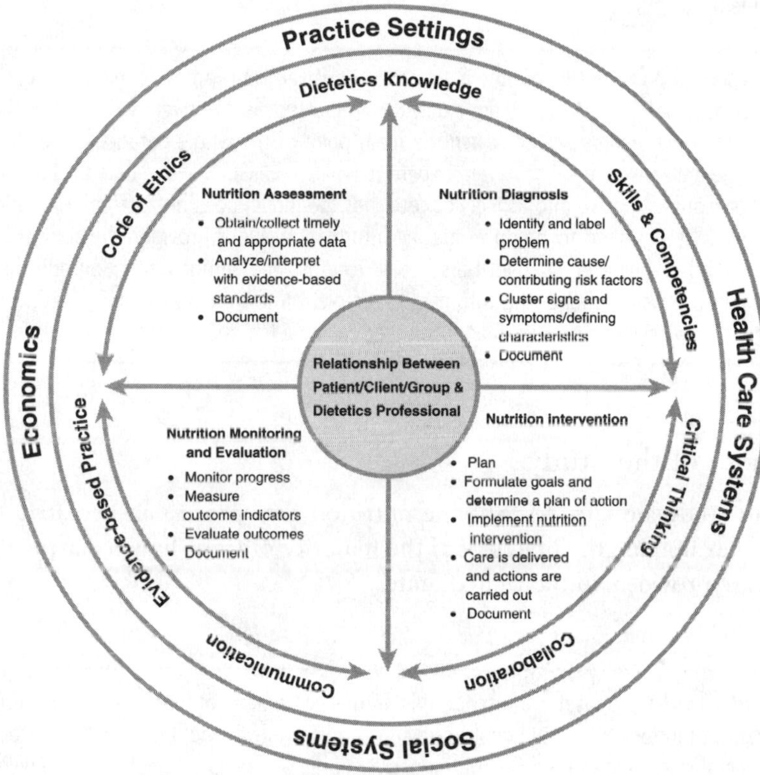

Figure 2 ADA Nutrition Care Process Model

SOURCE: Reprinted from the *Journal of The American Dietetic Association*, *V103(8)*: 1061–1072, copyright © 2003. Reprinted with permission of The American Dietetic Association.

The preceding paragraph presents several points that call for comment. First, the author has employed the third person "this researcher" rather than the direct first person "I" as a deliberate choice for self-reference. This grammatical style once was regarded as correct because it served to "distance" the person of the researcher from the process of doing a study. Personal opinions, dispositions, and values were deemed to be contaminants that had no proper part in formal inquiry. In contrast, to write that "I had an interest in understanding provider characteristics in relation to care . . ." would make it absolutely clear that personal interests did, in fact, shape the purpose and design of the study. Today, it has become more common in practice to accept that personal role for the investigator. In fact, writing a proposal or report in the third person is regarded by many as both unnecessary and cumbersome.

(Continued)

(Continued)

A second point worthy of note is that the "however," between the first and second sentences is left implicit and is not explicitly announced. Such indirect writing runs the risk that some readers will not catch the main point of the whole paragraph. Readers who mentally insert the "however," where it is left implicit, will understand that the author plans to inquire into aspects of care that the model does not presently include. Her study is designed to produce specific information about provider characteristics. Some readers, however, may not be so quick to grasp the author's intention and they will be at a disadvantage. The rule here is simple. Write in a way that is direct and explicit. Don't depend on your reader to detect what is left implicit.

Purpose of the Study

This study aims to describe the nutrition care process in pediatric HIV care by examining the practices of the majority of front-line pediatric nutrition care providers in the United States.

In the original proposal manuscript this formal statement of purpose occurs after 9 pages of introduction and literature review. An alternative would have been to place a general statement of purpose in the opening paragraph where it might then serve as a frame for the lengthy development of basic constructs that follows. Some authors also find it effective to include an assertion of the study's significance at or near the same early location in the proposal. Such decisions are a matter of expository style about which advisors sometimes have firm opinions. It is wise to inquire about preferences for the proposal's structural sequence, and to do so sooner rather than later.

In the following section the author presents the questions that guide the study. Note how she presents an overarching question and then sub-questions that are more specific. That strategy permits readers to understand the primary research focus in terms of more detailed questions. In that way, committee members will be relieved of the necessity of having to speculate about what exactly is intended by the more general statement of purpose.

Statement of the Question

1. What is the nutrition care process of front-line pediatric HIV providers?
2. What is included in the nutritional assessment of children living with HIV infection?

3. What are the current nutrition problems in pediatric HIV?
4. What interventions and strategies do pediatric HIV nutrition providers implement?
5. What outcomes do nutrition providers expect for the patient?
6. What are the characteristics of the interaction between the nutrition provider and the patient?

In the earlier introduction section the author provided detailed information that laid the foundation for the study's significance. Here she provides a paragraph that efficiently restates that rationale and leads gracefully into the review of literature that follows.

Significance of the Study

Nutrition care evolved with the dramatic changes in the medical management of children living with HIV infection in the United States. Health care providers working with children living with HIV infection need continuing education and training to keep abreast of emerging nutrition issues in HIV. An analysis of front-line nutrition providers practice offers a mechanism to identify interventions, strategies, and approaches that are necessary to manage the nutrition care of children living with HIV in the United States. This study will substantiate inclusion of content for continuing education and training for health care providers. Furthermore, description of the characteristics of the interaction between nutrition providers and their patients will further elucidate the central core of ADA's Nutrition Care Process Model and will provide evidence and contribute to our understanding of effective professional practice. The description of the characteristics of the interaction between patients and providers also may inform future research aimed at optimizing the patient- provider relationship to improve nutrition outcomes.

Related Literature

This chapter provides a basis for the research question and the reasons for the selection of the research strategy. It begins with a brief overview of nutrition and pediatric HIV infection and since medical management has evolved in pediatric HIV care, this section also describes this new era and the emerging nutrition issues and approaches to care. The next section provides a basis for selecting the study population, front-line nutrition providers. Finally, the last section, describes the job analysis literature that informed the research methods. This section includes the use of ADA's Nutrition Care Process Model, the framework for the deductive job analysis, and the critical incident technique, the strategy for inductive analysis of the characteristics of the interaction between the nutrition provider and children living with HIV infection.

As indicated above, the treatment of related literature is divided into three major topics. In the first of these, "Nutrition and Pediatric HIV Infection," we have retained the opening introductory paragraphs. The remaining review of literature consisting of 11 pages and 71 citations of reports and reviews has been deleted.

Nutrition and Pediatric HIV Infection

In 1983 James Oleske, MD, published the first report describing HIV infection in children (Oleske et al., 1983). Since Oleske's discovery of perinatally-acquired HIV virus, there have been numerous advances in prevention, detection and treatment of children living with HIV infection. Nevertheless, in the year 2000, in the United States, approximately 10,000 children were living with HIV/acquired immunodeficiency syndrome (AIDS) and worldwide 1.4 million children fifteen years old and younger were infected with HIV/AIDS (UNAIDS, 2000).

In children, growth is related to nutritional status, and appropriate growth rates are a marker of good health. The greatest complication faced by a child living with HIV infection is failure to gain weight, progressive weight loss, and malnutrition (Oleske, 1994). In the older long-term surviving child, this process eventually leads to the wasting syndrome that is frequently the beginning of a rapid downhill course ending in a painful and premature death of the child (Oleske, Rothpletz-Puglia, & Winter, 1996).

Preceding contributors to growth failure and wasting syndrome include appetite loss, abdominal pain, vomiting, and chronic diarrhea, all consequences of the multiple oral and gastrointestinal complications of HIV infection (Palumbo, Jandinski, Connor, Fenesy, & Oleske, 1990). These increased nutrient losses, in addition to altered metabolism and insufficient intake, place a person infected with HIV at high risk for malnutrition (Semba & Gray, 2001). Hence, it has been established that malnutrition is an important and potentially reversible cause of morbidity and mortality in patients living with HIV/AIDS (Kotler et al., 1989) and under-nutrition has been proposed as a cofactor of immune dysfunction in HIV (Ikeogu et al., 1997). As a result, nutrition care has become an integral aspect of the medical management for patients living with HIV/AIDS (American Dietetic Association, 2000; Centers for Disease Control and Prevention, 1998).

Following the detailed examination of literature dealing with nutrition and pediatric HIV infection, we have included here the closing summary paragraph for this first part of the section on related literature. Note how it serves to remind the reader of the investigation's potential significance.

These data support the need to update the nutrition care process in pediatric HIV to also include treatment for overweight children and children with metabolic changes. Additionally, when queried, the lead Title IV agencies agreed that their clinicians needed continuing education and training to address these new and emerging issues in pediatric HIV.

In the second major portion of the review of literature, the author introduces the population that will be of primary interest in the proposed study: "Front-Line Pediatric HIV Nutrition Providers." Here, we have included the entire text.

Front-Line Pediatric HIV Nutrition Providers

The American Dietetic Association (ADA) is the professional group that develops the dietary guidelines for medical nutrition therapy (MNT), pediatric HIV/AIDS protocol (Rothpletz-Puglia et al., 2000). The ADA is in the process of updating the MNT guidelines with an emphasis on evidence-based recommendations. However, the scientific evidence may be years away or may never be pediatric-specific.

The Pediatric AIDS Clinical Trials Group (PACTG) is the preeminent organization in the world for evaluating treatments for HIV infected children and adolescents, and for developing new approaches to the interruption of mother to infant transmission. The PACTG is a joint effort of the National Institute of Allergy and Infectious Disease (NIAID) and the National Institute for Child Health and Human Development (NICHD). Without the existence of the PACTG, studies in women and children with HIV infection would be severely limited or even nonexistent. That said, of the 28 research studies underway in the PACTG system, only two have nutrition components, while, of the 38 research studies open in the Adult AIDS Clinical Trials Group system, seven are nutrition-related. In general, nutrition research is not as well funded as pharmaceutical research, and in general, children are studied the least. Today, nearly 80 percent of all drugs used in children have not been specifically tested in children (Pediatric Aids Foundation, 2004).

Pediatric HIV research is no exception, and much of the medical and nutritional care that is provided is gleaned from the research and clinical practice guidelines for adults with HIV infection. Most of the current practice guidelines or articles written about the nutrition management of children with HIV are based on research in adults. For example, pediatric HIV nutrition practitioners recommend the use of psyllium to decrease cholesterol although no systematic research has been done in children. Due to the lack of research in children with HIV infection, pediatric HIV practitioners provide nutrition care that is informed by evidence-based adult literature and other more researched chronic pediatric diseases.

These new and emerging nutrition issues require that the pediatric HIV nutrition care process be updated, yet there is a lack of nutrition interventional research completed, open, or planned; it makes sense therefore to try to learn from front-line pediatric HIV nutrition providers. What are nutrition providers doing in practice, despite not having evidence-based guidelines? The collective knowledge of front-line providers has implications for understanding the process and characteristics of pediatric HIV nutrition practice in the era of highly active antiretroviral therapy.

> The third, and final, section dealing with related literature is, itself, divided into two topical parts. The first of these, "Job Analysis as Applied to this Research," introduces the process of obtaining and utilizing information about how jobs are performed and, particularly, how Flanagan's Critical Incident Technique (CIT) is used for that purpose.

Job Analysis

In industrial and organizational psychology, a job analysis is the process of obtaining information about how jobs are performed, and is one of the most widely used organizational data collecting techniques (Morgeson & Campion, 1997).The results of job analysis are used as the basis for developing performance appraisal systems, employee selection systems, career development, workforce planning, and curriculum and training design and development (Butler & Harvey, 1988).

We have deleted the four pages used to document job analysis but retained here both the author's introduction to the CIT and a table used to display how the stages of the CIT will be adapted to serve the purposes of the proposed study.

Flanagan's procedures for the critical incident technique include asking the respondent to report a critical incident that includes a description of the situation, action, and outcome. Critical incidents can be gathered through face-to-face interviews, self-administered questionnaires, telephone interviews, workshops, group interviews, systematic record keeping, and by direct observation (Flanagan, 1954, Anderson & Wilson 1997, Kemppainen, 2000). An *incident* is any observable human activity that is sufficiently complete in itself to permit inferences and predictions to be made about the person performing the act (Flanagan, 1954). The term *critical* refers to the fact that the behavior described in the incident plays an important or critical role in determining the outcome (Kemppainen, 2000).

Job Analysis as Applied to This Research

Job analysis research often includes the administration of a questionnaire, and the results of the questionnaire are sometimes used to inform training and development (McCormick, 1980). Therefore, job analysis techniques can be used to gather data about pediatric nutrition care and the results of the survey can be used to inform training and development for health care professionals. These relationships are displayed in Table 1.

The second topical part of this section presents several models that offer frameworks for use of CIT data in the job analysis proposed for this study. From those, the American Dietetic Association's "Nutrition Care Process Model" is selected for use in the proposed study.

Existing Models and Frameworks

Sometimes, jobs can be analyzed using an existing taxonomy, a deductive analysis which results in descriptions of a particular job within this existing framework. Sometimes there is no existing framework, or the model does not encompass all the behavioral aspects of the job. When there is no suitable existing taxonomy for a job, information about the job is collected from

Table 1 Procedural Stages in a Critical Incident Technique Study—Flanagan's (1954) Five Stages and Approach Adopted in Present Study

Flanagan's stages	*Approach in present study*
Stage 1: Establishing the general aim of the activity	Stage 1: To describe provider characteristics related to the patient provider relationship
Stage 2: Setting plans and specifications, e.g., who should be the observers (i.e., respondents), which situations should be observed and which activities should be noted	Stage 2: Respondents will be nutrition care providers within the Title IV network and these respondents will be asked to describe a memorable patient encounter.
Stage 3: Collecting the information – by direct observation retrospectively; by what means (e.g., interview); using only positive or both positive and negative incidents	Stage 3: Retrospective accounts collected by means of structured questionnaire items in a survey. Accounts could be positive or negative incidents.
Stage 4: Analyzing the information – Inductive classification, grouping of incidents into clusters, independent checking of categorization	Stage 4: Critical incident reports will be categorized by expert analysts and the findings will be validated by an independent set of reviewers.
Stage 5: Interpreting and reporting the findings	Stage 5: Derived categories will be verified within professional literature and discussed by the experts and independent reviewers. Critical incident reports with excellent inter-rater reliability will be used to describe exemplary nutrition practice and to illustrate the categories and their level of effectiveness.

subject matter experts. The job analyst then organizes this information to form constructs or categories of job behavior. Jobs are analyzed deductively or inductively, but both inductive and deductive approaches can be combined if the existing taxonomy is insufficient (Williams & Crafts, 1997).

> We have deleted the review of related literature that pertains to an examination of existing models, but included here the basic description of the ADA model and how the proposed methods of data collection will meet its demands.

Deductive job analysis often relies on existing taxonomies to analyze jobs in terms of predefined categories (Peterson & Jeanneret, 1997). Several models were evaluated for their use in this study, whose objective was to analyze the full range of nutrition care that is provided, including behavioral characteristics of nutrition care providers, and the implicit and exemplary aspects of nutrition care. In an effort to describe the full range of behavior and interaction in pediatric nutrition care, several models were examined for their relevance.

ADA's Nutrition Care Process Model

After examining the professional literature, this researcher found the American Dietetic Association's recently-published Nutrition Care Process Model (Lacey & Pritchett, 2003). The Nutrition Care Process is a standardized process for dietetic professionals and the Nutrition Care Process Model is a visual representation of this standardized process. The model reflects key concepts of each step of the nutrition care process and illustrates the greater context of nutrition practice. Figure 2, presented previously, provides a full depiction of the model.

The model is intended to depict the relationship among all the components as they overlap, interact, and move in a dynamic manner to provide the best quality nutrition care possible. The four steps in the nutrition care process are nutrition assessment, nutrition diagnosis, nutrition intervention, and nutrition monitoring and evaluation. In the outer rings, the model also identifies factors that influence the quality of nutrition care. Some of these factors include the practice setting, environmental impacts, and provider characteristics such as knowledge, skills, critical thinking, ethics, communication, and collaboration. The central core of the model depicts the relationship between the client and the nutrition provider, although specific attributes of this interaction are not described. Provider attributes that are related to positive outcomes are not depicted in any aspect of the model, and are not described in the article about the Nutrition Care Process Model (Lacey & Pritchett, 2003).

The purpose of the Nutrition Care Process Model is to represent the spectrum of nutrition care practices that can be provided by nutrition professionals (Lacey & Pritchett, 2003). The model is a useful tool for evaluating nutritionists' practice because it acknowledges the common dimensions of practice and provides structure to validate nutrition care and outcomes. This was the most comprehensive and relevant model to study nutrition practice, and was therefore used in this study as the framework for analyzing and describing the components of pediatric HIV nutrition care.

Conducting and Analyzing the Job Analysis

A job analysis identifies the work behaviors and associated tasks that are required for successful job performance. Conducting a job analysis involves two steps (Wheaton & Whetzel, 1997). The first step is to identify the most critical aspects of the job. This includes descriptions of what is done, how it is done and why it is done. In this research, the job-oriented aspects include the problems, strategies and interventions, barriers to success and expected outcomes. These job-oriented aspects of the nutrition care process are captured within the first part of the questionnaire for the present research. ADA's Nutrition Care Process Model (Lacey & Pritchett, 2003) informed the development of the variables or categories of interest. The Nutrition Care Process Model was also helpful in analyzing the components of the nutrition care process.

The next step is to identify the provider knowledge, skills, abilities, and personal characteristics needed to successfully perform critical job operations. This inductive approach usually results in specific knowledge, skills, abilities and other characteristics (KSAO's) that are integral to job performance (Dunnette, 1983; Williams & Crafts, 1997). These worker or nutrition provider-oriented (in the case of the research described here) aspects of the job analysis are captured within the latter part of this research survey, and the critical incident technique (Flanagan, 1954) is used to elucidate and evaluate these behavioral aspects of the nutrition care process. CIT is particularly useful for understanding the dimensions of professional roles and the interaction with patients and other clinicians (Byrne, 2001). A basic condition necessary for a functional description of an activity is a fundamental orientation in terms of the general aim of the behavior (Flanagan, 1954). The general aim of the critical incident report analysis will be to describe nutrition provider interaction with the patient.

> The section now concludes with a summary of how the literature supports both the question and the methodology of the proposed study. This portion of the review is reproduced here in its entirety. Note how the author uses the main points established in the related literature as a rationale for the primary focus of the proposed study.

Summary

There are a number of reasons for analyzing the nutrition care process in pediatric HIV infection. First, there is a lack of published research about the pediatric HIV nutrition care process, and there are several new and emerging issues in nutrition management such as lipodystrophy and overweight. A more subtle reason for a revisiting pediatric HIV nutrition practice is that HIV care has changed from being a primarily acute-care modality to a chronic-disease paradigm. This means that nutrition care can be less concentrated on critical illness and more participatory and focused on preventing other chronic diseases such as cardiovascular disease and diabetes. Finally, the most compelling reason for this research were the expressed needs of pediatric HIV clinicians within the Title IV network, who wanted more information about what other nutrition providers are experiencing and how they are handling the new and emerging clinical issues. An intended outcome and application of these research findings are the development of continuing education training for Title IV providers, and the research was designed to address the need for job training, as well as to describe the implicit and exemplary aspects of nutrition practice.

Job analysis serves as a needs assessment for job training of incumbents. Findings from job analysis serve to validate content for the training. Therefore job analysis techniques were used for this research. Job analyses can be deductive, inductive or mixed, depending on the existence of a comprehensive job taxonomy.

In this case, there is an existing framework for the nutrition care process called the Nutrition Care Process Model (Lacey & Pritchett, 2003). However, one aspect of this process, the central core, or the patient-provider relationship, is not fully explored within the model. As a result, a mixed approach of deductive analysis using the Nutrition Care Process Model for analyzing quantitative survey findings, and inductive analysis using the critical incident technique was used to elicit a more comprehensive picture of the nutrition care process in pediatric HIV. The findings from this study of the front-line nutrition providers' role will be the basis for the development of a training opportunity for health care providers that includes information about standards for the nutrition assessment of children living with HIV infection, the current issues in pediatric HIV, specific strategies and interventions, desired outcomes and the characteristics of the patient-provider interaction that impact on outcomes.

The next major section outlines the research methods to be employed in the study. As we recommended in Chapter 3, the author has conducted multiple pilot studies and reference to them is included in the following presentation. As the pilot studies proceeded, the author was in regular contact with members of her dissertation committee. Thus, she ensured that there would be no surprises in this portion of the proposal document.

Methods

There have been dramatic changes in the medical management of children living with HIV infection in the United States. New nutritional issues have emerged from these changes, and pediatric HIV medical program administrators have indicated that their healthcare providers need continuing education and training. To meet this need, an online survey containing several strategies for gathering these data will be distributed to all of the medical sites in the United States that provide care to children living with HIV infection.

> It is here, at a point that was 22 pages into the original proposal manuscript, that the reader first encounters an important fact about the proposed study. The primary agency for collecting data in this study will be an "online survey." As procedures for this form of data collection are explained, it becomes clear that this was a particularly appropriate solution to the problem of reaching a large and highly dispersed subject population. It would have been our choice to feature that element in the title, or, absent that notation, to at least make that aspect of the study apparent in the introduction.

The questions in the survey are intended to elucidate specific aspects of nutrition practice. In turn, that information will be used to substantiate specific content for continuing education and training. In addition, by examining data that describe the interaction between nutrition providers and their patients it should become possible to identify the aspects of exemplary nutrition care that heretofore have remained implicit in models of patient care. In sum, this section of the proposal describes the study population, the procedures used to develop the survey, and those that will be used to implement, collect, and analyze the survey data.

Description of the Population

The United States Department of Health and Human Services' (DHHS) Health Resources and Services Administration (HRSA), along with programs funded through the Ryan White Comprehensive AIDS Resource Emergency Act (CARE), work with cities, states, and local community-based organizations to provide services to individuals living with HIV. Grantees include health institutions, legal agencies, educational facilities, and community service organizations. The majority of CARE Act funds subsidize medical care and essential support services. Title IV of the CARE provides coordinated HIV services and access to research for women, infants, children, youths and families living with HIV/AIDS (United States Department of Health and Human Services, 2002).

In the year 2001, the Ryan White Act Title IV programs reported having 6009 HIV/AIDS infected clients younger than 13 years of age (*Title IV Report*, 2001). In December of 2001 the number of children younger than 13 living with HIV/AIDS in the United States, reported by the United States Centers for Disease Control and Prevention (CDC) was 5,409 (Center for Disease Control and Prevention, 2001). The discrepancy between the number of cases reported by the CDC and those served by the Title IV programs may be due, in part, to variation in reporting requirements, i.e., some states do not require mandatory HIV reporting.

These data indicate that programs receiving funding from the Title IV program provide care for the majority of children in the United States living with HIV infection. Therefore, the survey will be distributed to all the Title IV program sites as the most efficient way to reach healthcare professionals (the population) who provide care to children living with HIV infection in the United States.

The study population of healthcare professionals working in CARE Title IV funded settings included nutritionists, physicians, advanced practice nurses, nurses, social workers, physician assistants, and administrators. All of these healthcare professionals will be eligible to complete the online survey because many Title IV sites do not have funding for a nutrition specialist. Nutrition is integral to pediatric HIV medical care, so if a Title IV program does not have a nutrition specialist such as a registered dietitian, other healthcare professionals must fill this need. These healthcare professionals in Title IV programs provide care for HIV infected children from infancy throughout adolescence.

> The situation just described would attract the immediate attention of any dissertation committee. First, given the wide range of possible respondents, the identity of the person at the Title IV site who will assign the task of responding to the survey becomes a vital piece of information. Further it is important to describe the basis upon which that person will determine that assignment. Before reading ahead, you might wish to think about how you would plan proposed methods to deal with those two questions. Second, as the author has made clear, responses will be returned by individuals with widely varying professional backgrounds and very different responsibilities in their roles within the care program. What provisions should the author build into the proposed methodology to insure that data will be collected to allow analysis of that difference among and between respondents?

Instrument Development

Professional literature informed the development of the overarching research question regarding the need to describe pediatric HIV practice in a new era. Job analysis literature provided the basis for the strategies used within the survey. The ADA's Nutrition Care Process Model provided the framework for the development of categories within nutrition care to be examined. This model posits that the nutrition care process consists of four distinct but interrelated and connected steps: (a) nutrition assessment; (b) nutrition diagnosis; (c) nutrition intervention; and (d) nutrition monitoring and evaluation. This model also includes other dimensions of the nutrition care process such as the relationship between the professional and the patient. The Nutrition Care Process Model led to the development of specific research questions to elucidate key components or attributes of the nutrition care process. The methods for collecting information about each component of the nutrition care process within the survey are discussed below.

> As we alluded to in our previous comment, in the next paragraph the author describes how demographic information will be obtained concerning workplace context and provider roles. You will also note that in the text that follows the author uses past tense to describe those tasks that already have been completed—the development of an instrument and its pilot testing. While we often think of proposals as documents aimed at future research it is important that preliminary work is presented in the appropriate format. The extensive pilot work addressed here leaves little doubt that the instrument to be employed in the study will be adequate for the task of data collection.

Demographics

Demographic questions were designed to capture information about practice setting and nutrition provider characteristics. Questions include check boxes about work setting, geographic area (rural, urban, etc.), case load, time spent providing direct care, years of experience, and professional role.

Nutrition Assessment

This variable is related to the information gathered and analyzed by the nutrition provider for the evaluation of nutritional status. Several questions were designed to specifically relate to describing nutrition assessment. Two questions are open-ended: the first asks providers how staff members at their site determine which children need nutrition counseling, and the second inquires into how children with HIV infection are nutritionally assessed.

Nutrition Diagnosis

This variable is related to identifying nutrition problems and contributing risk factors. In an open-ended question format, respondents will be asked about the most common nutrition problems encountered. Risk factors will be identified with a question about factors that influence patients, using a Likert-type scale of 1–5.

Nutrition Intervention

This includes the action plan and strategies of nutrition care. The questions designed to gather data about this variable include asking respondents to report barriers to change faced by their patients and the specific strategies they use to help their patients overcome those barriers. Both of these are open-ended questions. An additional question will ask respondents to report particularly useful educational materials.

Nutrition Monitoring and Evaluation

This variable is an indicator of the progress and outcome of nutrition care. Expected outcome of care will be measured through use of a Likert-type scale of 1–5.

Patient–Provider Interaction

This variable is related to the characteristics of the interaction between nutrition providers and their patients. Critical incident reports will be the primary source of information about this dimension of nutrition care. In addition, some of the questions regarding specific components of the nutrition care process will be useful for describing the providers' role within the nutrition care process.

The primary aim of the critical incident technique in the research described in this dissertation is to elucidate the characteristics of the patient-provider relationship. The critical incident technique places the nutrition care process in context and offers a meaningful picture of salient aspects of patient care.

Critical incident reports present a series of questions prompting respondents to describe a memorable patient experience. A complete critical incident report includes a description of the situation, the action, and the outcome. In essence, this series of questions captures the entire nutrition care process for one patient encounter. Because the critical incident reports will contain all the components of the nutrition care process in an actual patient encounter, these data will also be used to validate the findings from the other parts of the survey.

Refinement of the Survey Instrument

Once the questions were developed to identify the steps and components of nutrition care, and the critical incident technique procedures were outlined, the questionnaire was formulated. Each survey question was evaluated for its relevance to the research questions. The instrument was given to three experienced researchers for comments. This resulted in a number of revisions (e.g., elimination of unnecessary questions). This streamlined and reduced the amount of time to complete the survey. During this process, the factors presented in Table 2 contributed to the determination of the relevance and purpose of each questionnaire item.

> The use of a table here concisely presents a great deal of information in an easy-to-interpret format. Readers immediately can see how sections within the instrument are related to the research questions.

Table 2 Relationship Between Survey Items and Research Questions

Research Sub-question #	*Concept*	*Related Questionnaire Item*	*Level of Measurement*
Demographics	Work setting	2 – 7	Nominal
Demographics	Professional role	8 – 14	Nominal
1	Assessment	15, 16	Nominal
2	Diagnosis	17, 24, 27, 28	Nominal
2	Diagnosis	18	Ordinal
4	Monitoring/ Outcomes	20	Ordinal
4	Monitoring/ Outcomes	31, 32	Nominal
3	Interventions	22, 23, 25, 29	Nominal
5	Patient–Provider Interaction	Critical incident report – 27 –34	Nominal

In the following paragraphs the author indicates how she made good use of both colleagues and her advisor in completing a preliminary review of her instruments. By doing so, she helped insure that the tools for collecting data were as refined as possible before pilot testing began. This reflects our advice in Chapter 3. Consult with others—and do so both widely and often.

The survey instrument was also evaluated by peers enrolled in the nutrition education doctoral program and by the dissertation sponsor. Their comments led to revision of the introductory letter to the survey, and several modifications in the wording of questions. Finally, the survey instrument was reviewed by the United States DHHS, HRSA HIV/AIDS Bureau (HAB). After this review, no further revisions were made and the research protocol was sent to the Institutional Review Board (IRB) at the University of Medicine and Dentistry of New Jersey (UMDNJ) (this researcher's place of work) and then to the Teachers College, Columbia University IRB for approval.

Survey Format

The data will be collected by an online survey (see Appendix A for the questions and format). Participants will be asked to access the survey from a direct link embedded in an invitation email (see Appendix B). The invitation will contain all the elements of consent as required by the two associated IRBs. A Web hosting company, Creative Research Systems, will post the survey on their server for a period of two months.

Pilot-Testing

Following IRB approval, the instrument was pilot-tested by five pediatric HIV nutritionists, one physician, a nurse practitioner, and a social worker, all of whom had a decade or more experience in pediatric HIV care. Each of these health care providers was asked to complete the online draft survey for purposes of validation, to evaluate comprehensibility, and to comment on the value and relevance of the data gathered. Two of the eight professionals were asked to complete the survey twice, to evaluate the reliability of the survey questions. The survey was modified to increase the space for several questionnaire items based on the pilot test. All of the healthcare providers reported that they thought the survey was useful, although the physician commented that he would have preferred fewer open-ended questions. Each questionnaire item produced the answer that the question was designed to elicit, and the two healthcare providers who answered the questionnaire twice answered the survey the same way both times.

Survey Implementation Protocol

Each of the 89 lead Title IV sites will receive a notification/invitation email (A copy of this message is contained in Appendix B) that describes the purpose of the study and clarifies the requirements for survey participants. The administrators of the lead Title IV sites will be asked to forward the invitation email to the person who provided the most nutrition education at their sites. This procedure is necessary because, while there is a comprehensive and updated contact list of Title IV program administrators, there is no comprehensive or current contact listing of healthcare professionals who provide nutrition education in these settings. All submitted surveys will be anonymous.

> Following the proposed procedures, the author will not have access to the names of individuals who complete the survey form. In effect, the participants will be anonymous. Each respondent's place of employment and professional role, however, will be recorded on the survey form and thus be available as part of the data set for demographic information.
>
> We now have answers to the questions raised earlier concerning how the distribution of invitations will be managed at the Title IV site, and by whom. The reason for doing it as described has been made perfectly clear—there simply was no alternative.

Participants will be given two months to complete the survey. Follow-up emails will be sent (Appendix C) to all of the potential respondents since there was no way to identify non-responders because of confidentiality. Participants will be offered a paper format of the survey via fax if they prefer this to completing the survey online. In addition, follow-up telephone calls will be placed to improve the response rate.

> The necessity of having to send a follow-up to each "potential respondent" at each site simply reflects how complex it can become to maintain confidentiality in a study such as this. The offer to provide a paper version for any respondent who requests one reflects the author's thoughtful and respectful attitude toward the participants.

Data Analysis

The main objectives of the data analysis will be fulfilled using descriptive statistics. Frequencies and prevalence rates will be calculated for demographic and other variables related to the nutrition care process. Fisher's exact test will be performed to evaluate differences in effectiveness according to nutrition education approaches. Two-tailed *p* values will be reported and $p < .05$ (two-tailed) will be set as the criterion for statistical significance. Cohen's kappa coefficient will be calculated for inter-rater agreement between two analysts who will extract data from the critical incident reports. For interpretation of kappa values this researcher will use Landis and Koch's (1977) suggested values of less than or equal to 0.4 indicating relatively poor agreement, values of 0.4–0.6 indicating moderate agreement, 0.6–0.8 good agreement, and 0.8–1.0 excellent agreement.

Each open-ended question will be coded, and counts entered into an Excel database (Microsoft Excel, 2000). Figures will be created within Excel, including those that contain frequencies. Survey Systems Software (version 8.0, Creative Research Systems, Petaluma, CA) will be used to generate descriptive statistics for the questions that are forced choice or Likert-type rating scales. Statistical Package for the Social Sciences (version 8.0, 1997, SPSS, Chicago, Ill) will also be used to generate descriptive statistics and Cohen's kappa.

A question that might be on the mind of committee members at this point is "what response rate would be acceptable for the analyses that are presented?" It would be a matter of choice and local custom as to whether to include that information here. If not presented in the proposal document it would be wise to be prepared for the question and the need to assure the committee that response rates and various forms of subject attrition will be presented and fully discussed in the final dissertation.

In the sections that follow, the author provides detailed information about procedures for systematically treating content from the open-ended questions, and completing the analysis of critical incident reports. Note how she has included information about reliability and validity in a manner that clearly indicates her understanding of the importance of those issues when dealing with these kinds of data.

Procedures for Open-Ended Questions

It is anticipated that many of the open-ended questions will require content analysis. Manifest content analysis, coding for and tallying of specific words or ideas, will be sufficient for most such questions. These counts will then be used to generate frequencies and prevalence rates for nutritional problems, assessment considerations, and nutrition education strategies. Two questions will require latent content analysis; that is, the process of identifying, coding, and categorizing the primary patterns in the data, within the context of the data (Mayan, 2001). These questions are related to patients' barriers to change and provider strategies for overcoming these barriers. It will be necessary to code for the context of the providers' suggested strategies for overcoming barriers. This process should lead to the development of conceptual categories for the barriers to success and related strategies.

Critical Incident Analysis Procedures

The critical incident series of questions will be analyzed according to Flanagan's methods as outlined by Anderson and Wilson's (1997) procedures. They begin with editing guidelines to ensure that the writer made the correct distinctions about where to record each part (situation, action, outcome). These procedures involve moving content to the appropriate sections, while keeping the text in the author's own words. Then, it will be necessary to make sure that the narrative was actually about the person writing the report. Incidents that provide insufficient information, or that do not include a plausible relationship between the person's actions and outcome, will be discarded.

> In the first sentence below, the author indicates that "experts" will participate in the next step of data analysis. An additional sentence about the experts' qualifications would be helpful to readers who might wonder about how expertise was certified.

Two expert analysts will review and categorize the incidents by provider-oriented attributes of the nutrition care process, and the interaction between the provider and their patient. Each analyst will sort the critical incidents independently, to achieve the full benefit of having more than one analyst. Each critical incident will be on a separate sheet of paper so that it can be sorted. Each analyst will have a complete set of reports. Both expert analysts will complete a critical incident analysis table, created by this researcher for the experts to use as a starting point for the analysis of the critical incident reports (Appendix D).

The matrix includes columns for the expert analysts to record the nutrition providers' knowledge, skills, abilities, personal characteristics, provider advice, and intended and actual outcomes. The matrix was developed to help the analysts code and categorize each critical incident report. The headings in the matrix were based on ADA's Nutrition Care Process Model and the characteristics that are integral to job performance (Dunnette, 1983; Lacey & Pritchett, 2003).

Note how in the above paragraph the author reminded the reader that this is based on the ADA model that was presented earlier. In the paragraphs below she shows how the matrix will be used. We have included the matrix in an appendix, as it was in the original proposal.

The matrix is a tool that will be used to characterize the nutrition providers. Once this is completed for each critical incident report, similar reports will be placed together, according to aspects of the provider's behavior in the critical incident. The analyst then will label each set of similar reports, with a brief abstraction of the behavior.

The next step will be to develop a common set of provider characteristics between the two analysts. The analysts will take turns explaining the rationale for how the categories were assigned. They will negotiate to derive a common set of provider characteristics based on the analysis of the critical incident reports. This also will involve an iterative process whereby the expert analysts will discuss their findings in relation to professional literature. After this, descriptions of the provider attributes will be written with enough detail about the defining characteristics so that the incidents can be categorized by independent reviewers who may not have expertise in pediatric HIV nutrition.

The provider characteristics will then be tested (validity of derived constructs) by an independent group of reviewers. This process is called retranslation by Anderson and Wilson (1997). During this part of the analysis of the critical incident reports, the independent reviewers will be asked to categorize each critical incident report using the descriptions and labels that the expert analysts have derived. The independent reviewers also will be asked to rate the effectiveness level of the incident on a scale of one to three. One is "not effective," two is "somewhat effective," and three is "very effective." Data from this set of reviewers will be analyzed by calculating the mean and standard deviation of the effectiveness ratings. In addition, the percent of reviewers who characterize each incident similarly will be calculated.

In the subsequent dissertation the author made excellent use of the critical incident reports by using respondents' own words to describe examples of the various nutrition education approaches. It would have been helpful to include an indication of that intent here—both to whet the appetite of the committee and alert them that this valuable information will be presented in the final document.

The author included a complete list of cited references. We have included only the first few to indicate format and placement in the proposal document.

References

Amaya, R. A., Kozinetz, C. A., McMeans, A., Schwarzwald, H., & Kline, M. W. (2002). Lipodystrophy syndrome in human immunodeficiency virus-infected children. *Pediatric Infectious Disease Journal, 21,* 405–410.

American Academy of Pediatrics. (2003). *Pediatric nutrition handbook* (5th ed.). Washington, DC: American Academy of Pediatrics.

American Association of Diabetes Educators. (1998). *A core curriculum for diabetes educators* (1st ed.). Chicago: American Association for Diabetes Educators.

American Dietetic Association. (1998). *Medical nutrition therapy across the continuum of care* (2nd ed.). Chicago: American Dietetic Association.

Appendixes

The author included all the appendices indicated throughout the proposal. They were:

A. Survey Instrument
B. Invitation email
C. Follow-up email
D. Critical Incident Analysis Matrix

We have included the critical incident matrix here to indicate the extent and complexity of planning that was accomplished prior to data collection.

Appendix D. Critical Incident Analysis Matrix

% agreement	
Actual outcome	
Intended outcome	DIRECT NUTRITION OUTCOME (knowledge gain behavior change, food or nutrient intake change, improved nutritional status) CLINICAL AND HEALTH STATUS OUTCOMES (lab values, growth parameters, BP, risk factor profile changes, signs and symptoms, clinical status, infections, complications) PATIENT CENTERED OUTCOMES (quality of life, satisfaction, self-efficacy, self-management, functional ability) HEALTH CARE UTILIZATION AND COST OUTCOMES (medication changes, special procedures, planned/unplanned clinic visits, preventable hospitalizations, length of hospitalization, hospice care)
Linked to CI Yes or No	
Advice Provided (from last two questions asked)	
Personal characteristics (ethics, values, "bed-side" manner—listening, empathy, advocacy etc., think rationally, think creatively)	
Ability to (critically think, make decisions, collaborate, communicate, problem solve)	
Skill in (competencies such as tube feed management etc.)	
Knowledge of (medical nutrition therapy, HIV disease, pediatrics, etc.)	

PROPOSAL 4

Funded Grant

How Higher Education in Prison Affects Student-Inmates' Social Capital and Conceptions of Self

This proposal is typical of those requesting dissertation funding from a source within the student's college or university. It was developed in response to a request for proposals (RFP) issued under a research training grant (RTG) in the doctoral candidate's institution. One function of the RTG was to provide funding for student research that could be used alone or as a supplement to support from other grants. The RFG had the following guidelines for completion of the grant proposal:

Cover Sheet (1 page)

- Student's name
- Social security number
- Program and degree (e.g., music education, Ed.D.)
- Tentative dissertation title

AUTHORS' NOTE: This proposal, used with permission, was prepared in 2005 by Jed B. Tucker in response to a request for proposal at Teachers College, Columbia University. The author is now completing his doctoral degree in Applied Anthropology at the same institution.

- Dissertation sponsor's name
- Mailing address
- E-mail address
- Phone number(s)

Overview of the Study (maximum 5 single-spaced pages)

- Background and significance of the study
- Research questions and/or hypotheses
- Summary of the method

Statement of Professional Goals (maximum 1 page)

This should include a brief narrative of what the student plans to do after graduation. It should include a brief description of how the student's current research fits into future research plans.

Budget (maximum 1 page)

- Listing of research-related expenses and the total amount requested
- A rationale for the expenses

In addition, the proposal had to have been approved by both the dissertation committee and by the Institutional Review Board.

Imposition of a page limit for the document is common for internal grants since review committees often receive many such proposals. Graduate students who respond promptly to the request and convey the required information in the space permitted will have a definite advantage in the selection process. The author's cover sheet has been omitted here.

Background and Significance of this Study

With five percent of the world's total population, the 2.1 million incarcerated individuals in the United States represent 25% of the world's prisoners (Shiraldi 2003). Historically, that represents the largest number of individual prisoners and highest proportion of the population to be imprisioned within any country in the world (Elsner 2005). While new convictions continue to fuel prison growth, recidivism (i.e. re-incarceration post-release) has emerged as its sustaining force. Sixty-eight percent of ex-prisoners nationally are re-arrested within three years of release, 52% return to prison (Bureau of Justice Statistics 2002). These numbers, along with the well-documented negative *effects* of incarceration (Rhodes 2004), challenge the efficacy of the criminal justice system's use of prisons to "correct."

A growing body of evidence concerning "Correctional Practices That Work" presents some hope for change (Shrum 2004). In particular, the dramatic and unmatched success of higher education programs in prisons in lowering recidivism rates (Anderson et. al. 1988; Blackburn 1981; CESA 1990; Chappell 2004; Clark 1991; Fine et. al. 2001; Harer 1994, 1995; Johnson 2001; Lawyer et. al. 1993; Lockwood 1998; O'Neill 1990; Shumacker et. al. 1990; Steurer et. al. 2001; Tracy et. al. 1994) suggests an effective intervention in the phenomenon of inmate recycling.

Because of the sensitive nature of this study, in the description below (and throughout the proposal) we have used pseudonyms for the institutions and locations.

Until 1994, there were over 350 college programs in prisons throughout the country. That year, however, the federal Violent Crime Control and Law Enforcement Act was adopted by Congress. Among other restrictions, that legislation removed prison inmate's eligibility for Pell Grants, effectively ending college in prison thereafter. Nevertheless, a private initiative by Local College starting in 2001, restored higher education to two men's prisons in the state, Large Correctional Facility (LCF) and Other Correctional Facility (OCF). (Local College will be opening its third college program, the first in a women's prison, in the Fall of 2005).

While the research on the impact of higher education programs in prisons completed prior to their closing was encouraging, it also was inconclusive because so little of the inquiry was directed at examining exactly *how* those educational programs functioned to produce desirable outcomes. The proposed research will begin to address that omission through an ethnographic study of the college program at LCF, a maximum-security prison where I have taught for two years. With the largest student body (*n*=75) of either of the extant Local College programs, LCF offers the best possibility to investigate the far-reaching impact of higher education programs in prisons.

The next section continues to develop background that directly supports the importance of the questions.

Research Questions

An underlying assumption in the dominant literature about the effects of higher education programs in prisons—primarily from the disciplines of criminology, criminal psychology and the sociology of crime—is that recidivism is a justifiable measure of rehabilitation.

In almost all of this literature (in all but one of the studies referred to above), the rehabilitative potential of higher education programs (as well as most other forms of intervention) is measured by rates of re-incarceration. If the ex-prisoner is not re-incarcerated during a specified period (customarily, 3 years) following exit from the program, he/she is assumed to have been constitutionally transformed and the program is deemed successful. Implicit in this understanding of rehabilitation is the assumption that only problematic individuals (referred to as anti-socials, psychotics, career criminals, those exhibiting low self-esteem, and many more such descriptive labels) go to prison, which is why the "successful" rehabilitation program is one that can be presumed to have "corrected" the offender's "deficiency."

In contrast to that traditional model of incarceration and rehabilitation, penologist Shadd Maruna (2001) has questioned the essentialist portrayal of prisoners as defectives. Through his comparative study of recidivist crime "persisters" versus non-recidivist crime "desisters," (offenders who have stopped all illicit activities) he has undermined some of our most basic assumptions about crime, punishment, and reform. Desisters, he argues, do not *become* straight, nor does one *become* criminal; both labels describe status positions, and therefore, both may require the on-going work of intervention. That understanding of criminal behavior also has invited questioning of the dominant paradigm for prisoner rehabilitation.

Two facts about incarceration in North America today further disturb what I refer to as the dominant explanation: (1) the incarceration explosion of the last thirty years has resulted from changes in legislation and administration *not* changes in human behavior (Currie 1998, ch.1; Gilmore 1998/9; Mauer 1999, ch.8); (2) of the 52% of ex-prisoners nationally (40% in the state) who are re-incarcerated within three years of release, half (70% in the state) result from "technical parole violations," not new crimes (BJS, *Recidivism* 2002; DOCS, *1999 Releases*). This means that re-incarceration rates, the litmus test for rehabilitation programs, often reveal nothing about the activities or psychology of incarcerated individuals.

By presenting this challenge to the dominant paradigm's explanation of the relationship between higher education and recidivism, I do not mean to suggest that such a relationship may not, in fact, exist. Restricting the impact of higher education in prisons to its effects on recidivism, measured in terms of re-incarceration, threatens to obscure other potentially critical outcomes. This study will explore the alternative hypothesis that college participation enriches the inmate's social networks with *social capital,* a resource inherent in a social collective, or group (i.e., not an individual), that helps the individual members of that group "get things done." Examples of social capital are things like instrumental knowledge (where to get a license, who to call in an emergency, etc.), cultural knowledge (what dialect to use, appropriate dress in court, etc.), and emotional or other non-remunerated support (unpaid babysitting, counseling, general assistance).

The strong social ties constitutive of networks with high levels of social capital are a proven critical factor in the post-release success of ex-offenders (Wolff et al. 2003). This study also aims to explain how the content of academic courses, and the entire learning process, encourages students to re-examine their past experiences and understandings of themselves.

Here the author presents the purposes of the proposed study. Even given the restrictions on length, he appears to have made a reasonably compelling case for the potential importance of the study.

In sum, this study will explore one of the unintended consequences of higher education in prisons by (1) describing the changes in student-inmates' social networks (all the individuals with whom the subject has meaningful social ties) and the qualities of those new networks, and (2) analyzing student-inmates' personal narratives diachronically over a one-year period for their own re-interpretations of their life experiences. If significant changes can be established in these two areas, we will be much closer to an empirical explanation of the relationship between college participation and lowered recidivism rates.

Theoretical Framework: Why Narratives and Networks?

Here, the proposal offers more information about the theoretical frameworks that are commonly employed in this type of research. This is intended to rationalize the choice of methodology that follows.

Narratives

This study will analyze the personal narratives, or life stories of student-inmates (see: Bruner 2002; Greimas 1990; Josselson et. al. 1993; McAdams 1985; Ochs et. al. 2001; Ricouer 1984 for a general theoretical discussion of narratives). At the surface level, personal narratives, or life stories, are considered repositories of information about the individual's social environment and life experiences. At a deeper level, life stories reveal, if not help construct, the narrator's perspective (see: Rosenthal 1993 for a discussion of the two "levels" of narratives).

As Ricouer (1984) explains, the life story is based on a preconceived understanding of experience, which changes in the narrative re-telling due to the imposition of a particular structure. This process has been described as "giv[ing] shape to experience" (Ochs et al. 1996:20), or as "construct[ing], reconstruct[ing], in some ways reinvent[ing] yesterday and tomorrow" (Bruner 2001: 93). Narratives help make our experiences intelligible by providing a smooth chronology to life's haphazard events (Goffman 1974; Labov 1972; Polanyi 1989; Ricouer 1984).

There are no hard and fast rules about where "new" narrative structures come from, but importantly, they are acquired or learned through cross-fertilization from new experiences. This means that while it cannot be determined why one "text" will impact upon the interpretation of another (Derrida 1972), one's personal narrative changes as a result of new interpretations of new (or old) activities in the individual's life. Because this study is concerned with how the college program impacts the life of the student-inmates, I will focus on how the subjects' re-interpretations of their lives are affected by incorporating the content of the course material and by becoming members of new social networks.

In addition to its impact on the individual's point-of-view, college participation establishes the context for new relationships, and formal schooling infuses the entire social group with social capital (Bourdieu 1984). Social capital, in turn, is a critical factor in the post-release success of ex-offenders (Wolff et. al. 2003). By increasing the social capital of the collective student body, individual members benefit from the enhanced resources of the group (Coleman 1990). This process is explored in the following section.

Social Networks

In addition to insights into the narrator's point of view, the personal narratives will provide information about the individual's social network (all the individuals with whom the subject has meaningful social ties) over his lifespan. A significant rupture in all the students' social networks occurred upon incarceration. This study will attempt to account for the more recent rupture resulting from participation in the college program.

There are two questions about social networks that are of interest for this study: (1) How are strong and supportive interpersonal relations maintained? and, (2) What is their value? Classic (Mauss 1950[1924]), and somewhat more contemporary (Stack 1974) anthropological studies have shown that social ties are formed through the informal exchange of goods or services. Informal exchange refers to a giving/receiving transaction that is not remunerated immediately. The *delayed reciprocity* of these exchanges necessitates a long-term social bond, characteristic of strong social ties. To be successful in the Local College program requires students to work together. They must learn to cooperate and depend upon one another. The support and assistance the students provide one another can be thought of as one example of the "informal exchange" that builds strong social ties.

The degree of social capital of the student's networks will be measured by the "strength" and "diversity" of its social ties, and as the "ability of actors to secure benefits by virtue of membership" in the particular social network (Portes 1998:6; see also Coleman 1990; Kadushin 2004; Lin 2001; Putnam 2000). The strength of a social tie is a "combination of the amount of time, the emotional intensity, the intimacy (reciprocal confiding), and the reciprocal services which characterize the tie" (Granovetter 1973; see also Burt 1992). The diversity of a network generally refers to the "range" in social rank (job position, economic status, position of authority) of the individuals who the subject knows (Erickson 2001; Kadushin 2004). The focus of data collection during this study will be each of the above categories.

Based on analyses of the distribution of incarceration in the state by neighborhood (see: Clear et. al. 2003; Fagan 2004), combined with relevant ethnographic research on the daily life of residents in those very neighborhoods (Bourgois 1995; Sharff 1998; LeBlanc 2004), it is likely that the networks of current inmates lacked social capital (as defined here). For example, Bourgois describes being solicited as the *de facto* cultural broker for the subjects of his study whenever they needed to navigate the "rules and regulations of legal society" (1995: 28). Some of the deficits of social capital Bourgois describes (see. ch.'s 1 and 4) are those very ones which formal education can overcome (Bourdieu 1984). The limitations resulting from the low social capital of the student-inmates' social networks on the outside will be examined through the personal narratives and follow-up interviews.

Summary of Research Design

This study will take place at Large Correctional Facility, a maximum-security prison in Erehwon that houses 1,200 men, 75% of whom come from a nearby metropolitan area. The 75 men enrolled in the Local College program will be the subjects of this study.

The design of this study consists of 3 sections:

1. Narratives: the analysis of demographic and personal perspective data.

The 15 new student-inmates who are accepted into the program in the next enrollment term will write a personal narrative before their first semester. By not specifying or restricting the content of the narratives, the subject's choices will indicate personal relevance (Rosenthal 1993). After completing the first semester in the college program, and then again after the first year, these same individuals will be asked to write another narrative. Up to 50 veteran student-inmates (second and third year students) will also write personal narratives.

The narratives will be quantitatively coded individually according to "theme, tone, style, motivation, and characterization" (see: Maruna 2004:50). The body of narratives will then be compared diachronically and synchronically for changes in personal perspective.

I will conduct follow-up interviews with the subjects after writing each narrative to discover possible topics that were omitted, forgotten, or deemed irrelevant. Comparing this supplemental information with the narrative data will allow for comparative analyses of topic relevance.

2. Social networks: a descriptive account of how social networks are shaped by the inmates' participation in the educational program.

At the outset, I will meet with a core group of 6 to 8 student-inmates to elicit some of the relevant characteristics of a generalizable inmate demographic "matrix." The variables discussed will inform a preliminary interview schedule (see attachment 1 for sample interview questions). This section will utilize structured interviews (peer-to-peer, when appropriate) focusing on informal socializing experiences prior to the period of incarceration, as well as interviews with willing family members of inmates, focusing on the impact of the college program on family relations.

The interviews, among other techniques such as questionnaires, will provide information about the *strength* and *diversity* of the inmates' social networks prior to incarceration. The next sub-section will focus on network formation inside the prison.

The geographic "concentration of incarceration" (Fagan 2004) means that newly admitted inmates often recognize an acquaintance on the inside. Thus, network formation inside a prison is partially influenced by customs from the outside. Nonetheless, social ties inside a prison are subject to the particular constraints of that environment (Goffman 1961), such as cell location, work assignments, and program activities. This section aims to isolate the impact of the higher education program on social group formation. By asking subjects to describe their new socializing experiences and, in particular, the reciprocal responsibilities associated with their roles as students, it will be possible to determine how the college program shapes their social networks.

3. Post-release follow-up: assessment of long-term impact of college participation.

Though few in number, I believe the experiences of the several student-inmates who will be released during the period of this study will be useful for considering the long-term impact of college participation. Traditionally, attempts to explore the post-release experience of ex-inmates have been hindered by difficulties of tracking. This problem will be diminished with this population thanks to a recently implemented Local College program designed to assist ex-inmates with pursuing their academic careers post-release. Their continued connection to Local College provides the unusual opportunity for a meaningful follow-up component to this research.

It is likely, given the success of the proposal, that the review committee assumed that the author (and his advisors) would attend to the matter of developing a protocol for analysis of the data. A possible alternative would have been to offer a considerably condensed version describing the proposed method of analysis.

In the next section, the author restates his argument for the possible utility of findings from the study. He ends by simply asserting that the study has the potential to provide answers to the important questions posed at the outset.

Conclusion

What might be learned from the life stories of student-inmates? One clue comes from Maruna's (2001) analysis of the narratives of criminal *desisters* (long-term offenders who have stopped all illicit activities). Unlike *persisters*, they imagined their past illicit activities as aberrations to who they really are, as opposed to expressions of an unalterable identity. Desisters often took this a step further by using their past actions as a springboard for "some newfound calling," expressed in their interest to "give something back to society" (Maruna 2001: 9, 88). By acknowledging their actions and incorporating them into a "success narrative," desisters formulated a sustainable rejection of illicit activity. Maruna's study found these unique qualities in longtime offenders who were already living a newfound self. This study may provide evidence of how formal education can play a part in making this transformation take place.

As you see below, the author provided an attachment with the sample interview questions. This was an effective way to provide further information about the method without using space within the main proposal. If you were completing a proposal such as this it would be important to make certain that the rules do not prohibit attachments or appendices. In this case, using an attachment was an acceptable strategy, but in some cases it might not be permitted, or, if excessive, would be penalized during the review process.

Attachment 1

Sample Question Guide for Participant Interviews

1. Do you have any siblings?
   If so, did you have a good relationship with your siblings growing up?
2. Who did you spend most of your time with as a child?
3. How many "good" friends did you have as a child?, as an adolescent?, as a young adult? . . .
4. What kinds of activities did you do with your friends on a regular basis?
5. Have you had the same group of friends/acquaintances for a long time? If not, what precipitated the change?
6. How would you describe yourself in comparison to your childhood friends?
7. Where did you spend most of your time growing up, when you weren't in school?
8. What was your favorite thing to do with your parents/guardians?, friends?, or other relatives?
9. What was school like for you?
10. Were you involved in any other organized activities such as a community sports team, music lessons, summer camp, religious activities, etc.?
11. What are some of the more memorable events of your childhood?
12. As an adult, what were/are most of your friends doing for a living?
13. Do most of the people you consider to be a "good" friend live in your area?, or are they far away?
14. Before you were incarcerated, what was a typical day like for you?
15. Did you spend most of your time alone?, or with specific company?
16. Have you been able to keep in touch with some of your old friends since your incarceration?
17. Did you know anyone on the inside when you were first incarcerated?
18. What is it like trying to create a social group in here?
19. Have you maintained the same group of friends/acquaintances during your time on the inside?

20. Have you met any new people, including perhaps a significant other, since being incarcerated? If so, who introduced you?

21. Have you met any new people, who you would now consider a friend, since your participation in the college program or any other "program"? If so, do you study together?, Help one another with assignments?

My Professional Goals

At this point the author prepared a one-page overview of his professional goals. Because of its personal nature we have deleted that statement here. We should note, however, that the author directly responded to the RFP by briefly addressing how his graduate work led into the topic of interest and how he will use this study as a springboard for future research, including logical follow-up studies.

Budget and Narrative

In this section of the proposal the author provides a budget that is aligned with the maximum request permitted by the RFP ($3,000). Note how in the narrative he directly addresses funding priorities of the RFP and efficiently lets the review committee know that this grant, if awarded, would supplement other funding for the study. In addition, In the brief space provided, and as we suggested in Chapter 9, he provides a rationale for the expenses that are included in the budget.

Budget and Narrative

Itemized expenses for 1 year of research	*Cost*
Research materials	
Laptop computer	$1,400
Digital recording device*	$140
Transcription services and/or software	$500
Removable storage media (hard drive, zip disks, DVDs)	$300
Batteries, printing paper, notepads	$200
Photocopying	$300
Misc. office supplies	$200
TOTAL	$3,040

*already purchased

I will be the sole researcher working on this project. Therefore, all costs are associated with my own research expenses over the period of data collection. With respect to the stipulations of the Spencer RTG, as indicated above, the grant money requested will not be directed toward living expenses (i.e., a stipend). This proposal has already received positive notice of funding from two other grants that will cover those costs.

Because the principle research techniques will be interview and document analysis, the expense of research materials is limited to a recording device, storage media, and a laptop computer. The computer will be used to administer questionnaires at the research site, to store the digitally-recorded interviews, and for my own writing. There are computers at the site, but they cannot be used for this purpose because no transferable recording media can be removed from the research site.

The high cost of photocopying is due to the need for frequent and multiple copies of progress reports, required by the prison administration.

Not included here are the references cited in the proposal. In the original proposal the reference list was two and a half pages long. The RFP, however, was clear that the reference list was not included in the page count. Thus, the author was able to include all of the references without having to reduce the length of the main body of the proposal.

We should note that the RFP also asked for a support letter from the dissertation advisor. If you encounter a similar requirement, it will be important to notify your advisor promptly so that she or he will have time to draft a letter affirming support for both the importance of the topic and the adequacy of the proposed methodology. In addition, if you are able to provide your advisor with a draft of the proposal and then get feedback about the content and writing you will have an opportunity to do revisions.

Appendix A

Annotated Bibliography of Supplementary References

Information that might be helpful in preparing a research proposal can be found in publications of several different types. For example, valuable direction can be obtained from some textbooks that focus primarily on other topics. Among such references are books that deal with the conduct of research. Some of them do contain sections or chapters on the proposal, but because the central concern of the text is with the technical specification of design and method, the treatment of the proposal generally is brief and superficial. Standard references of this type, such as Gall, Gall, and Borg (2007), are available in most college libraries, and we have not included examples in this appendix. If desired, colleagues, advisers, and research professors can help the novice locate a general research method text that will best serve his or her particular proposal.

Another, and very different, category of potentially helpful books are those designed to assist in obtaining a grant—for support of research or (more commonly) a development project. In many of these, the focus is on how to locate funding sources, how to match proposed projects with agency agendas, and how the proposal evaluation process works. We have included several such guides in the annotations below, but again, because the primary focus of these books often is not on the development of a research proposal, readers will do well to examine library or bookstore copies before purchase.

A small additional category of publications consists of documents prepared by professional and scholarly organizations as guides in the area of research ethics, as manuals for document preparation and format, or as general advice for those embarking on careers in research. An appropriate selection of these should be on the desk of everyone who does research, and the novice will find that the standard references among them, such as the *Publication Manual of the American Psychological Association* (American Psychological Association, 2001), quickly become dog-eared with use. We have not attempted to include annotations of such documents for each disciplinary specialization, but some of the most helpful generic examples are abstracted here.

Finally, a growing number of texts are now available that, like this one, are directed specifically to the needs of people faced with the task of preparing a research proposal—whether for academic or grant solicitation purposes. Some of these presume investigation in certain areas (education, medicine, or social services, for example), but others are directed to a more general audience. Most are oriented toward empirical/quantitative studies, but some, like Marshall and Rossman's *Designing Qualitative Research* (2006) or Maxwell's *Qualitative Research Design: An Interactive Approach* (2005), serve investigators who intend to employ other paradigms. The political and social circumstances that are unique to graduate study are made central in some of these texts, whereas the logic of formulating design and method (or the mechanics of writing the formal document) are given greater attention in others. Accordingly, we have tried to include a wide selection of the proposal books now on the market. The best choice for many novices would be to combine several of those that serve your particular needs—for use in combination with the present text (of course).

American Psychological Association. (2001). *Publication manual of the American Psychological Association* (5th ed.). Washington, DC: Author.

This paperbound manual contains a comprehensive set of guidelines for the particular format commonly called "APA," as well as substantial sections that deal with other aspects of the process of scholarly writing. The manual is the standard reference for those students and professors in the social and behavioral sciences for whom APA is either a choice or a requirement of publication. The major topics covered are (a) content and organization of the manuscript, (b) expression of ideas (style, grammar, and reduction of bias), (c) APA editorial style, (d) manuscript preparation and sample paper, and (e) manuscript acceptance and production (submission, review, and proofreading). The sections on style and grammar provide a very condensed version of those grammatical rules that editors find more frequently violated by both beginning writers and more seasoned writers. The guidelines for reduction of bias (gender, race, disability, and sexual orientation) in language will be particularly valuable for proposal writers. The section devoted to handling of materials from on-line information sources and other electronic media reflects standards as of the date of publication. Given the rapidity of changes in this area, authors should seek more recent guidelines. The APA format and citation style are now accepted or required by a large number of universities and journals. If you are going to be using APA in the future, the proposal represents the perfect vehicle for mastering the details of that format.

Becker, H. S. (1986). *Writing for social scientists: How to start and finish your thesis, book, or article*. Chicago: University of Chicago Press.

This is not another style and form manual. The book is a collection of nine elegantly simple essays dealing with the problems of writing about research (a 10th chapter on confronting the psychological risks of writing is contributed by sociologist Pamela Richards). Becker attends to the day-to-day work demands of producing a document—whether a research proposal, dissertation, or journal report. These include the familiars of procrastination, writing successive drafts, editing your own work, getting help, using (or not using) a word processor, and how to write clear and attractive prose within the confines of academic conventions.

It takes a special kind of confidence to write a book about good writing, particularly if that book also has to give helpful advice to people who do not know how to do it. The author does this with such grace and gentle encouragement that many readers will be reminded of another slim volume—Strunk and White's (2000) *The Elements of Style*. Becker indeed is an established scholar with a formidable reputation for his writing style; he succeeds here by remembering exactly why writing is not easy for most people. If you do not have confidence in your prose voice or the established work habits required to accomplish a piece of extended writing, this is the book for you.

Coley, S. M., & Scheinberg, C. A. (2000). *Proposal writing* (2nd ed.). (Sage Human Services Guide, Vol. 63). Thousand Oaks, CA: Sage.

This lively little paperback is directed specifically at the task of preparing grant proposals (applications to agencies and foundations for funding of any activity—including research). Intended primarily for use by students and first-time grant writers, it is an ideal companion for Chapters 8 and 9 in the present text. The format is a straightforward and highly pragmatic walk through the process of proposal development from conceptualization to follow-up activities after the proposal is submitted. The authors even attend to such sophisticated topics as management of multisource funding, strategies for obtaining continuation funding, use of technology in proposal writing, and activities that enhance possibilities for future grants.

Clear and concrete guidelines and examples are provided at every step, including a particularly useful appendix containing a specimen grant proposal accompanied by full critique and evaluation as they would be prepared by an agency reviewer. By attending to proposals for a wide range of human service activities, the text does not focus exclusively on the particular case of locating and obtaining funding for research. Accordingly, some "translation" by the reader may be required to adapt the authors' advice to the demands of proposals in which inquiry is the central element. Our copy of the second edition also lacks a subject index. Notwithstanding those limitations, if soliciting funds for research is the purpose of your proposal, and if agencies or foundations are among the sources you want to tap, this guide may represent a useful addition to your reference shelf.

Committee on Science, Engineering, and Public Policy, National Academy of Science, National Academy of Engineering, and Institute of Medicine. (1995). *On being a scientist: Responsible conduct in research* (2nd ed.). Washington, DC: National Academy Press.

This 27-page booklet was a project of the National Academy of Science, National Academy of Engineering, and the Institute of Medicine's joint Committee on Science, Engineering, and Public Policy. It provides an overview of ethical principles as they apply to the process of research. As a statement reflecting the position of the sponsoring organizations, this is essential reading for those beginning research careers, as well as a source of continuing guidance for experienced investigators. A wide variety of topics is covered, including the social foundations of science, conflicts of interest, human error, negligence and misconduct in science, allocating credit and authorship practices, and responding to violations of ethical standards. Easy to read, with helpful sidebars to supplement the main text, this booklet is well worth the modest price (at the time of this writing, $5.00 for downloading as a PDF file from the Academy Press website) and the investment of reading time.

Creswell, J. W. (1998). *Qualitative inquiry and research design: Choosing among five traditions*. Thousand Oaks, CA: Sage.

In this substantial book (403 pages), the author undertakes an ambitious and unusual task. He has set out to illustrate in detail how the choice of a particular "type" of qualitative inquiry (also called a qualitative research "tradition") will shape the design—and, thus, the subsequent proposal for a study. That is not a purely theoretical matter, because just as is true within other research paradigms, all those who propose to do a qualitative study must answer the question "What type of qualitative study?" The considerable array of alternatives that, at least in theory, are available to the novice can be overwhelming. Creswell has selected five (biography, phenomenology, grounded theory, ethnography, and case study) and proceeds to explain how each component of the study design (definition of the study problem, collection of data, analysis, and preparation of the report) is shaped by the selection of one of the qualitative genres. Previously published study reports representing each of the five types are reprinted in appendices, as is a glossary of terminology unique to each type of research.

The section most closely related to our own text (Chapter 2, "Designing a Qualitative Study") offers only a general overview of the proposal process; nevertheless, if you are puzzling about which kind of qualitative research best fits your problem or question, capabilities, resources, and personal dispositions, then this text should help you to make an informed decision. That the detail contained in your proposal will be powerfully shaped by that choice is convincingly demonstrated in this book.

Creswell, J. W., & Plano-Clark, V. L. (2006). *Designing and conducting mixed methods research*. Thousand Oaks, CA: Sage.

Many of you will be tempted (probably for good reasons) to propose studies that mix either methods from the qualitative and quantitative traditions, or the paradigmatic models themselves. In Chapter 5, we present both our support and our reservations concerning such choices when made by a novice. Whatever your stage of development, however, if mixed methodology is to be considered, then this book, with its brisk prose style, wonderfully transparent set of starting definitions, and carefully selected exemplars of design, analysis, and interpretations, should be at the top of your reading list. If you couple this with the much more directly "how to" text by Thomas (below) you should have sufficient background to determine whether a mixed method design will serve your proposed study.

Creswell and Plano-Clark start right where many proposal authors are located—at the point of wondering exactly what mixed methodology is, what considerations have to be entertained in selecting or rejecting it as a study design, and, in the end, whether it really fits their research purpose. By dealing with those very practical decisions up front, they have placed their priority, we think, exactly where it belongs.

The authors then move on to treat the practical problems of actually executing a mixed method design: describing it in a proposal, collecting and analyzing data, and writing the report. Finally, they address exactly the questions that are most frequently raised about mixing methods from different paradigms (which leads us to recommend that many of you will want to move to Chapter 9 as soon as you have finished the Preface and opening chapter of introduction). As with other Sage Publications textbooks authored by John Creswell, there is a distinct pragmatic flavor to his advice. Keeping mixed method research projects manageable while ensuring quality is a theme throughout the book, and certainly one that ought to appeal to graduate students everywhere.

Fink, A. (2005). *Conducting research literature reviews: From paper to the Internet* (2nd ed.). Thousand Oaks, CA: Sage.

This book is designed for those who wish either to assess current knowledge in a particular area of research or to produce a literature review themselves. The former, of course, is relevant to your preparation for writing the proposal. As we have indicated here in Chapter 4, however, the latter is not the appropriate model for most proposals. In this paperbound text, the attention given to electronic search procedures, as well as to strategies for screening what is found, is both up to date and relevant to any area of inquiry. The examples and exercises in this second edition make the book particularly appropriate for use in proposals dealing with the areas of health, medicine, and social services.

Although some sections in this comprehensive text, notably those dealing with the methodological adequacy of retrieved reports, go beyond what most readers will require at the proposal stage of development, we think the book can serve both as an excellent resource for organizing the effort to identify and digest related literature found through use of the Internet, and for writing the synthesis of results that must be prepared for the proposal.

Galvan, J. L. (2004). *Writing literature reviews* (2nd ed.). Glendale, CA: Pyrczak Publishing.

Although Galvan had several audiences in mind when he prepared this second edition of his popular text (including graduate students proposing theses and dissertations, grant writers, and students facing term paper assignments), it is clear that it continued to be the genuine beginners who were foremost in his mind. If you already have experience with preparation of reviews and feel ready for some advanced coaching, there are other resources. Fink (2005), annotated above, might serve you better. The present book is a step-by-step guide written in straightforward prose that will be completely accessible to any student.

There are guidelines here for selecting a topic, retrieving studies, analyzing reports, synthesizing results, writing draft reviews, and editing the final product. The text is replete with examples and exercises, and the four sample literature reviews are excellent models of brief reviews that lead smoothly into proposals for a new research question. In this attractively produced paperback the author accomplishes only what he proposed to do. The resulting text introduces the beginner to the mysteries of doing a sound and serviceable review—and gives a firm push toward writing it up in ways that will serve the proposal process.

Girden, E. R. (2001). *Evaluating research articles: From start to finish* (2nd ed.). Thousand Oaks, CA: Sage.

This book is specifically intended for graduate students who have finished an introductory/intermediate course in statistics and who are at least familiar with the protocols of research design. It will not serve to replace textbooks in those areas, but for many who are preparing to write their proposals it will be the perfect bridge between the abstractions of academic courses and the realities of crafting a sound design and winnowing the research literature that must serve as its foundation.

The purpose of the book is to train both serious research consumers and beginning researchers in the task of critically reading a research report. Two examples are drawn from the literature to illustrate each of 11 different research designs. The first example is dissected by the author, while the second is presented as an exercise in critical evaluation for the reader (with answers for each critique question provided in the rear of the book). The stress throughout is on what particular elements make for the construction of sound designs—and supportable conclusions.

The research designs include both quantitative and qualitative formats: case studies, narrative analysis, surveys, quasi-experimental studies, correlation studies, several formats for experimental research, and quantitative designs involving regression analysis, factor analysis, and discriminant analysis. The author introduces each design with a list of "caution factors" that point the critical reader toward aspects that are most likely to be troublesome. Her own analyses pursue those points with complete explanation for her own evaluations.

Our advice here is simple: (a) if you are going to make use of one of the designs included in this text, and (b) if you are prepared for some tutorial exercises that go beyond the introductory and elementary, this book may serve you in ways that make the difference between adequacy and excellence in your analysis of the related literature and, ultimately, in the design you propose.

Glatthorn, A. A., & Joyner, R. L. (2005). *Writing the winning thesis or dissertation* (2nd ed.). Thousand Oaks, CA: Corwin Press.

Although not indicated in the title, this step-by-step guide was prepared explicitly for graduate students in the field of education. Despite that focus (most of the illustrative material involves school settings), first-time proposal authors will find a great deal of practical advice that is relevant to the generic problems of formulating research plans in any professional field.

Twelve chapters are devoted to activities that occur prior to actual execution of the study. In two of them, the author urges first the development of a preproposal (a "prospectus"), and then its subsequent use in conferences with advisers (advice that we heartily second). Additional chapters take the reader through developing, writing, and formally presenting the proposal. Added in this second edition are chapters dealing with peer collaboration during preparation of the proposal, and using technology in prudent and skillful ways—both of which respond to opportunities and problems that most graduate students will immediately recognize.

As the authors themselves observe, this book was not prepared as an exercise in sophisticated scholarship. It is a handbook intended to give technical assistance to people who have never performed the tasks they now face—identifying a research topic, designing a study, and preparing the written proposal. Although purists may wince at the suggestion that novices have to learn how to "sound academic" when they write about their study, some acquaintance with the prose styles found in published research reports will suggest that the observation contains more than a grain of truth. In any case, Glatthorn and Joyner are far more concerned with helping the novice achieve clarity than they are with giving advice about creating the right image.

As experienced dissertation advisers, the authors have accumulated a substantial fund of suggestions that, though they often deal with routine matters, are at the heart of getting the work done. Such homey things as keeping duplicate records, maintaining a planning chart, and settling the first person versus third person question early in the writing process are better attended to than neglected. This modest paperback will allow anyone to profit from what hard experience has taught others about preparing research proposals.

Krueger, R. A., & Casey, M. A. (2000). *Focus groups: A practical guide for applied research* (3rd ed.). Thousand Oaks, CA: Sage.

The textbook marketplace now offers a variety of books dealing with the use of focus group techniques for social research. A number of these are identified in our discussion of the technique in Chapter 5. Some are intended to develop the theoretical basis for the design, while others serve the more practical end of explaining how to use focus groups to generate data that will answer research questions. This book is unabashedly one of the latter as the authors devote their efforts entirely to the work of planning and executing a focus group study. Accordingly, the sequence of treatment runs from planning, through moderating groups, and on to analysis of data and reporting findings.

With the advantage of revisions through multiple editions, the writing is completely transparent and free of jargon, lively and full of good humor, and yet also serious about the importance of doing sound inquiry and respecting the people who offer their stories as data. Produced with a utilitarian spiral binding, lovingly illustrated with original drawings, organized with the insertion of graphic icons to signal different functions in the text (procedural cautions, tips for good practice, general principles, checklist reminders, etc.), and replete with examples, this text is precisely what the title purports—a guide for applied research.

If you need the detail of a thorough theoretical framework, or want to examine the history and development of focus group technique in the social sciences, you will need to go elsewhere. If you need the "how-to" advice of veteran focus group researchers, we think this should be your first stop.

Locke, L. F., Silverman, S. J., & Spirduso, W. W. (2004). *Reading and understanding research* (2nd ed.). Thousand Oaks, CA: Sage.

Our motives in suggesting this text actually are not entirely self-serving. For the beginner, reading and making sense out of research reports will occupy a major part of the time and effort expended in preparing to write a proposal. Knowing how to do that work from the outset can make the whole process far more efficient—and rewarding. Further, having at least rudimentary skills in estimating the quality of research represented in a report (and, thus, the degree of confidence warranted for its findings) will serve the novice well, both when conversing with advisers and when preparing the related literature portions of the proposal. Those three functions—reading, understanding, and making critical assessments—are the central topics of our "other" book.

Written in a style similar to the text you now have in hand, *Reading and Understanding Research* is designed as a guide for the novice investigator, whether used as a textbook in a formal course or as a do-it-yourself handbook for independent study. The text contains step-by-step study and practice procedures as well as a variety of recording forms and checklists designed to support beginners during their first encounters with actual reports. The book is intended for use with a variety of paradigmatic formats (both quantitative and qualitative) as they are employed in the social sciences and fields of applied study.

Among the sections that will be of particular interest to proposal writers are those devoted to (a) an overview of the types of research encountered in reports, (b) a discussion of the sources of credibility in published studies, (c) identification of reasons to suspend trust in the findings of an investigation, (d) a list of basic questions to ask when reading research with a critical eye, and (e) directions for "explaining" reports as a means of improving your understanding of what they contain. Appendices include a beginner's guide to reading statistics (in reports), as well as summary explanations of actual examples of both research reports and research reviews.

Marshall, C., & Rossman, G. B. (2006). *Designing qualitative research* (4th ed.). Thousand Oaks, CA: Sage.

This remains one of the few books devoted exclusively to the topic of formulating plans for qualitative research (also see Maxwell, below). Although much updated with details that reflect the contemporary research environment, this new edition retains the direct and pragmatic flavor that characterized the earlier versions. Put simply, the authors treat the proposal document as a framework for organizing the many activities that go into thinking through and presenting a research design.

Although this is not a how-to-write-it guidebook for preparing the proposal, neither is it a standard how-to-do-it research methods text for beginners. The book is intended to provide clear, practical guidance for anyone at the starting point of a qualitative study, whether novice or experienced investigator. The target throughout is thorough and systematic consideration of the major problems in designing a study—from identifying and defining the problem to defending the proposed method of investigation.

Discussion of the design process is cast as the development of an argument that will convince funding agency reviewers or dissertation committee members of the importance and feasibility of a qualitative study—and of the author's competence to bring it to a successful completion. Each step is illustrated with brief (one to three pages) vignettes that underscore the nature of the problems encountered and the solutions available. The twin themes of maintaining design flexibility and marshaling supportive evidence are used throughout to weld the individual parts of the process into an integrated whole.

The text is sensitive to the many alternative approaches to qualitative study, including feminist research, critical ethnography, and qualitative evaluation formats. Graduate students will find the chapter devoted to defending the value and logic of qualitative research particularly helpful. The authors understand that some of their readers will have to argue their case before reviewers who either are unfamiliar with the paradigm or are uneasy about the adequacy of qualitative methods. Their commentary on this problem avoids defensive posturing by staying firmly positive and pragmatic. Also of particular value for first-time authors of qualitative proposals will be the unique chapter on estimating and managing time and resources—a topic almost invariably ignored or underestimated by the beginner.

It is impossible for us to imagine a qualitative thesis or dissertation proposal that would not profit from a close reading of this volume, as early in the process as possible. Teamed with our own text, this book offers a solid foundation for preparing a successful proposal.

Maxwell, J. A. (2005). *Qualitative research design: An interactive approach* (2nd ed.) (Sage Applied Social Research Methods Series, Vol. 41). Thousand Oaks, CA: Sage.

Although this small paperback has only one full chapter devoted to writing the proposal for a qualitative study, in a real sense the entire book is about preparing to write that proposal. The author simply regards the research "design" as the logic of a study and the research "proposal" document as a vehicle for communicating that logic to a particular audience.

As a result of the author's "interactive approach," the reader is engaged in a series of provocative questions and lively writing exercises that are precisely those required to create a research plan: (a) Why are you doing this study? (b) What do you think is going on? (c) What do you want to understand? (d) What will you actually do? and (e) How might you be wrong? The last question in that sequence deals with the problem of reassuring proposal readers about the validity of the study. Because that is one of the most commonly asked questions in the presentation/defense of a qualitative proposal, such attention gives particular strength to the text. A full proposal for a qualitative study is contained in an appendix—complete with the author's commentary on each section.

Among the changes in this new edition is an expanded discussion of personal goals as a legitimate factor in research design, a topic that is rarely addressed in research texts. Also contained here is a valuable discussion of how a conceptual framework (often a troublesome construct for novice investigators) is built and how it must be connected to the research paradigm within which the study will be situated.

This is not a compendium of methods for data collection and analysis, nor is it a thorough explication of the philosophical and theoretical underpinnings of the qualitative paradigm. Instead, it is an overview of what qualitative researchers must actually do in thinking through the study they wish to propose.

Patten, M. L. (2005). *Proposing empirical research* (3rd ed.). Glendale, CA: Pyrczak Publishing.

The subtitle for this text is "A guide to the fundamentals" and it is exactly that. The book contains step-by-step instructions for students who are writing their first research proposal. Although the approach is generic in both tone and coverage, it is clear that research in the social and behavioral sciences is the intended target. The level of difficulty is pitched perfectly for an undergraduate senior seminar or an introductory research course for master's degree candidates. The general absence of jargon and unessential technical terminology means that, although set up with student exercises and class discussion topics, the text also can be used as a self-study guide. Along that line, if you are facing the threatening rigors of a first research course, Patten's guide will allow you to embark on that journey with confidence and a solid head start.

Nine model proposals allow you to see exactly what such documents look like. Each is based on a study report already published in a research journal. Patten and the author of each study simply worked backwards to create a proposal that would fit the final form of the report. Although such artificial products might lack the verisimilitude provided by the false starts, superfluous language, and general floundering about of the novice, these proposals are completely competent, transparent, compact, and tightly reasoned. They also involve a variety of designs including surveys, experiments, quasi-experiments, qualitative inquiry, and mixed method research.

The book is produced in the attractive paperbound format used for the entire research series issued by Pryczak Publications. If you have not previously encountered one of the Pryczak introductory guides for research, this is a good place to start. Just remember that the intention here is "introduction and overview." If you persevere through all 208 pages you will have acquired basic literacy in the system language of inquiry, understand what a proposal must accomplish, and, whether you find research attractive or not, you will surely have some sense that writing one is something you actually could do. Not an inconsiderable set of outcomes for any textbook.

Penslar, R. L. (Ed.). (1995). *Research ethics: Cases and materials*. Bloomington, IN: Indiana University Press.

This book fills a gap that has long existed in the literature on the ethical problems that arise in planning, conducting, and reporting research. The editor's intention in this text, which is designed as a casebook, was to present materials that would lead research workers to think more deeply about the ethical problems presented by the human circumstances accompanying formal inquiry. The book includes chapters on both ethical theory and the teaching of research ethics. Case materials are grouped into (a) natural sciences, (b) behavioral sciences, and (c) humanities. Case examples are followed by questions designed to stimulate individual reflection or group discussion. Touching on such issues as scientific misconduct, authorship, animal and human subjects, deception, faculty-student relationships, intellectual property rights, and other areas of ethical concern, the text both extends and reinforces our own discussion in Chapter 2. The proposal is the appropriate point at which to begin your consideration of ethics, and *all* studies require such consideration.

Rossman, G. B., & Rallis, S. F. (2003). *Learning in the field: An introduction to qualitative research* (2nd ed.). Thousand Oaks, CA: Sage.

Given the overlap of authorship, you might expect to find that the task of preparing a proposal is described here in the same broad terms as were used in Marshall and Rossman's book on designing qualitative studies (see above). Although the lengthy chapter devoted to the planning and proposing process indeed does summarize some of the material from that other text, we believe there still are good reasons to consult *Learning in the Field* as part of your preparation for writing a proposal.

We think that many readers will find the authors' strategy of describing how three fictional graduate students gradually developed very different proposals is an effective teaching device. Through imaginary conversations, the reader is made aware of how experience gradually leads people with differing interests and research goals toward different types of inquiry—and proposal content. Because all proposals must argue for the logic of links between researcher, research goal, and research method, this text can assist beginning investigators to gradually forge those connections on the anvil of their own experiences.

Based on both extensive feedback from users of the first edition and the personal teaching experiences of the authors, the book has been substantially expanded and reorganized. These changes range, on the one hand, from a much-augmented presentation of methods for data analysis to, on the other hand, the simple use of "communities of practice" to describe what can happen when small groups of peers work together on the development of research proposals. Along the range that stretches between those two examples rest all of the other thoughtful efforts that Rossman and Rallis have employed to craft a more transparent and useful text. If you are considering a qualitative study, we think you will find that their investment as authors will fully justify your own as a reader.

Sales, B. D., & Folkman, S. (Eds.). (2000). *Ethics in research with human participants*. Washington, DC: American Psychological Association.

This edited volume, a project of an American Psychological Association task force, deals with many issues related to ethical treatment of human subjects. The book begins with a chapter titled "Moral Foundations of Research with Human Participants" that is less a discussion of morals and more a prelude to ethical decision making in research. This chapter establishes a theme for the entire book by proposing that research ethics involve much more than simply following rules, and by cautioning that there often will be difficult tensions that must be resolved among various moral principles.

Throughout this collection the authors also emphasize the responsibility of the investigator for making ethical decisions that are appropriate to the particular context of his or her study. In the final chapter, titled "Identifying Conflicts of Interest and Resolving Ethical Dilemmas," the twin themes of moral complexity and personal responsibility are brought together within a simple dictum for what constitutes responsible behavior. Ethical conduct in research involves making ethically informed decisions.

Between the first and last chapter are two sections that focus on "Ethical Issues across the Research Process" and "Ethical Concerns within the Research Community." The section on the research process has chapters on planning research, recruitment of participants, informed consent, and privacy and confidentiality. The section on ethics and the research community goes beyond what most books of this type cover and addresses additional responsibilities to participants, authorship and intellectual property, and ethical training.

We believe that this collection, in combination with our own treatment of ethics in Chapter 2 and the several other resources dealing with responsible science included in this appendix, will provide a fully sufficient foundation for writing a proposal that meets any reasonable test for ethical research. Profound wisdom is not required, but a proposal that includes thoughtful anticipation of the ethical concerns described in this book is the necessary first step toward meeting your responsibility for doing ethical research.

Schumacher, D. (1992). *Get funded!* Newbury Park, CA: Sage.

This is the third edition of *Proposals That Work* that has offered an annotation for this small guide to the writing of grant proposals. That longevity reflects the fact that Schumacher continues to offer material that is not otherwise available in such inexpensive and compact form. Advertised as a "practical guide for scholars seeking research support from business," it begins with an explanation of the corporate environment, gives suggestions on how the researcher can discover needs and interests within that culture, outlines strategies for convincing corporate managers that a proposed study might contribute to their objectives, and ends with a practical discussion of ways of organizing such privately sponsored research once it has been funded. Throughout, the emphasis is on thinking in terms of a partnership with business—with the necessary reciprocity that entails.

Declining governmental support for research in some areas dictates that many investigators will have to widen their search for funding. As the author suggests, building partnerships with business may well be the wave of the future. Those who wish to test the corporate waters should find this text an excellent place to begin.

Schwandt, T. A. (2001). *Dictionary of qualitative inquiry* (2nd ed.). Thousand Oaks, CA: Sage.

In keeping with its title, this book can serve as a desktop resource for looking up the meaning of terms used in qualitative research reports. In addition, however, because many of the definitions contain explanations of how qualitative researchers think and do their work, many readers will find themselves browsing through the pages as though the dictionary were a textbook. That the author makes both kinds of use easy through graceful and inviting prose is a perfect example of what is meant by "added value."

There are several sorts of readers who may want this reference on their bookshelf. One group is made up of people who want to read reports of research in which the approach was qualitative, but find that the specialized terminology makes it an unpleasant (or impossible) chore. A second group of customers for Schwandt's lexicography will be the many graduate students who are struggling to master the rudiments of the qualitative paradigm, whether as a research course requirement or as preparation for a thesis or dissertation proposal. For either purpose this will be a helpful supplement to the standard texts on qualitative research.

The book includes 304 alphabetically ordered terms, including 100 that did not appear five years ago in the first edition, which tells you something very important about what is going on in the practice of qualitative inquiry. The "definitions" given really are commentaries on the meaning of words in this particular research language (and sometimes run to the length of a small essay). Cross-references are given in bold face making it easy to follow a thread of meaning across several levels of terminology.

The author does have a point of view about social inquiry, although that partisanship is well disguised. Nonetheless, we think you will find that his treatments of central constructs in the qualitative research tradition come across as honest, economical, and richly informative.

Shadish, W. R., Cook, T. D., & Campbell, D. T. (2002). *Experimental and quasi-experimental designs for general causal inference.* Boston: Houghton Mifflin.

A follow-up to Campbell and Stanley's classic monograph on this topic (1963) and to Cook and Campbell's later more extensive explication of that work (1979), this text provides a comprehensive discussion of validity in research design for everything from true experiments to all forms of quasi-experiments. Included as part of the discussion are examples drawn from studies conducted in field setting such as schools, community centers, hospitals, and other sites where it would be difficult to control some of the variables. In that way, the authors address many topics that range beyond the comfortable confines of laboratory experiments.

The focus throughout is on the mechanisms of inquiry that influence the validity of data. Even if you are going to propose a study employing one of the traditions within the qualitative paradigm for which the construct of validity may assume a different function, we think that you must start with basic literacy concerning all forms of empirical research—whatever their paradigmatic foundation.

Thomas, R. M. (2003). *Blending qualitative & quantitative research methods in theses and dissertations.* Thousand Oaks, CA: Corwin Press.

Our cautions and reservations about research proposals involving mixed methodologies are presented in Chapter 5 and form the background for this annotation. Whatever concerns we may have, the construct of mixed methods obviously is here to stay. Mixed method studies are being published in virtually every outlet for social and behavioral research. It is inevitable that proposals for mixed method theses, dissertations, and grant-supported research projects would be a part of that broad movement.

In this plump little paperback Professor Thomas follows one rule with tenacious consistency. He believes that it is far better to show readers how things work than to talk (or write) about how things work. Surely, more than half of the 240 pages in this text are devoted to carefully selected examples—many of them extracted from theses or dissertations.

It is important for you to understand what this book is not. It does not provide a detailed resource for learning how to plan, conduct, and report research—mixed method or otherwise. It also is not a prescription for the "correct" way to combine quantitative and qualitative approaches to inquiry. Instead, the book describes a wide range of data collection methods and appurtenant design formats, including an inventory of their advantages and limitations, and then leaves the reader to select the combination (mixture) that best suits the particular needs of his or her study.

We particularly laud Thomas's inclusion of an entire section for illustration of replicated studies as a legitimate (or even desirable) form of mixed method research. Thomas has his own way of categorizing types of research and it may take you some time to become comfortable with its use. Various forms of correlational study, for example, are found with case studies and surveys in a chapter entitled "Present Status Perspectives." If you will persevere with thumbing through the pages, however, we think you will find it well worth the effort.

Voght, W. P. (1999). *Dictionary of statistics and methodology: A nontechnical guide for the social sciences*. Thousand Oaks, CA: Sage.

This nifty little paperback provides nearly 2000 nontechnical definitions of statistical concepts and methodological terminology in a simple alphabetical format that makes everything easy to find. The goal of the author was to enable people to read research reports containing accounts of procedures with which they are not familiar. All definitions are written in crisp, transparent English and none depend on formulas for their explanation. If nothing else, this allows readers to bypass the usual chore of searching through a standard textbook (often without a glossary or adequate index) to locate an accessible explanation of a research concept or procedure.

Virtually all of us, present authors included, have once learned about terms and techniques that through lack of use become rusty and thus require occasional refreshment. Examples such as "fixed-effects model," "coefficient of determination," "floor effect," "purposive sampling," and "R-squared test," can be read intelligently given a simple working definition (or the gentle shove of a reminder). Cross-referencing to related definitions is nicely provided as part of many entries. For example, the entry for "paradigm" lists "schema" and "model" as associated constructs that may properly flesh out the definition of the original entry. Finally, this lexicon includes a list of frequently encountered symbols used in research reports, as well as a list of other dictionaries and reference works (from a variety of disciplines) that offer greater depth of information.

References

Altman, L. K., & Broad, W. J. (2005, December 20). Global trend: More science, more fraud. *New York Times*. Retrieved December 20, 2005 from http://www.nytimes.com

American Educational Research Association. (1992). Ethical standards of the American Educational Research Association. *Educational Researcher, 21*(7), 23–26.

American Historical Association. (1998). *Grants, fellowships, and prizes of interest to historians* (1998–99 ed.). Washington, DC: Author.

American Psychological Association. (2001). *Publication manual of the American Psychological Association* (5th ed.). Washington, DC: Author.

American Psychological Association. (2002, October 8). *Ethical principles of psychologists and code of conduct*. Retrieved July 2, 2006 from http://www.apa.org/ethics/code2002.html

Bailey, C. A. (1995). *A guide to field research*. Thousand Oaks, CA: Pine Forge Press.

Bauer, D. G. (2003). *The "how to" grants manual: Successful grantseeking techniques for obtaining public and private grants* (5th ed.). Westport, CT: Praeger.

Bazeley, P., & Richards, L. (2000). *The NVivo qualitative project book*. London: Sage Ltd.

Becker, H. S. (1986). *Writing for social scientists: How to start and finish your thesis, book, or article*. Chicago: University of Chicago Press.

Bell, R. (1992). *Impure science: Fraud, compromise, and political influences in scientific research*. New York: Wiley.

Bogdan, R., & Biklen, S. (2003). *Qualitative research for education: An introduction to theory and method* (4th ed.). Boston: Allyn & Bacon.

Boote, D. N., & Beile, P. (2005). Scholars before researchers: On the centrality of the dissertation literature review in research preparation. *Educational Researcher, 34*(6), 3–15.

Bronowski, J. (1965). *Science and human values*. New York: Harper & Row.

Bryant, M. T. (2004). *The portable dissertation advisor*. Thousand Oaks, CA: Corwin Press.

Campbell, D. T., & Stanley, J. C. (1963). *Experimental and quasi-experimental designs for research*. Chicago: Rand McNally.

Carr, W., & Kemmis, S. (1986). *Becoming critical: Education, knowledge and action research*. London: Falmer.

Carspecken, P. F. (1996). *Critical ethnography in educational research: A theoretical and practical guide* (Routledge Critical Social Thought Series). New York: Routledge.

Coffey, A. J., & Atkinson, P. A. (1996). *Making sense of qualitative data: Complementary research strategies.* Thousand Oaks, CA: Sage.

Coley, S. M., & Scheinberg, C. A. (2000). *Proposal writing* (2nd ed.) (Sage Human Services Guide, Vol. 63). Thousand Oaks, CA: Sage.

Committee on Science, Engineering, and Public Policy, National Academy of Science, National Academy of Engineering, and Institute of Medicine. (1995). *On being a scientist: Responsible conduct in research* (2nd ed.). Washington, DC: National Academy Press.

Cook, T., & Campbell, D. (1979). *Quasi-experimentation: Design and analysis issues for field settings.* Boston: Houghton Mifflin.

Creswell, J. W. (1998). *Qualitative inquiry and research design: Choosing among five traditions.* Thousand Oaks, CA: Sage.

Creswell, J. W. (2003). *Research design: Qualitative, quantitative, and mixed method approaches* (2nd ed.). Thousand Oaks, CA: Sage.

Creswell. J. W., & Plano-Clark, V. L. (2006). *Designing and conducting mixed methods research.* Thousand Oaks, CA: Sage.

Datta, L. (1994). Paradigm wars: A basis for peaceful coexistence and beyond. In C. S. Reichardt & S. F. Rallis (Eds.), *The qualitative-quantitative debate: New directions for program evaluation* (pp. 53–70). San Francisco: Jossey-Bass.

Delamont, S. (2001). *Fieldwork in educational settings* (2nd ed.). London: Falmer.

Denscombe, M. (2003). *The good research guide for small scale research projects* (2nd ed.). Buckingham, UK: Open University Press.

Denzin, N. K., & Lincoln, Y. S. (Eds.). (2002). *The qualitative inquiry reader.* Thousand Oaks, CA: Sage.

Denzin, N. K., & Lincoln, Y. S. (Eds.). (2005). *The SAGE handbook of qualitative research* (3rd ed.). Thousand Oaks, CA: Sage.

Eisenhart, M. A., & Howe, K. R. (1992). Validity in educational research. In M. D. LeCompte, W. L. Millroy, & J. Preissle (Eds.), *The handbook of qualitative research in education* (pp. 643–680). San Diego: Academic Press.

Eisner, E. W. (1981). On the differences between scientific and artistic approaches to qualitative research. *Educational Researcher, 59*(4), 5–9.

Eisner, E. W., & Peshkin, A. (Eds.). (1990). *Qualitative inquiry in education: The continuing debate.* New York: Teachers College Press.

Emerson, R. M., Fretz, R. I., & Shaw, L. L. (1995). *Writing ethnographic field notes.* Chicago: University of Chicago Press.

Fanger, D. (1985, May). The dissertation, from conception to delivery. *On Teaching and Learning: The Journal of the Harvard-Danforth Center, 1,* pp. 26–33.

Farrell, A. (Ed.). (2005). *Ethical research with children.* New York: Open University Press.

Fetterman, D. M. (1998). *Ethnography: Step by step* (2nd ed.) (Sage Applied Social Research Methods Series, Vol. 17). Thousand Oaks, CA: Sage.

Fink, A. (2005). *Conducting research literature reviews: From paper to the Internet* (2nd ed.). Thousand Oaks, CA: Sage.

The Foundation Center (2005a). *The foundation directory* (2005 ed.). New York: Author.

The Foundation Center (2005b). *The foundation directory: Part 2* (2005 ed.). New York: Author.

The Foundation Center (2005c). *The foundation directory supplement* (2005 ed.). New York: Author.

The Foundation Center (2005d). *Foundation grants to individuals* (14th ed.). New York: Author.

The Foundation Center (2005e). *National directory of corporate giving* (11th ed.). New York: Author.

Freebody, P. (2003). *Qualitative research in education.* Thousand Oaks, CA: Sage.

Gall, M. D., Gall, J. P., & Borg, W. R. (2007). *Educational research: An introduction* (8th ed.). Boston: Allyn & Bacon.

Galvan, J. L. (2004). *Writing literature reviews* (2nd ed.). Glendale, CA: Pryczak Publishing.

Girden, E. R. (2001). *Evaluating research articles: From start to finish* (2nd ed.). Thousand Oaks, CA: Sage.

Gitlin, A. (Ed.). (1994). *Power and method: Political activism and educational research.* New York: Routledge.

Glatthorn, A. A., & Joyner, R. L. (2005). *Writing the winning thesis or dissertation* (2nd ed.). Thousand Oaks, CA: Corwin Press.

Glesne, C. (2006). *Becoming qualitative researchers: An introduction* (3rd ed.). Boston: Allyn & Bacon.

Golden Biddle, K., & Locke, K. D. (1997). *Composing qualitative research.* Thousand Oaks, CA: Sage.

Green, J. C., & Caracelli, V. J. (Eds.). (1997). *Advances in mixed method evaluation: The challenges and benefits of integrating diverse paradigms.* San Francisco: Jossey-Bass.

Greenbaum, T. L. (1998). *The handbook for focus group research* (2nd ed.). Thousand Oaks, CA: Sage.

Greenbaum, T. L. (2000). *Moderating focus groups: A practical guide for group facilitation.* Thousand Oaks, CA: Sage.

Guba, E. G., & Lincoln, Y. S. (1989). *Fourth generation evaluation.* Newbury Park, CA: Sage.

Guston, D. H. (1993). Mentoring and the research training experience. In Panel on Scientific Responsibility and the Conduct of Research, Committee on Science, Engineering, and Public Policy, National Academy of Sciences, National Academy of Engineering, and Institute of Medicine (Eds.), *Responsible science: Ensuring the integrity of the research process* (Vol. 2, pp. 50–65). Washington, DC: National Academy Press.

Holliday, A. (2002). *Doing and writing qualitative research.* Thousand Oaks, CA: Sage.

Huberman, A. M., & Miles, M. B. (2002). *The qualitative researcher's companion.* Thousand Oaks, CA: Sage.

Information Today, Inc. (2005). *Annual register of grant support* 2006 (39th ed.). Medford, NJ: Author.

Jacob, E. (1987). Qualitative research traditions: A review. *Review of Educational Research, 57,* 1–50.

Jacob, E. (1988). Clarifying qualitative research: A focus on traditions. *Educational Researcher, 17*(1), 22–24.

Jacob, E. (1989). Qualitative research: A defense of traditions. *Review of Educational Research, 59,* 229–235.

Jensen, B. E., Martin, K. A., & Mann, B. L. (2003). Journal format versus chapter format: How to help graduate students publish. *Measurement in Physical Education and Exercise Science, 7,* 43–51.

Johnson, R. B. (1997). Examining the validity structure of qualitative research. *Education, 118,* 282–292.

Johnson, R. B., & Onwuegbuzie, A.J. (2004). Mixed methods research: A research paradigm whose time has come. *Educational Researcher, 33*(7), 14–26.

Jorgensen, D. L. (1989). *Participant observation* (Sage Applied Social Research Methods Series, Vol. 15). Newbury Park, CA: Sage.

Kidd, P. S., & Parshall, M. B. (2000). Getting the focus and the group: Enhancing analytical rigor in focus group research. *Qualitative Health Research, 10,* 293–308.

Kimmel, A. J. (1988). *Ethics and values in applied social research* (Sage Applied Social Research Methods Series, Vol. 12). Newbury Park, CA: Sage

Kraemer, H. C., & Thiemann, S. (1987). *How many subjects? Statistical power analysis in research.* Newbury Park, CA: Sage.

Krueger, R. A., & Casey, M. A. (2000). *Focus groups: A practical guide for applied research* (3rd ed.). Thousand Oaks, CA: Sage.

Kroll, W. (1993). Ethical issues in human research. *Quest, 45,* 32–44.

Kuhn, T. S. (1996). *The structure of scientific revolutions* (3rd ed.). Chicago: University of Chicago Press.

Kvale, S. (1994). Ten standard objections to qualitative research interviews. *Journal of Phenomenological Psychology, 25,* 147–173.

Kvale, S. (1995). The social construction of validity. *Qualitative Inquiry, 1,* 19–40.

Lather, P. (1991). *Getting smart: Feminist research and pedagogy with/in the postmodern* (Routledge Critical Social Thought Series). New York: Routledge.

LeCompte, M. D., Milroy, W. L., & Preissle, J. (Eds.). (1992). *The handbook of qualitative research in education.* San Diego: Academic Press.

Lincoln, Y. S. (1995). Emerging criteria for quality in qualitative and interpretive research. *Qualitative Inquiry, 1,* 275–289.

Lincoln, E. S., & Guba, E. G. (1985). *Naturalistic inquiry.* Beverly Hills, CA: Sage

Locke, L. F., Silverman, S. J., & Spirduso, W. W. (2004). *Reading and understanding research* (2nd ed.). Thousand Oaks, CA: Sage.

Lofland, J., Snow, D. A., Anderson, L., & Lofland, L. H. (2005). *Analyzing social settings: A guide to qualitative observation and analysis* (4th ed.). Belmont, CA: Wadsworth.

Macrina, F. L. (2005). *Scientific integrity: Text and cases in responsible conduct of research* (3rd ed.). Washington, DC: ASM Press.

Madison, D. S. (2005). *Critical ethnography: Method, ethics, and performance.* Thousand Oaks, CA: Sage.

Marshall, C., & Rossman, G. B. (2006). *Designing qualitative research* (4th ed.). Thousand Oaks, CA: Sage.

Mauthner, M., Birch, M., Jessop, J., & Miller, T. (2002). *Ethics in qualitative research.* Thousand Oaks, CA: Sage.

Maxwell, J. A. (1992). Understanding and validity in qualitative research. *Harvard Educational Review, 62,* 279–300.

Maxwell, J. A. (2005). *Qualitative research design: An interactive approach* (2nd ed.). (Sage Applied Social Research Methods Series, Vol. 41). Thousand Oaks, CA: Sage.

Mays, N., & Pope, C. (2000). Assessing quality in qualitative research. *British Medical Journal, 320,* 50–52.

Merriam, S. B. (2001). *Qualitative research and case study applications in education* (2nd ed.). San Francisco: Jossey-Bass.

Merriam, S.B. (Ed.). (2002). *Qualitative research in practice: Examples for discussion and analysis.* San Francisco: Jossey-Bass.

Miceli, M. P., & Near, J. P. (1992). *Blowing the whistle.* Lanham, MD: Lexington Books.

Miles, M. B., & Huberman, A. M. (1994). *Qualitative data analysis* (2nd ed.). Thousand Oaks, CA: Sage.

Milinki, A. K. (Ed.) (1999). *Cases in qualitative research: Research reports for discussion and evaluation.* Los Angeles, CA: Pyrczak.

Morgan, D. L. (1997). *Focus groups as qualitative research* (2nd ed.) (Sage Qualitative Research Methods Series Vol. 16). Thousand Oaks, CA: Sage.

Morse, J. M. (Ed.). (1997). *Completing a qualitative project: Details and dialogue.* Thousand Oaks, CA: Sage.

Morse, J. M., & Richards, L. (2002). *Readme first for a user's guide to qualitative methods.* Thousand Oaks, CA: Sage.

Morse, J. M., Swanson, J. M., & Kuzel, A. J. (Eds.). (2001). *The nature of qualitative evidence.* Thousand Oaks, CA: Sage.

National Academy of Science, National Academy of Engineering, and Institute of Medicine. (1997). *Advisor, teacher, role model, friend: On being a mentor to students in science and engineering.* Washington, DC: National Academy Press.

O'Brien, P. K. (1995). The reform of doctoral dissertations in humanities and social studies. *Higher Education Review, 28,* 3–19.

Oliver, P. (2003). *The student's guide to research ethics.* Philadelphia: Open University Press.

Oryx Press. (2005a). *Directory of biomedical and health care grants* (19th ed.). Westport, CT: Author.

Oryx Press. (2005b). *Directory of grants in the humanities* (19th ed.). Westport, CT: Author.

Oryx Press. (2005c). *Directory of research grants 2005* (annual). Westport, CT: Author.

Palgrave Macmillan (Ed.). (2005). *The grants register* 2006 (23rd ed.). New York: Author.

Panel on Scientific Responsibility and the Conduct of Research. (1992). *Responsible science: Ensuring the integrity of the research process* (Vol. 1). Washington, DC: National Academy Press.

Patten, M. L. (2005). *Proposing empirical research: A guide to the fundamentals* (3rd ed.). Glendale, CA: Pyrczak Publishing.

Patton, M. Q. (2001). *Qualitative research and evaluation methods* (3rd ed.). Thousand Oaks, CA: Sage.

Payne, D. (2005, March 21). Researcher admits faking data. *The Scientist.* Retrieved July 2, 2006 from http://www.the-scientist.com/article/display/22630/

Penslar, R. L. (Ed.). (1995). *Research ethics: Cases and materials.* Bloomington, IN: Indiana University Press.

Piantanida, M., & Garman, N. B. (1999). *The qualitative dissertation.* Thousand Oaks, CA: Corwin Press.

Przeworski, A., & Salomon, F. (1995). *The art of writing proposals.* New York: Social Science Research Council. Retrieved July 2, 2006 from http://www.ssrc.org/fellowships/art_of_writing_proposals.page

Puchta, C., & Potter, J. (2004). *Focus group practice.* Thousand Oaks, CA: Sage.

Reece, R. D., & Siegel, H. A. (1986). *Studying people: A primer in the ethics of social research.* Macon, GA: Mercer University Press.

Reichardt, C. S., & Rallis, S. F. (Eds.). (1994). *The qualitative-quantitative debate.* San Francisco: Jossey-Bass.

Ribbens, J., & Edwards, R. (Eds.). (1997). *Feminist dilemmas in qualitative research.* Thousand Oaks, CA: Sage.

Riessman, C. K. (Ed.). (1994). *Qualitative studies in social work research.* Thousand Oaks, CA: Sage.

Roberts, C. M. (2004). *The dissertation journey.* Thousand Oaks, CA: Sage.

Roberts, G. C. (1993). Ethics in professional advising and academic counseling of graduate students. *Quest, 45,* 78–87.

Rossman, G. B., & Rallis, S. F. (2003). *Learning in the field: An introduction to qualitative research* (2nd ed.). Thousand Oaks, CA: Sage.

Rubin, H. J., & Rubin, I. S. (2005). *Qualitative interviewing: The art of hearing data* (2nd ed.). Thousand Oaks, CA: Sage.

Safrit, M. J. (1993). Oh what a tangled web we weave. *Quest, 45,* 52–61.

Sales, B. D., & Folkman, S. (Eds.). (2000). *Ethics in research with human participants.* Washington, DC: American Psychological Association.

Schlachter, G. A., & Weber, R. D. (2006a). *Money for graduate students in the social & behavioral sciences, 2005–2007.* El Dorado Hills, CA: Reference Service Press.

Schlachter, G. A., & Weber, R. D. (2006b). *Money for graduate students in the biological & health sciences, 2005–2007.* El Dorado Hills, CA: Reference Service Press.

Schlachter, G. A., & Weber, R. D. (2006c). *Money for graduate students in the physical & earth sciences, 2005–2007.* El Dorado Hills, CA: Reference Service Press.

Schlachter, G. A., & Weber, R. D. (2006d). *Money for graduate students in the arts & humanities, 2005–2007.* El Dorado Hills, CA: Reference Service Press.

Schlachter, G. A., & Weber, R. D. (2006e). *How to pay for your degree in education & related fields, 2006–2008*. El Dorado Hills, CA: Reference Service Press.

Schmidt, F. L. (1996). Statistical significance testing and cumulative knowledge in psychology: Implications for training of researchers. *Psychological Methods, 1,* 115–129.

Schumacher, D. (1992). *Get funded!* Newbury Park, CA: Sage.

Schwandt, T. A. (2001). *Dictionary of qualitative inquiry* (2nd ed.). Thousand Oaks, CA: Sage.

Seidman, I. (2006). *Interviewing as qualitative research* (3rd ed.). New York: Teachers College Press.

Shadish, W. R., Cook, T. D., & Campbell, D. T. (2002). *Experimental and quasi-experimental designs for general causal inference*. Boston: Houghton Mifflin.

Sieber, J. E. (1992). *Planning ethically responsible research*. Newbury Park, CA: Sage.

Simons, H., & Usher, R. (Eds.). (2000). *Situated ethics in educational research*. London: Routledge Falmer.

Sprague, R. L. (1998). The voice of experience. *Science and Engineering Ethics, 4,* 33–44.

Stake, R. E. (1995). *The art of case study research*. Thousand Oaks, CA: Sage.

Stanley, B. H., Sieber, J. E., & Melton, G. B. (Eds.). (1996). *Research ethics: A psychological approach*. Lincoln, NE: University of Nebraska Press.

Strauss, A., & Corbin, J. (Eds.). (1997). *Grounded theory in practice*. Thousand Oaks, CA: Sage.

Strauss, A., & Corbin, J. (1998). *Basics of qualitative research* (2nd ed.). Thousand Oaks, CA: Sage.

Strunk, W., Jr., & White, E. B. (2000). *The elements of style* (4th ed.). Boston: Longman.

Swazey, J. P., Anderson, M. S., & Lewis, K. S. (1993). Ethical problems in academic research. *American Scientist, 81,* 542–553.

Tashakkori, A., & Teddlie, C. (Eds.). (2003). *Handbook of mixed methods in social and behavioral research*. Thousand Oaks, CA: Sage.

Taubes, G. (1995). Plagiarism suit wins: Experts hope it won't set a trend. *Science, 268,* 1125.

Thomas, J. (1992). *Doing critical ethnography*. Newbury Park, CA: Sage.

Thomas, R. M. (2003). *Blending qualitative and quantitative research methods in theses and dissertations*. Thousand Oaks, CA: Corwin Press.

Thompson, B. (1996). AERA editorial policies regarding statistical significance testing: Three suggested reforms. *Educational Researcher, 25*(2), 26–30.

Thompson, B. (1997). Rejoinder: Editorial policies regarding statistical significance tests: Further comments. *Educational Researcher, 26*(5), 29–32.

Thompson, B., & Kieffer, K.M. (2000). Interpreting statistical significance test results: A proposed new "what if" method. *Research in the Schools,* 7(2), 3–10.

Thornton, S. J. (1987). Artistic and qualitative scientific approaches: Influence on aims, conduct, and outcome. *Education and Urban Society, 20,* 25–34.

United States General Services Administration. (Semiannual). *Catalog of federal domestic assistance*. Washington, DC: Author.

Van Maanen, J. (1988). *Tales of the field: On writing ethnography.* Chicago: University of Chicago Press.

Voght, W. P. (1999). *Dictionary of statistics and methodology: A nontechnical guide for the social sciences.* Thousand Oaks, CA: Sage.

Webb, C., & Kevern, J. (2001). Focus groups as a research method: A critique of some aspects of their use in nursing research. *Journal of Advanced Nursing, 33,* 798–805.

West, L. J. (1992). How to write a research report for journal publication. *Journal of Education for Business, 67,* 132–136.

Wolcott, H. F. (2001). *Writing up qualitative research* (2nd ed.) (Sage Qualitative Research Methods Series, Vol. 20). Thousand Oaks, CA: Sage.

Yin, R. K. (2002a). *Applications of case study research* (2nd ed.) (Sage Applied Social Research Methods Series, Vol. 34). Thousand Oaks, CA: Sage.

Yin, R. K. (2002b). *Case study research: Design and methods* (3rd ed.) (Sage Applied Social Research Methods Series, Vol. 5). Thousand Oaks, CA: Sage.

Yin, R. K. (Ed.). (2004). *The case study anthology.* Thousand Oaks, CA: Sage.

Zeni, J. (Ed.). (2001). *Ethical issues in practitioner research.* New York: Teachers College Press.

Index

About the Authors

Lawrence F. Locke is Professor Emeritus of Education and Physical Education at the University of Massachusetts at Amherst. A native of Connecticut, he received his bachelor's and master's degrees from Springfield College and a Ph.D. from Stanford University. He has written extensively on the production and utilization of research on teaching and teacher education. He has authored a number of books designed to assist non-specialists with the tasks of reading and understanding research. At http://www.UnlockResearch.com, his service website offers monthly annotations of research reports, reviews of research journals and textbooks, and guides for physical educators who seek ways to apply research in professional practice. He makes his home in Sunderland, MA, but with his wife, Professor Lorraine Goyette, he spends much of each year writing, running, and exploring the Beartooth Mountains at Sky Ranch in Reed Point, MT. At both locations, he can be contacted at lflocke@hotmail.com.

Waneen Wyrick Spirduso is the Mauzy Regents Professor in the Department of Kinesiology and Health Education at The University of Texas at Austin. She is a native of Austin and holds bachelor's and doctoral degrees from The University of Texas and a master's degree from the University of North Carolina at Greensboro. Her research focuses on the effects of aging and the mechanisms of motor control. She has been a prolific contributor to the research literature and has authored textbooks related to research methods and aging. She has taught research methods and directed student research for more than three decades and has received numerous research grants from the federal government and foundations. She plays golf and rows, and lives with her husband, Craig Spirduso, in Austin, TX. Her website is http://www.edb.utexas.edu/coe/depts/kin/faculty/spirduso/index.html

Stephen J. Silverman is Professor of Education at Teachers College, Columbia University. He is a native of Philadelphia and holds a bachelor's degree from Temple University, a master's degree from Washington State University, and a doctoral degree from the University of Massachusetts at Amherst. His research focuses on teaching and learning in physical education and on methods for conducting research in field settings. He has authored numerous research articles and chapters, and is coauthor of a number of books. He has served as editor of two research journals, is an experienced research consultant, has directed graduate students, and has, for many years, taught classes in research methods, statistics, and measurement. He enjoys running, following politics, and aquatic sports, and lives with his wife, Patricia Moran, on the Upper West Side of Manhattan. His website is: http://www.tc.columbia.edu/faculty/ss928